WORLD
GOVERNMENT

WORLD GOVERNMENT

GENERAL EDITOR

Dr Peter Taylor

New York
OXFORD UNIVERSITY PRESS
1994

CONSULTANT EDITOR
Professor Peter Haggett, University of Bristol

Professor John Agnew, Syracuse University, New York
Italy and Greece

Dr Gerald Blake, University of Durham
The Middle East; Northern Africa

Dr Terry Cannon, The University of Greenwich
China and its neighbors

Professor J. Clark Archer, University of Nebraska
The United States

Dr Colin Clarke, Jesus College, Oxford
Central America and the Caribbean

Dr James Cotton, National University of Singapore
Southeast Asia

Dr Brian Fullerton, University of Newcastle upon Tyne
The Nordic Countries

Dr Barry K. Gills, University of Newcastle upon Tyne
Japan and Korea

Dr Alan Jenkins, Oxford Brookes University
China and its neighbors

Professor R.K. Johnston, University of Sheffield
The British Isles; Australasia, Oceania and Antarctica

Dr Eleonore Kofman, Middlesex University and the University of Caen
France and its neighbors; Spain and Portugal

Dr Janet Momsen, University of Newcastle upon Tyne
Central America and the Caribbean; South America

Professor Joanna Regulska, Rutgers State University of New Jersey
Eastern Europe

Dr Barry Munslow, University of Liverpool
Central Africa; Southern Africa

Dr Ian Neary, University of Newcastle upon Tyne
Japan and Korea

Dr Phil O'Keefe, University of Newcastle upon Tyne
Central Africa; Southern Africa

Professor John O'Loughlin, University of Colorado, Boulder
Central Europe

Professor Fred Shelley, University of Southern California
The United States

Dr G.E. Smith, Downing College, Cambridge
Northern Eurasia

Dr Peter Taylor, University of Newcastle upon Tyne
Governing the World; The British Isles; The Indian Subcontinent

Professor H. van der Wusten, University of Amsterdam
The Low Countries

Professor Colin Williams, Staffordshire University
Canada and the Arctic; Spain and Portugal

AN EQUINOX BOOK

Copyright © Andromeda Oxford Limited 1990

Planned and produced by
Andromeda Oxford Limited
9–15 The Vineyard, Abingdon,
Oxfordshire, England OX14 3PX

Published in the United States of America by
Oxford University Press, Inc.,
200 Madison Avenue,
New York, N.Y. 10016

Revised and updated 1994

Oxford is a registered trademark of
Oxford University Press

Library of Congress
Cataloging-in-Publication Data

A catalog record for this book is available from
the Library of Congress, Washington, D.C.

Volume editor	Susan Kennedy
Designers	Jerry Goldie, Rebecca Herringshaw
Cartographic manager	Olive Pearson
Cartographic editor	Alison Dickinson
Picture research manager	Alison Renney
Picture researchers	Libby Howells, David Pratt
Project editor	Candida Hunt
Art editor	Steve McCurdy

ISBN 0-19-521096-4

Printed (last digit): 9 8 7 6 5 4 3 2

Printed in Spain by Fournier Artes Graficas SA, Vitoria

INTRODUCTORY PHOTOGRAPHS
Half title: *Prague demonstrations 1989 (Gamma/Frank Spooner Pictures/Gilles Bassignac)*
Half title verso: *East German troops (Rex Features/Van Morvan)*
Title page: *Voting in El Salvador (Rex Features/Wesley Bocxe)*
This page: *Keeping the peace in Lebanon (Magnum/Gilles Peress)*

Contents

PREFACE

THE WORLD IS IN THE MIDST OF A DYNAMIC PHASE OF POLITICAL CHANGE. The old order – the Cold War – that dominated world politics for more than forty years has collapsed. The new world politics that will replace it is still beyond our view. However, we may confidently predict that this period of change will be succeeded by a new stability or order, for history shows us that over the centuries world politics have evolved in an alternating "stop–go" pattern of change and equilibrium.

It is an exciting period to be living through, and a challenging one, too, for old assumptions are being discarded by politicians and professional analysts alike as they are forced to consider new ways of looking at the world. How can we make sense of our current dynamic political world? The demise of the global politics of Cold War will have far-reaching implications for the way the countries of both East and West are governed.

The aim of this volume is to unravel the processes that are shaping the world of politics today in order to understand the possibilities that lie ahead. It does this by looking in detail at individual government within the countries and regions that make up our world.

There are three main themes. First, there are questions of formal state power. How have state boundaries been established? What sort of constitutional arrangements are present? Next it considers how states are actually governed: what is the basis of a government's power and how stable is its possession of it? Does it recognize one, two or more political parties – or none at all? Are the people free to express their political choice? Through what kind of electoral system? Finally there is the question of how states relate to each other. Why are some neighboring states economic partners while others are politically hostile? Where does conflict exist, and how is it contained? What overall global patterns can we find?

We can only begin to glimpse the changes that the future has in store for us if we have knowledge of these themes. Many new global scenarios have been predicted: a united Europe will reach from the Atlantic to the Ural Mountains; a power bloc that includes both the United States and Japan will emerge in the countries fringing the Pacific; democracy will be rekindled in China as the "old guard" die; Japan will ascend to world leadership.

Some of these things may happen, others not. All are possibilities, some may even be probabilities. This volume will have succeeded if it helps the reader to sort out in his or her mind the nature of our political world and to make a reasoned assessment of what the future holds in store.

Professor Peter Taylor
UNIVERSITY OF NEWCASTLE UPON TYNE

The heroes of communism, paraded in Havana, Cuba

Antinuclear demonstrators in Japan (*overleaf*).

GOVERNING THE WORLD

What is a State?

THERE ARE IN THE WORLD TODAY NEARLY 190 countries that are recognized as independent, each with a defined territory, a people, and a government that is responsible for making and implementing laws within that territory. Most important of all, these "states", as they are called, possess sovereignty.

Sovereignty has nothing to do with size. Luxembourg, with an area of 2,586 sq km (999 sq mi) and a population of 336,000, is a sovereign state, but Texas, with an area of 692,407 sq km (267,339 sq mi) and a population of over 14 million, is not. Luxembourg possesses the authority to govern its citizens without interference from any outside power. Although Texas has its own government able to make certain policy decisions, it is subject to the authority of the government of the United States, of which it is a constituent element or unit.

Sovereignty defines a state as a member of the international community of states. It is like a membership card to an exclusive club. Sovereign states can sign treaties, swap ambassadors and join the United Nations. Luxembourg can do all these things, but Texas cannot. Its international affairs are formally conducted through the federal government situated in Washington DC. Similarly, a colony or dependency conducts its affairs through the government of the external power to which it belongs. Hong Kong, a British colony, is due to be reintegrated with the government of mainland China in 1997. All the negotiations agreeing to its transfer were carried out between the British government and that of the People's Republic of China.

It is an essential condition of sovereignty that it is recognized by the international community of states. It cannot simply be proclaimed. There has been no international recognition for the independent Turkish state in northern Cyprus set up after Turkey's invasion of Cyprus in 1974. Although Lithuania, Estonia and Latvia declared independence in 1991, and Croatia and Slovenia seceded from Yugoslavia, only when the European Community recognized them did they become part of the states system.

States express their sovereignty in different ways. They all have a national flag and a national anthem. Most have their own currency, postage stamps, even their own airline. They have a capital city, in which the diplomats of other states represent their country's interests in relations with the host country. All states provide passports for their citizens when traveling abroad. Most join the world association of states, the United Nations (UN).

A meeting place for sovereign states
The UN has its headquarters in New York. It was set up after World War II to strive for international peace, security and cooperation between its members. The General Assembly of the UN, in which all decisions are taken, embodies the principle of equality between sovereign states: each member has one vote in the assembly, with the exception of the Soviet Union, which retains three seats as part of the deal for setting up the UN in 1945. This means that the United States, with its huge wealth and resources, and China, with its vast population, have the same voting weight in assembly divisions as Djibouti or the Solomon Islands, each of which has a population of less than half a million.

DIPLOMACY

Diplomacy has been described as the management of international relations by negotiation. Modern diplomatic procedures date back to the 15th century in Western Europe, when Italian city states maintained permanent missions in each other's territory. Today all sovereign states keep a diplomatic service to conduct foreign policy beyond their borders.

Diplomats have several functions. First, they are the means by which governments communicate with each other. A diplomat's prime object is to represent his country's interests and influence the host country in its favor. Diplomats also protect their country's citizens abroad and represent their country at ceremonial events. They obtain information and advise in the making of foreign policy toward the country they are assigned to.

Day to day diplomacy is conducted through the network of embassies and legations across the world. Special meetings of diplomats may be held to discuss particular issues, such as peace conferences, and the UN and its associated organizations provide a permanent forum for diplomatic exchange.

Diplomats must be able to work without fear of retribution by the host country. Hence the custom of treating all embassies as the sovereign territory of their country, and of granting immunity to diplomats from local laws.

The world political map Although all sovereign states are equal in the eyes of the UN, vast differences exist between them in terms of size and population, and in degrees of wealth and poverty.

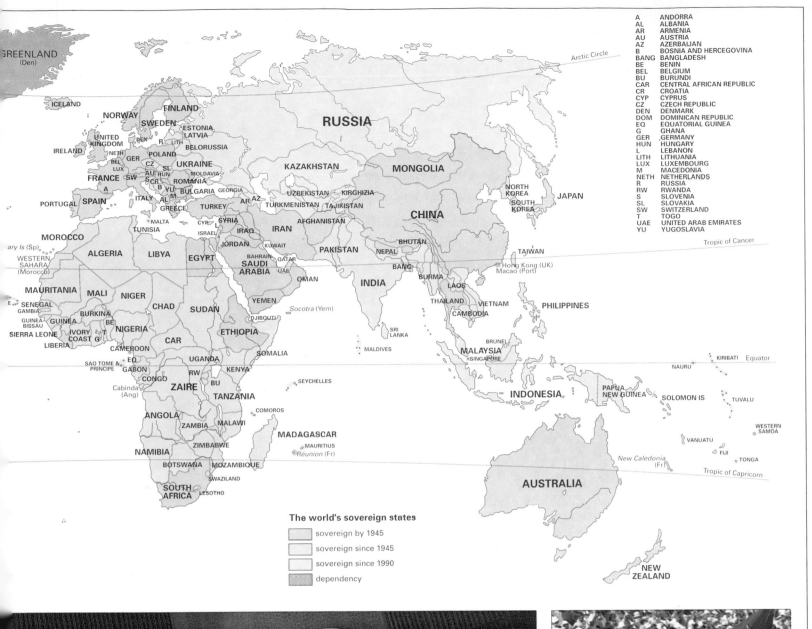

ANDORRA
AL ALBANIA
AR ARMENIA
AU AUSTRIA
AZ AZERBAIJAN
B BOSNIA AND HERCEGOVINA
BANG BANGLADESH
BE BENIN
BEL BELGIUM
BU BURUNDI
CAR CENTRAL AFRICAN REPUBLIC
CR CROATIA
CYP CYPRUS
CZ CZECH REPUBLIC
DEN DENMARK
DOM DOMINICAN REPUBLIC
EQ EQUATORIAL GUINEA
G GHANA
GER GERMANY
HUN HUNGARY
L LEBANON
LITH LITHUANIA
LUX LUXEMBOURG
M MACEDONIA
NETH NETHERLANDS
R RUSSIA
RW RWANDA
S SLOVENIA
SL SLOVAKIA
SW SWITZERLAND
T TOGO
UAE UNITED ARAB EMIRATES
YU YUGOSLAVIA

The world's sovereign states

- sovereign by 1945
- sovereign since 1945
- sovereign since 1990
- dependency

Flags and sporting teams are part of the means that states use to declare their membership of the international community of states. Here a team from China parades behind its national flag.

The General Assembly of the UN in session. Not all states are members. Switzerland chose not to join; North and South Korea have not been admitted to membership. China's seat, originally held by the Nationalist government in Taiwan, was taken over by the People's Republic of China in 1971.

11

Territorial Definitions

EVERY SOVEREIGN STATE POSSESSES AND controls a territory within recognized boundaries. There are three main types of interstate boundaries. Physical boundaries follow major natural landscape features, such as a range of mountains like the Andes, separating Argentina and Chile, or the Pyrenees, which divide Spain and France, or a river such as the Rio Grande, which forms the border between Mexico and the United States.

Geometric boundaries were often drawn when an area was largely unknown and uncharted: there are many such arbitrary boundaries in the Sahara desert. Often they run along a line of latitude or longitude. The 49th parallel, separating Canada and the United States, is the best known of these.

Finally, there are cultural boundaries, often created as the result of conflict. The boundaries of Europe were redrawn after World War I and modified after World War II. Communal violence that broke out between Muslims and Hindus in India after the withdrawal of British rule led to the creation of a separate Muslim state in Pakistan.

A state's claim to territory may be diputed for a number of reasons. A claim to a piece of territory may be made on the grounds of territorial integrity or geographical proximity. This is the basis on which Argentina claims the Falkland Islands/Malvinas from Britain, and Spain claims the British colony of Gibraltar. Such claims are frequently maintained in defiance of the wishes of the majority of the inhabitants.

Sometimes a state may claim prior settlement or possession of an area. Hungary lays claim to Transylvania, lying inside Romania, on the grounds that it settled the area over a thousand years ago. Closely allied to such historical arguments are cultural claims to a territory, which are associated with the idea of a nation-state.

This is the principle that a state's boundaries should match the territory of a nation or a group of people with a shared ethnic or linguistic heritage. It emerged in Europe during the 19th century, and lay behind the struggle to recover Greece's independence from the Turkish Ottoman empire after 1821.

Self-determination

Associated with it is the doctrine of self-determination, which asserts that the people of a national territory or group have the right to decide their own political status. It was on this basis that the political boundaries of Europe were redrawn after World War I, when the vacuum left by the fall of the Austro-Hungarian empire and the tsarist empire in Russia led to the creation of a number of independent states in eastern and southeastern Europe.

The carving up of Africa between a number of European powers in the 19th century created territorial boundaries that cut across ethnic and linguistic territories. Postcolonial states are living with the consequences of these decisions. Somalia, for example, has never recognized colonial boundaries, and claims part of Kenya as well as the whole of the Ogaden region in eastern Ethiopia, which are predominantly inhabited by Somalispeaking people.

The UN charter recognizes self-determination as one of the underlying principles leading to friendly relations between states, but it is rarely honored, except in support of nationalist groups opposing colonial rule, since to do so provides an example for undermining the territorial integrity of other states. External pressure may, on the other hand, lead to the partitioning of a nation to make two separate states. This happened in Germany, Korea and Vietnam as a result of Cold War pressures. In such cases both sides continue to recognize each other as part of the same nation, and reunification remains an eventual goal. Reunification occurred in Vietnam in 1975 and Germany in 1990.

NONTERRITORIAL BOUNDARIES

State sovereignty extends beyond the surface land territory established by territorial boundaries. Rights over all subterranean resources reach downward as a "cone" to the center of the Earth, but extending sovereignty upward is more difficult. A 1919 convention gave states the right to prevent their territory being overflown. The space treaty of 1967 tried to define an upper limit to this sovereignty, but an exact distance was not agreed. The operational ceiling of conventional aircraft normally defines the sovereign space of the state below. Everything above this level, such as satellites, are considered free from sovereignty restrictions.

More important is the extension of the sovereignty of coastal states over their adjacent seas. Originally territorial waters were fixed at 5 km (3 nautical mi). The 1982 United Nations Convention on the Law of the Sea (UNCLOS) extended territorial waters to 22 km (20 nautical mi). Within this zone states have absolute sovereignty except for the "innocent passage" of ships. A new exclusive economic zone of 370 km (200 nautical mi) was established, giving a state rights to all fishing and seabed resources within it, as well as the further right to exploit resources on their continental shelf where that extends beyond 370 km (200 nautical mi). The United States disagrees with this provision and has declined to ratify the new Law of the Sea.

A heavily fortified frontier US troops patrol the demilitarized zone that separates North and South Korea along the line of the 1953 ceasefire agreement. The country was partitioned in 1945 along the 38th parallel, with a communist government in the North and a pro-Western regime in the South, leading to war between them (1950–53). Reunification had still not been attained 40 years later.

A river boundary The length of the Rio Grande as it runs for 1,995 km (1,240 mi) between El Paso and the Gulf of Mexico acts as the frontier between Mexico and the state of Texas in the United States. This decorative plaque at Laredo, Texas straddles the line of division along the center of the river. The frontier is difficult to patrol effectively, and is easily crossed by immigrants.

SOME MAJOR TERRITORIAL DISPUTES SINCE 1945

Territory in dispute	Claimants	Conflicts
Kashmir	India/Pakistan	1947–49
Palestine/Israel	Palestine Arabs/Israel	1948–49,1967
North Assam	China/India	1962
Sarawak (Borneo)	Malaysia/Indonesia	1962–66
Ussuri	China/USSR	1969
Cyprus	Greece/Turkey	1974
Western Sahara	Morocco/Mauritania/Polisario Front	since 1975
Ogaden	Ethiopia/Somalia	1977–78
Northern Vietnam	China/Vietnam	1979
Shatt al Arab waterway	Iraq/Iran	1980–88
Malvinas/Falkland Islands	Argentina/Britain	1982
Kuwait	Iraq/Kuwait	1990–1
Bosnia	Croatia/Serbia/Bosnia	1991–

The Two Faces of Nationalism

THE DOCTRINE OF NATIONALISM ASSUMES that every individual belongs to a nation, and that every nation requires a sovereign state as its homeland in which to express its particular culture. It evolved in the 19th century and has come to dominate the politics of the 20th century. At its heart lies an essential ambiguity: as a force for good, nationalism liberates peoples from foreign oppression, but as a force for evil it leads to conflict between them. The two major wars of the 20th century are a clear illustration of this: World War I was partly caused by the suppression of Serbian nationalism, World War II by the excessive expression of German nationalism by the Nazis.

Membership of a nation confers identity to its people and gives meaning to their lives. Each nation has its particular aspirations, but common to them all is a tradition of national heroes and national myths harking back to a "golden age" when the nation was free and strong. The stone dwellings of Zimbabwe, the center of a great empire in east Africa that predated European colonization by several hundred years, provided in-spiration to black nationalists fighting white settler rule in Southern Rhodesia and gave the country its name once independence was achieved.

The ideology of nationalism is so strongly imbued in people's minds that the nation is sometimes thought to be synonomous with state. Even the United Nations is misleadingly named, since it is not an organization for nations, but for states. States that have a multiplicity of nationalities among their citizens, such as Canada, are members, but nations without states, such as the Kurds or Palestinians, are not.

The world is not as neat as the doctrine of nationalism supposes. It can be very difficult to define precisely cultural groups that may be designated "nations", but however it is done, the result of the calculation is always many more potential nations than there are states. According to one classification, there are as many as 1,300 potential nations throughout the world. Since there are less than 200 states, it follows that national self-determination for each of them would produce a world of mini-states.

A	ANDORRA
AL	ALBANIA
AR	ARMENIA
AU	AUSTRIA
AZ	AZERBAIJAN
B	BOSNIA AND HERCEGOVINA
BANG	BANGLADESH
BE	BENIN
BEL	BELGIUM
BU	BURUNDI
CAR	CENTRAL AFRICAN REPUBLIC
CR	CROATIA
CYP	CYPRUS
CZ	CZECH REPUBLIC
DEN	DENMARK
DOM	DOMINICAN REPUBLIC
EQ	EQUATORIAL GUINEA
G	GHANA
GER	GERMANY
HUN	HUNGARY
L	LEBANON
LITH	LITHUANIA
LUX	LUXEMBOURG
M	MACEDONIA
NETH	NETHERLANDS
R	RUSSIA
RW	RWANDA
S	SLOVENIA
SL	SLOVAKIA
SW	SWITZERLAND
T	TOGO
UAE	UNITED ARAB EMIRATES
YU	YUGOSLAVIA

National pride A British youth shows his support for his country's forces fighting in the Falklands. Patriotism can easily lead to excessive nationalism as one race or group strives for superiority over others.

A nation without a state Palestinians demonstrate against the Israeli occupation of the West Bank. The 1994 withdrawal of Israeli forces was a step toward the creation of a Palestinian homeland.

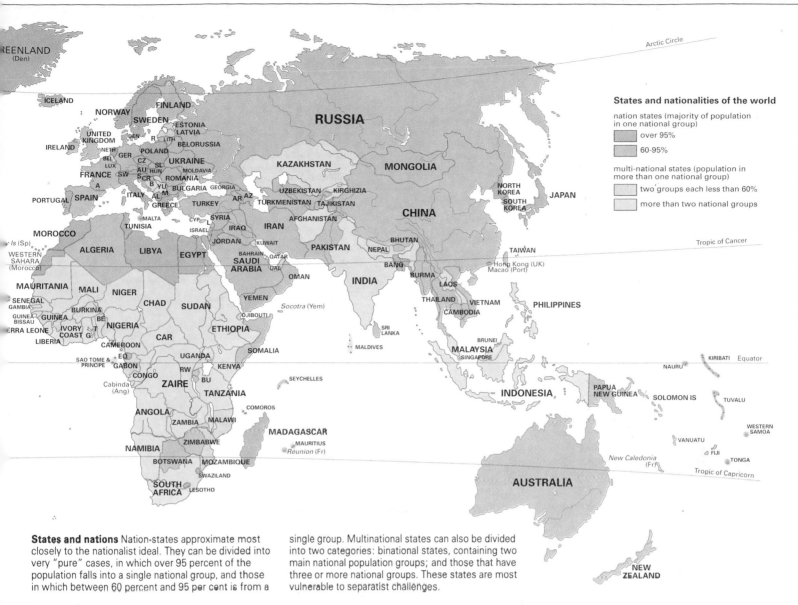

States and nationalities of the world

nation states (majority of population in one national group)

over 95%

60-95%

multi-national states (population in more than one national group)

two groups each less than 60%

more than two national groups

States and nations Nation-states approximate most closely to the nationalist ideal. They can be divided into very "pure" cases, in which over 95 percent of the population falls into a single national group, and those in which between 60 percent and 95 per cent is from a single group. Multinational states can also be divided into two categories: binational states, containing two main national population groups; and those that have three or more national groups. These states are most vulnerable to separatist challenges.

THE RIGHTS OF INDIGENOUS PEOPLES

Many of the indigenous peoples who occupied vast tracts of land in North and South America and in Australasia before colonization were brought close to the edge of extinction by the activities of European farmers and settlers. They have recently begun to find a new political voice in their struggle for ethnic survival. The United Nations Commission on Human Rights working group provides a focal point for their demands. Population estimates are only approximate, but there are possibly 2.5 million in the United States, 800,000 in Canada, 400,000 in New Zealand, and 200,000 in Australia. Only some 200,000 survive in tribal groups in the Amazon basin of South America.

In all these states they are a growing political force that can no longer be ignored. Strong movements have developed in Australia and New Zealand to demand the redress of the wrongs imposed by European settlement and to seek the restitution of land rights, and similar protests have been voiced among the Amerindians and the Inuit people of Canada and the United States. Recently in Brazil there has been united action by the Indians to protect their remaining lands from commercialization.

The reality is that nearly all states contain a mix of ethnic groups and nationalities. In only a handful of them is more than 95 percent of the population from one national group. Most have two or more national groups. Some nations are dispersed over several states. The Kurds are in six states, though mainly in Iran, Iraq and Turkey. The Palestinians are scattered among several Arab states as well as Israel; Armenians are found in 11 states. Some 16 states make up the Arab nation, and three states and a colony (Hong Kong) the Chinese nation. These examples show the complexity of the relationship between state and nation.

Challenging the state

Nationalist minority groups such as the Basques and Catalans in Spain and the Québécois in Canada began to demand separate status during the 1960s. Political protest took a number of forms, ranging from campaigns of violence waged against agents of the state to the formation of political parties to win support through the ballot box. None of them proved successful in their primary goal of separate sovereign status, but most won large-scale concessions in the form of regional self-government.

Violence has often emerged as the territorial arrangements imposed on states at independence have proved unable to satisfy the nationalist ambitions of ethnic minorities. This has been the case in many parts of Africa, in Southeast Asia (Burma, Indonesia, Malaysia, the Philippines) and in India and Sri Lanka.

The growth of nationalist movements in the Soviet Union (especially the Baltic Republics and Central Asia) in the late 1980s precipitated its dissolution once communism collapsed there in 1991. In Eastern Europe, emerging ethnic rivalries caused the federal states of Czechoslovakia and Yugoslavia to break up, and in the latter case led to bitter war.

The Creation of New States

OF ALL THE CHANGES IN WORLD POLITICS IN the second half of the 20th century one of the most remarkable has been the rapid increase in the number of new states taking their seats in the United Nations. In 1945 the number of member states was 51; by the 1980s the number had risen to 159. This growth came about as the result of the breakup of the major European colonial empires after World War II. The rapid movement toward colonial liberation produced more than a hundred new states in Africa, the Caribbean, Southeast Asia and the Pacific between the independence of the Philippines, achieved in 1946, and that of Namibia in 1989.

The throwing off of external domination is the aspiration of peoples in most dependent states. The first country to achieve it in the modern world was the United States, which declared its independence from Britain in 1776. The success of its revolution served as a model and inspiration for others: by 1826 the Spanish and Portuguese empires in Central and South America had collapsed, creating 17 new states.

Elsewhere in the world, particularly Africa, the 19th century saw the consolidation of European imperial rule. Although Britain had granted independence to its "white" colonies of Australia, Canada, New Zealand and South Africa by 1931, the main surge toward decolonization did not take place until after World War II, which saw the end of Europe's domination of world power.

The occupation by Germany of Belgium, France and the Netherlands caused severe upheavals to their colonies, and Germany's ally, Japan, overran French Indochina and the Dutch East Indies, as well as seizing the Far East possessions of Britain, the world's major imperial power. European prestige was not to recover from these military humiliations.

After 1945 the Soviet Union and the United States, both of whom were hostile to old-style colonialism, emerged as the two dominant world powers. Both of them gave their agreement to the Atlantic Charter Declaration (1941), which committed the Allies to supporting self-determination throughout the world. In the postwar period Jordan, Syria and Palestine (which immediately proclaimed itself the Jewish state of Israel) gained their independence in the Middle East; India, Pakistan, Burma, Ceylon (Sri Lanka) and Indonesia achieved it in Asia; and Libya, Morocco, Tunisia and Sudan in North Africa.

During the 1950s a new world bloc of states began to emerge. It deliberately distanced itself both from Europe and from the two new superpowers. The first meeting of independent Asian and African countries, held at Bandung, Indonesia in 1955 pledged itself to supporting anticolonialism throughout the world. A second wave of decolonization swept through west and central Africa, starting with Ghana (formerly the British colony of the Gold Coast) in 1957. Some 17 African states attained independence in 1960 alone. Jamaica and Trinidad were the first of the Caribbean states to be granted independence, in 1962.

This rapid increase of postcolonial states transformed the United Nations by providing it with a new shift of balance. Decolonization and development issues came to dominate the agenda as Third World countries asserted their political strength. Independence movements in all the remaining colonies of Africa, except Western Sahara (under Moroccan rule), were successful by 1989 when Namibia finally achieved its freedom from South African occupation.

A citizen of France in French Guiana, South America, one of a handful of dependent territories that are left around the world. The majority of them are islands or groups of islands in the Caribbean and the Pacific; mostly they are self-governing.

THE CHRONOLOGY OF DECOLONIZATION

Belgium
Britain
France
Spain
Italy
Netherlands
USA

Year	Country	Belgium	Britain	France	Spain	Italy	Netherlands	USA
1946	Jordan		•					
	Syria			•				
	Philippines							•
1947	Bhutan		•					
	India		•					
	Pakistan		•					
1948	Burma		•					
	Ceylon (Sri Lanka)		•					
	Palestine (Israel)		•					
1949	Indonesia						•	
1951	Libya		•			•		
1954	Cambodia			•				
	Laos			•				
	North Vietnam			•				
	South Vietnam			•				
1956	Morocco			•				
	Sudan		•					
	Tunisia			•				
1957	Ghana		•					
1958	Guinea			•				
1960	Cameroon			•				
	Chad			•				
	Central African Republic			•				
	Congo			•				
	Dahomey (Benin)			•				
	Gabon			•				
	Ivory Coast			•				
	Malagasy Republic (Madagascar)			•				
	Mali			•				
	Mauritania			•				
	Niger			•				
	Nigeria		•					
	Senegal			•				
	Somalia					•		
	Togo			•				
	Upper Volta (Burkina)			•				
	Zaire	•						
1961	Cyprus		•					
	Kuwait		•					
	Tanganyika (Tanzania)		•					
1962	Algeria			•				
	Burundi	•						
	Jamaica		•					
	Rwanda	•						
	Trinidad and Tobago		•					
	Uganda		•					
	Western Samoa	•						

More than 100 dependent states acquired sovereign status between 1946 and 1989. One other new state was created during this period when Bangladesh (formerly the province of East Pakistan) seceded from Pakistan in 1971. A number of British dependencies in the Pacific had been administered by Australia or New Zealand.

Belgium
Britain
France
Spain
Portugal
Netherlands
South Africa

Year	Country	Belgium	Britain	France	Spain	Portugal	Netherlands	South Africa
1963	Kenya		●					
	Malaysia (including Singapore)		●					
	Zanzibar (part of Tanzania)		●					
1964	Malawi		●					
	Malta		●					
	Zambia		●					
1965	Gambia		●					
	Maldives		●					
	Singapore (from Malaysia)							
1966	Barbados		●					
	Botswana		●					
	Guyana		●					
	Lesotho		●					
1967	Southern Yemen		●					
1968	Equatorial Guinea				●			
	Mauritius		●					
	Nauru		●					
	Swaziland		●					
1970	Fiji		●					
	Tonga		●					
1971	Bahrain		●					
	Qatar		●					
	Sierra Leone		●					
	United Arab Emirates		●					
1973	Bahamas		●					
1974	Grenada		●					
	Guinea-Bissau					●		
1975	Angola					●		
	Cape Verde					●		
	Comoros			●				
	Mozambique					●		
	Papua New Guinea		●					
	Saõ Tomé and Principe					●		
	Surinam						●	
	Western Sahara				●			
1976	Seychelles		●					
1977	Djibouti			●				
1978	Dominica		●					
	Solomon Islands		●					
	Tuvalu		●					
1979	Kiribati		●					
	St Lucia		●					
	St Vincent		●					
1980	Vanuatu		●					
	Zimbabwe		●					
1981	Antigua		●					
	Belize		●					
1983	St. Kitts-Nevis		●					
1984	Brunei		●					
1989	Namibia							●

The creation of a new state The British flag is lowered in April 1980 as Southern Rhodesia becomes the new state of Zimbabwe. A white-controlled government had illegally declared independence in 1965, precipitating a nationalist war for black majority rule.

The legacy of empire can be seen in many political institutions and systems of government that have taken root around the world. These lawyers, in traditional 18th-century wigs and gowns, are in the former British colony of Sierra Leone in west Africa.

The new membership of the UN brought a significant shift of balance in the political alignments of the organization. In 1951 Asian and African countries represented less than a quarter of the voting members; by 1970 they dominated the assembly.

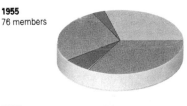

1945
51 members

1955
76 members

1965
120 members

1985
158 members

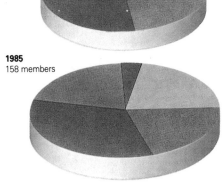

Membership of the United Nations

North and South America
Australasia and Oceania
Asia
Africa
Europe

Ruling the People

Most states have a constitution, a code of rules that embodies the supreme law of the state. A few states operate through an "unwritten" constitution. This means the rules of the state are contained partly in laws and partly in custom and convention. Britain and Israel both have constitutions of this kind. The majority of states, however, possess a written constitution, a document that enshrines all the principles of government on which that state is based. The United States' is the first written constitution, ratified in 1789.

Constitutions define the relationship between the three main branches of government: the executive power (the president, or prime minister and cabinet), which makes and carries out policy; the legislature (often called the parliament or national assembly), which approves policy and passes laws; and the judiciary, or courts of law, which interprets the constitution and the laws of the state. Also defined are the rights that the citizen has in relation to the government. These include civil liberties such as the freedom of speech and religion, and political rights – the right to vote or to join a trade union, for example.

The constitution additionally lays down the way the country is to be governed. First of all it establishes whether there are one or two layers of government. In a federal or confederal system the country is divided into states or regions. The national or federal government makes policies that concern the whole country; at the second level, state assemblies are empowered to pass laws on domestic affairs relating to themselves alone. Federal systems are found most often among the larger states of the world: Australia, Brazil, India, the United States. In unitary systems (Britain, France, Japan, Sweden) power is concentrated at the center, and the devolution of responsibility to the regional authorities is discretionary.

Systems of government

The head of government is the chief executive of the state – either a president or a prime minister – who has ultimate responsibility for making and implementing all decisions. In some states the president is both head of state and head of government; in others, the head of state may be a formal role only – either a constitutional monarch (Britain, Sweden) or an elected non-executive president (India, Germany). Executive power is vested in the prime minister, who is responsible to the legislature. Some systems are mixed. In France the president selects the prime minister and retains certain executive functions, such as foreign policy.

In most countries the legislature is a representative body elected by the people. Most consist of a single assembly, or chamber, and are often called unicameral. A minority have a two-chamber (bicameral) assembly. Generally the second (sometimes called the upper) chamber has the role of revising and checking legislation, but power remains with the first, or "popular" chamber. Second chambers are often unelected.

Governments vary greatly in their goals and the means of achieving them. Most strive for economic growth, but some seek to intervene directly to redistribute wealth more evenly. In the richer countries liberal democratic governments seek some consensus on the degree of redistribution through party elections. In the

A personality cult During China's Cultural Revolution, Mao Zedong, Communist Party chairman, was the subject of a personality cult; his thoughts, contained in *The Little Red Book*, were made official doctrine.

Dismantling communism The red star of communism is taken down from the front of a building in Prague, Czechoslovakia. The rapid collapse of communist power throughout Eastern Europe in the last months of 1989 brought chaos and uncertainty as those countries sought new forms of government.

France's national assembly is the dominant chamber of the two-chamber legislature. It was established during the Revolution in 1789 and has since provided a model for many other state legislatures. It has given the terms "right wing" and "left wing" to politics, since conservatives traditionally sat on the right side of the assembly and radicals on the left.

King Fahd of Saudi Arabia is one of the world's last absolute monarchs. There is no written constitution and no legislature: the king rules, in accordance with traditional Islamic law, by decree.

THE WORLD'S SURVIVING MONARCHIES

By far the largest number of countries in the world are republics – the head of state is nonhereditary, usually a president. The world's surviving hereditary monarchies are of two kinds: **constitutional**, where the king or queen has no executive powers, and is symbolic head of state only, with a constitutionally defined formal function, and **traditional**, where the king or queen still controls government. Many of Britain's former colonies still accept the British monarch as symbolic head of state.

Constitutional monarchies

Belgium, Britain, Cambodia, Denmark, Japan, Liechtenstein, Luxembourg, Monaco, Netherlands, Norway, Spain, Sweden

Traditional monarchies

Bahrain, Bhutan, Brunei, Jordan, Kuwait, Lesotho, Morocco, Nepal, Oman, Qatar, Saudi Arabia, Swaziland, Thailand, Tonga, United Arab Emirates, Western Samoa

world's remaining communist countries the redistribution goals are a formal part of the constitution, and the means are more authoritarian, through one-party control of government.

In the Third World there is an even wider range of government systems, including a number of surviving traditional monarchies. Huge material inequalities produce unstable government, and to counter this one-party systems, often centered on the personality of a strong popular leader, or authoritarian, often military regimes committed to preserving the status quo, have emerged. Such leaders are able to override or ignore constitutions. Hence many of the rights guaranteed in constitutions across the world today exist on paper only.

Parties and Party Systems

IN LIBERAL DEMOCRACIES CONTROL OF THE policy-making branches of government is competed for by political parties. These are voted into office at general elections on the basis of their program of political promises. Modern political parties first began to emerge in Europe during the 19th century when the franchise was extended and the existing political elites had to compete for the support of the new classes of voters. They did this by claiming to represent the public interest rather than their own factional interests.

Soon the newly enfranchised voters began to demand a voice in politics for themselves. The broadly socialist parties that formed toward the end of the 19th century to defend workers' interests against those of their employers have continued to be an important force in European politics. Other interest groups also formed parties. Conflict over religious education led to the creation of church parties; agrarian parties reflected the traditional interests of farmers and landowners in conflict with new industrial interests.

Thus the multi-party system emerged; it has continued to dominate politics in Western Europe, and in most of the countries of the world that were settled by Europeans, throughout the 20th century.

The broad pattern is one of socialist or social democratic parties on the left, offering policies of wealth redistribution and state welfare programs, competing with conservative or Christian democratic parties on the right and liberal and radical parties in the center. The system of proportional representation that is used to elect governments in the majority of these countries allows a large number of issue-related parties to be represented as well. Recently the most successful of these have been the Green parties; they can be influential when coalition governments are being formed.

The system of party politics that emerged in both the democracies of North America – Canada and the United States – differed significantly from the European pattern. Here a split in the working class on ethnic and religious lines that reflected the pattern of immigration from Europe in the 19th century inhibited the development of a strong socialist party, leaving a vacuum on the left of the political spectrum. The result has been much greater pragmatism in politics, and there are few apparent differences between the political programs of the two main parties. Party support is based on sectional, or regional, interests rather than ideology.

A	ANDORRA
AL	ALBANIA
AR	ARMENIA
AU	AUSTRIA
AZ	AZERBAIJAN
B	BOSNIA AND HERCEGOVINA
BANG	BANGLADESH
BE	BENIN
BEL	BELGIUM
BU	BURUNDI
CAR	CENTRAL AFRICAN REPUBLIC
CR	CROATIA
CYP	CYPRUS
CZ	CZECH REPUBLIC
DEN	DENMARK
DOM	DOMINICAN REPUBLIC
EQ	EQUATORIAL GUINEA
G	GHANA
GER	GERMANY
HUN	HUNGARY
L	LEBANON
LITH	LITHUANIA
LUX	LUXEMBOURG
M	MACEDONIA
NETH	NETHERLANDS
R	RUSSIA
RW	RWANDA
S	SLOVENIA
SL	SLOVAKIA
SW	SWITZERLAND
T	TOGO
UAE	UNITED ARAB EMIRATES
YU	YUGOSLAVIA

ONE-PARTY SYSTEMS

A number of countries in the world have one-party systems. In many supposedly multi-party states, such as Indonesia, Japan, Mexico and Singapore, though opposition parties are allowed to exist, one party dominates to such a degree that it is effectively the only party of government. However, in a true one-party state only one party is recognized by law. In some, it is the Communist Party, but the number of communist countries fell to a handful following the rejection of communist rule in Eastern Europe in 1989 and the Soviet Union in 1991.

Most one-party states are to be found in the Third World. Often the ruling party is the party that achieved independence. The Revolutionary Party of Tanzania (CCM), for example, became the official party of Tanzania in 1977. As the Tanganyika African National Union (TANU), it won independence in 1961 and dominated politics thereafter.

One-party rule in the Third World is usually the result of chronic economic crisis, which creates a need for strong,

unified leadership. Liberal democratic countries react in a similar way when faced with crisis or recession by forming national or coalition governments. The withdrawal of Soviet aid in the early 1990s, coupled with economic pressure from Western countries, led many one-party states in Africa to move toward multi-party systems. In 1990 less than 9 percent of states in sub-Saharan Africa possessed genuine pluralistic systems. By 1992 the figure was far higher.

All one-party states face the problem of accommodating opposition. It can either be repressed, or attempts be made to absorb it within the party. For example, more than one candidate from the party may be allowed to stand in a district. A further effect of one-party rule is that the relationship between the party and the government may become confused. Generally the party is supposed to define policy and the government to execute it. In practice, though, the two functions often merge, producing inbuilt conservatism.

Party symbols In countries where illiteracy is widespread political parties are often represented by symbols, which appear on ballot papers next to the candidate's name. India, the world's largest democracy, has many political parties, each with its own symbol. This hammer and sickle represents the Left Front Party.

Convention fever The elephant has long been the emblem of the Republican Party in the United States, where the two parties – the other being the Democratic Party – meet in national party conventions every four years to choose their candidate for the presidential election. In other systems, party conferences or assemblies are held to thrash out details of policy and win consensus for election programs from the party members.

Different styles of government
A number of states that declare themselves to have multi-party systems do so only in name – their civilian governments are under military influence, and opposition parties are severely restricted.

Systems of government

- multi-party state
- one-party state
- no political party

- dependency

- • federal state
- ★ military influence on government

A socialist platform Prime Minister Felipe Gonzalez campaigns for reelection in Spain. The red rose is the symbol of Western Europe's socialist parties. In most liberal democracies the socialist–conservative split dominates the political party spectrum: socialist parties offer nationalization and social welfare programs; they are opposed by conservative parties, which support a free market economy. In many countries the election system means that coalition government is the norm, with the balance of power being held by the center parties.

Other party systems

In most other parts of the world competitive party systems have failed to develop. This has led either to more fragmented and divisive politics, or to one-party systems. Some traditional regimes, such as the desert kingdoms of the Middle East, have never allowed parties to develop, while military regimes frequently ban them.

Even where competitive party politics is allowed, the parties do not perform an integrative function as they do in the liberal democracies. In South America, for example, the party system proved unable to adapt to a more democratic form of government, and self-serving elites continued to control politics. In reaction to this populist parties, of which the best known is the Peronist Party in Argentina, emerged. They claimed to represent the workers against the landowning aristocracy, and won popular support with welfare programs and higher wages. Elsewhere, notably in Chile, a strong socialist party developed. These trends in turn provoked a right-wing backlash, often resulting in military coups.

Electing the Government

Most people in the world will cast a ballot at some time in their lives. Elections are an essential part of the machinery of liberal democracies, but they also figure in other forms of government. In communist and single-party states the ruling party's candidates are endorsed through popular elections, and military dictatorships often hold referendums to justify their rule. Elections are excluded only in some of the traditional regimes, such as the Middle East oil autocracies, which still resist the idea that the people have a right to take part in choosing the government.

From being a privilege of the few, voting rights were gradually extended in most Western countries during the 19th century by the removal of property qualifications and other bars to voting such as race or religion. Most elections throughout the world are now conducted on the basis of universal suffrage, or the right of all adults to vote regardless of age, sex or race. The most notorious exception to the last of these rules is the white regime of South Africa, which excludes the country's blacks, who comprise 70 per cent of the population, from voting in national elections.

Women finally won the vote in most Western countries in the first quarter of the 20th century, with Australia leading the way in 1901, though Spanish women were not enfranchised until 1977. Women in a number of Islamic countries are still unable to vote. Women's representation in national assemblies remains very low: in more than half the world's legislatures it is less than 10 per cent.

Electoral corruption and abuse may be controlled in a number of ways. The most important of these is the secret ballot, first introduced into Western European parliamentary systems in the 19th century. The drawing of electoral district boundaries in a way that favors one party over another (malapportionment or gerrymandering) is difficult to eradicate. Other abuses, such as bribery or intimidation of voters, also occur, particularly in some Third World countries where representative politics is in its early stages or – as in Central and South America – is closely controlled by the military.

ELECTORAL SYSTEMS

In 1987 the *Economist* identified only 39 countries in the world in which parliamentary elections were regular, freely contested and genuinely competitive. The return of a number of countries in South America to civilian rule, as well as the move toward multi-party systems in Eastern Europe, means that this number is likely to grow in the 1990s, though other countries such as Fiji may well have been eliminated.

Plurality system

Bahamas, Barbados, Botswana, Britain, Canada, Fiji, India, Italy, Jamaica, Japan, New Zealand, Papua New Guinea, Solomon Islands, Trinidad and Tobago, United States

Preferential system or PR

Australia, Austria, Belgium, Colombia, Costa Rica, Cyprus, Denmark, Dominican Republic, Ecuador, Finland, France, Greece, Iceland, Ireland, Israel, Luxembourg, Netherlands, Norway, Portugal, Spain, Sweden, Switzerland, Venezuela, Germany

The machinery of democracy A locked ballot box of spoiled voting papers is carried from the count during the Philippines presidential election in 1986. Although the defeated President Marcos attempted to falsify the results, international observers had been on hand to make sure the election was properly conducted and expressed the true will of the people.

Careful scrutiny of electors is necessary in all elections to make sure that people do not vote twice, or send others to vote in their place. This Chilean voter has been required to produce his identity card and is thumbprinted before registering his vote in the 1988 national referendum. Most states in the world have universal suffrage; in a few of them voting is compulsory.

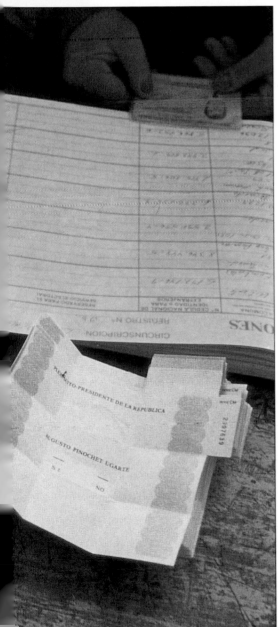

REFERENDUMS

Elections are mostly a form of indirect democracy, in that they produce representatives who make policy on behalf of the people. In complex modern societies it is usually considered impossible for the people to make policy directly. Yet referendums – when the entire electorate is asked to vote on a specific issue – are used in some countries to supplement, or occasionally constrain, the decision of elected representatives.

National referendums are rare and are usually used to settle constitutional questions such as Britain's entry into the European Community in 1973.

Local referendums are more common, and are particularly important in Switzerland and in several states in the United States.

By contrast, referendums have also been a popular device among dictators as a means of justifying their rule. This use of direct democracy has less to do with public participation in policy-making than with exploitation of the popular fear of instability and violence. But not all referendums produce the result the government campaigns for: in 1988 the Chilean people voted against President Pinochet's request for a further term in office.

Different voting systems

In liberal democracies elections are generally organized in one of two ways. In the plurality, or "first past the post", system used mostly in the English-speaking democracies – Britain, Canada, New Zealand and the United States – representatives are elected by obtaining the highest numbers of votes in their district.

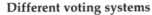

Knocking on doors to give out leaflets to voters has for years been the familiar pattern of election campaigns in liberal democracies, but these methods are becoming outmoded. Election campaigning has been revolutionized by television, and advertising agencies and "image experts" are replacing old-style campaign managers.

If there are more than two candidates in any district, the winner may well poll less than 50 per cent of the total votes cast.

Other voting systems seek to avoid "minority winners". One way is to require electors to rank candidates in order of preference. If nobody wins 50 percent of the vote in the first preference count, lower preferences are counted. Alternatively, a second, "run-off" ballot may be used if no candidate secures a majority first time round. However, most liberal democracies use a form of proportional representation (PR). This uses multi-member districts so several candidates are elected, reflecting the variety of voters. The most common method used is to vote for party lists. The aim of all types of PR systems is to ensure that the number of seats gained by each of the parties should be closely proportional to their nationwide vote.

Supporters of PR argue that it is the fairest electoral system as the legislature directly reflects the distribution of party voting. Opponents maintain that it produces weak coalition governments, as the system makes it more difficult for any one party to achieve a majority.

Elections in single-party states are a simpler affair. In the former Soviet Union, for example, electors were required to strike out all names except that of the candidate for whom they were voting. Until 1989, all candidates had to be Party members. Although multiple candidates were allowed in theory, in practice electors were given the choice of only one candidate. Only a tiny percentage (0.06 percent in 1984) failed to endorse the Party's nominees.

Unstable Governments

POLITICAL INSTABILITY, MARKED BY THE frequent coming and going of governments, often in circumstances of violence and upheaval, affects countries in many parts of the world. It does not necessarily always produce extreme political change: Italy, for example, had 43 governments between 1945 and 1985, but these frequent changes took place within a relatively stable political system. Bolivia, on the other hand, has had more than 200 governments in its history of under 200 years, and many of these have been accompanied by radical alterations to Bolivia's political structure.

Since 1945 Western industrial countries with liberal democratic forms of government have generally been very stable. The communist states of Eastern Europe also experienced a long period of relative stability until the upheavals of the late 1980s. It is in the Third World that political instability has been most common, though it is not the invariable rule. Kenya, Mexico and Tanzania with their one-party governments, Costa Rica, India and Jamaica with their multi-party systems, and Saudi Arabia with its intensely traditional autocratic regime are all examples of stability in the Third World.

The causes of instability

Individual circumstances determine every case of political instability, but some general causes can be found. Ethnic or national conflicts produce highly volatile politics, which may easily erupt into civil war; rivalries between the Hausa and Ibo peoples of Nigeria, for example, led to a damaging conflict between 1967 and 1970 when an attempt was made to create a separate Biafran state. However, ethnic mix does not always produce political instability, as the example of the United States shows.

The prosperity of a country is also an important factor; it often follows that the poorer the country, the greater the probability of unstable government. The effects of modernization create higher expectations among the people that the economy cannot meet. Extreme inequalities of wealth can also lead to a sense of deprivation and social frustration among most of the community, which creates the conditions for instability. This chain of events is typical of the precarious governments in South America, where only 11 governments were reelected in the 43 elections held between 1945 and 1986.

A portrait of the Shah, Iran's dynastic ruler, ablaze during the revolt against his Westernizing policies that led to his fall in 1979. An Islamic republic was set up amid disorder as strict Muslim laws were imposed.

In Haiti and the Philippines, stability was only gained through the iron rule of dictators, but once they were forced to flee by the surge of popular protest against them, the countries were plunged into political chaos. Ideological revolutions, such as took place in Cuba in 1959, Iran in 1979, and Nicaragua in the same year, can also lead to political instability. This is the result of imposing a new form of government suddenly on a society previously molded by completely different power structures.

Outside influences may affect internal stability. The dependence of most Third World countries on First World markets tends to produce inherently weak governments, since the economy is effectively beyond their control. These governments have little to offer their citizens, and are consequently unpopular and prone to defeat. Governments of richer countries can organize their own economies and "buy off" dissent by ensuring the material wealth of their people; poorer countries, however, often see coercion as the only method of controlling discontent.

Domestic politics may also be destabilized by the direct interference of unfriendly external powers. Usually it is done covertly and remains undiscovered. However, the United States is acknowledged to have used such tactics in Central and South America, particularly in Chile, El Salvador and Nicaragua, as was the Soviet Union in many African states.

MILITARY REGIMES

When political instability is present the army is often regarded as the only sector of the state able to maintain order. About half the countries of the Third World have military governments at any one time – an indication of their turbulent politics. Sometimes military governments alternate with civilian ones; the military takes control after the civilian government has failed, but fails in its turn. Ghana has had three elected civilian governments and three military governments since independence, a sure sign of deep-rooted instability.

At the simplest level a military regime installs itself when a minority of dissatisfied army officers and soldiers seizes power. This was the nature of Idi Amin's coup in Uganda in 1971. Pinochet's coup in Chile in 1973 was more politically motivated; the military swung into action to prevent the elected (Marxist) civilian government carrying out its policies. Other military governments claim to be above politics, and justify their seizure of power on the grounds that the country must be rescued from the disastrous policies of factious civilian politicians. Nigeria and Turkey have had military coups of this type.

Some military governments style themselves revolutionary, sweeping away anachronistic government; for example, Egypt in 1952, Peru in 1968, Libya in 1979, and Ethiopia and Portugal in 1974. Military regimes are inherently neither conservative nor radical, and may range in ideology from extreme antisocialism to anticapitalism. Their power, however, always derives from force, or the threat of it.

Under the rule of the army Instability has been a constant feature in South America's political history, and nearly all the states of the region have had some experience of military government this century.

A supporter of "people's power" in the Philippines tells President Marcos and his wife Imelda to go in 1986. His successor, Corazón Aquino, had to overcome resistance from the army and other quarters.

POLITICAL ASSASSINATIONS

The heads of state in the following list have all been murdered in office. Most assassinations are the work of religious or political fanatics, though conspiracy theories at the time will often attempt to implicate outside influences. Although political assassination may produce short-term chaos and confusion, order and stability are restored quickly.

1948	Mohandas Gandhi	nationalist leader, India
1951	Liaquat Ali Khan	prime minister, Pakistan
1958	Anastasio Somoza	president, Nicaragua
1959	Solomon Bandaranaike	prime minister, Sri Lanka
1961	Rafael Trujillo	dictator, Dominican Republic
1963	John F. Kennedy	president, USA
1966	Hendriks Verwoerd	president, South Africa
1973	Salvador Allende	president, Chile
1975	Faisal ibn Abdul Aziz	king, Saudi Arabia
1978	Mohammad Daud	president, Afghanistan
1979	Park Chung Hee	president, South Korea
1981	Zia ur-Rahman	president, Bangladesh
1981	Anwar Sadat	president, Egypt
1984	Indira Gandhi	prime minister, India
1986	Olof Palme	prime minister, Sweden
1987	Thomas Sankara	president, Burkina
1991	Rajiv Gandhi	prime minister, India
1992	Mohammed Boutief	president, Algeria
1993	Ranasinghe Premadaza	president, Sri Lanka
1993	Melchior Ndadaye	president of Burundi
1994	Juvenal Habyarimana	president of Rwanda

Government against People

ALL STATES CLAIM THE RIGHT TO ENSURE that laws are obeyed. They may do this by imposing fines or other punishments, restricting liberty or, ultimately, by the use of force. The army, the police and the judiciary may all legitimately use force to uphold the law. Ideally they should function independently of the particular government in power, but many governments take over the established machinery of law and order to further their own ends, and some are able to stay in power only by massive coercion of their own people.

Crushed by the power of the state A lone figure places himself in front of a line of tanks as they roll into Beijing's Tiananmen Square in June 1989. The Chinese army's brutal suppression of pro-democracy demonstrations was an extreme example of a state's untrammeled use of power to coerce its own people.

Political repression
Repressive governments ensure their stranglehold over the people by curtailing political freedom and denying individual, or civil, rights. In no-party or one-party states all opposition groups are repressed by not being able to form legitimate political parties. Apartheid in South Africa systematically repressed a whole section of the community.

Governments may prevent the legal activity of hostile groups simply by banning them. Communist countries have done this to trade unions; for example, the Polish Solidarity union was banned in 1981. Dissenting individuals may incur specific penalties. During the Brezhnev years in the Soviet Union (1964–82) dissidents were committed to mental

institutions or banished to internal exile, the most famous example being Andrei Sakharov (1921–89), the Nobel Peace Prize winner who was exiled to Gorky in 1980 following his criticism of the Soviet occupation of Afghanistan.

Other regimes may restrict freedom by confining people to their homes; the South African government commonly placed antiapartheid campaigners under house arrest. People may be blacklisted from employment because of their beliefs. The notorious anticommunist campaign that was led by Senator Joseph McCarthy in the United States during the 1950s resulted in many thousands of people losing their jobs.

State violence
In some unstable states violence may become endemic. Violent repression ranges from attacks on peaceful demonstrators to torture, murder and even genocide. It may happen in all parts of the world, both rich and poor, but it is most prevalent in poorer regions, where governments tend to be unpopular, having little to offer their people in the way of

Police brutality in South Africa Thousands of people in this large squatter camp were made homeless when their houses were bulldozed in 1986. The government's massive program of forced removals was part of its apartheid policy to establish separate living areas for blacks and whites.

Suppressing political dissent Police move in to arrest a protester in front of Chile's presidential palace as water cannon are used to disperse the crowds, who had gathered to call for an end to the military rule of General Pinochet in 1988.

goods and services. In a situation of civil unrest, the need to restore law and order is often used to justify further repression. Bangladesh, Burma, Israel and Venezuela are just some countries that imposed martial law under these circumstances in the 1980s.

Taken to extremes, political violence develops into a reign of terror. This happened in Uganda under Idi Amin in 1971–78, but was most widespread in Central and South America during the 1960s and 1970s, when the army and police in a number of states found

"subversion" everywhere and punished it ruthlessly. Torture, "disappearances", assassinations and secret death squads became the fabric of everyday politics.

Brutal repression in certain circumstances escalates into the horrifying attempts to exterminate a whole racial group. The Armenians under Turkey before 1915, the Jews under Nazi Germany, the Cambodians under the Khmer Rouge, and the Kurdish minority in Iraq have been victims of this most extreme form of state violence.

Because every state is sovereign in its own territory, governments are able to coerce the population with little fear of outside interference. It is the business of the people themselves to rid themselves of a tyrant, though sometimes a neighboring state may take the law into its own hands. In 1978 Tanzanian troops helped an internal movement overthrow Idi Amin in Uganda, and in 1989 the United States moved to overthrow General Manuel Noriega in Panama. Usually such interference by an outside state is highly controversial.

All members of the United Nations are party to the Universal Declaration of Human Rights. But the continuing catalog of atrocities committed by ruling elites against their own people shows that its provisions remain more violated than observed. We are far from inhabiting a world in which no individual need fear his or her own government.

The Human Rights Movement

THE UNIVERSAL DECLARATION OF HUMAN Rights, agreed by the United Nations in 1948, provided for the first time in international law all individuals with rights that transcend a state's sovereign rights. But there is a difference in the way human rights are interpreted by Western and non-Western cultures. This difference has been bridged in the United Nations by recognizing two sets of rights: the civil and political rights that derive from the Western political tradition, and the economic and social rights that are of particular concern to much of the rest of the world.

Debate continues on the relative importance of the two forms of human rights. Western countries maintain that civil and political rights are more fundamental, in particular the individual right to life and security. Third World critics reply that civil and political rights are meaningless unless the first social and economic right, the right to subsistence, or to be free from hunger, is assured. For the majority of the world's population this is far from being the the case.

Human rights politics

The debate over human rights has sometimes been cynically drawn into international politics. Human rights were used as a pawn in the Cold War, when the West staged human rights offensives against communist regimes, denouncing their generally inferior record in observing civil and political rights.

The 1975 agreement of the Helsinki Conference, to which 35 countries were signatories, is a good illustration of some of the contradictions in this area of international politics. The Soviet Union was willing to sign the agreement because for the first time Western governments recognized existing European boundaries, and implicitly accepted the Soviet Union's domination of Eastern Europe. In return, Western governments inserted into the agreement provisions to set up groups to monitor human rights in East and West. But the Soviet interpretation of the agreement emphasized the principle of non-interference in the internal affairs of a country, while the Western interpretation held that human rights override national boundaries.

Watchdogs and campaigners

A huge worldwide movement exists to champion human rights and publicize their violation. Major international organizations such as the International Commission of Jurists, the International Committee of the Red Cross and Amnesty International are important watchdogs for civil and political rights, and international charities such as Oxfam and War on Want monitor economic and social rights. The United Nations regularly monitors human rights abuses, and in 1987 established its Convention against Torture and Other Cruel, Inhuman or Degrading Treatment or Punishment. International movements, such as the antiapartheid sports boycott of South Africa, have put powerful pressure on governments that deny human rights.

Despite all the work of human rights activists, precise information on human rights abuse is very difficult to obtain. No

---

AMNESTY INTERNATIONAL

Amnesty International is the world's foremost human rights organization. Launched in London in 1961, by the late 1980s it had over 700,000 members in more than 150 countries. It campaigns for freedom of conscience everywhere, publicizing curtailments of freedom of speech and religion, exposing cruel and degrading treatment of dissidents, and calling for the release of prisoners of conscience who have not used or advocated violence, and for fair and prompt trials for others. It has a worldwide network of "adoption groups", which each take on a limited number of prisoners of conscience and barrage the offending government with letters of protest until the prisoner is released.

Amnesty International is independent of all states and is funded entirely by individual donations and subscriptions. Its annual report documents the human rights record of every country in the world, and it also operates an "Urgent Action Network" through which members respond to immediate risks of arbitrary arrest, torture or execution. During 1987 373 Urgent Action appeals were issued on behalf of prisoners in 82 countries.

government will provide statistics on its own violations, and countries with poor human rights records render assessment more difficult by exercising strict censorship of information. Reports, which are frequently compiled at very great risk to those involved, provide estimates of the extent of human rights violations throughout the world.

A repressive government may some-times be pressured into moderating its behavior by the weight of international opinion. Damage sustained to Chile's international reputation by the continued flood of publicity about its abuse of human rights forced General Augusto Pinochet to take the first steps toward the restoration of democracy there in the late 1980s. However, China's ruthless supp-ression of the democracy demonstrations in Tiananmen Square, carried out in front of the world's television, and Nicolae Ceausescu's brutal silencing of dissent in Romania, both in 1989, were stark re-minders that there will always be some elites whose determination to remain in power overrides all other considerations.

The world against apartheid Crowds attend a pop concert in London held to celebrate the 70th birthday of Nelson Mandela, the black South African nationalist leader who was then still in prison in South Africa after more than 25 years. International protest of this kind to bring pressure to bear on governments that deny human rights is a comparatively recent development.

PHYSICAL REPRESSION

In 1987 Amnesty International reported abuses by the following countries to the United Nations Commission on Human Rights.

Country	Reports of torture or fear of torture	Summary or arbitrary executions	"Disappearances"
Brazil			•
Burma		•	•
Burundi			•
Cambodia		•	•
Central African Rep.	•		
Chad	•	•	•
Chile	•		•
China			•
Colombia	•	•	•
El Salvador	•		•
Ethiopia			
Equador	•		•
Guatemala	•	•	•
Haiti	•	•	•
Honduras	•	•	•
Indonesia	•	•	•
Israel	•		
Iran	•		•
Iraq		•	•
Kenya	•	•	•
Lebanon			•
Mexico	•	•	•
Namibia			•
Nepal	•		
Pakistan			•
Paraguay			•
Peru	•		•
Philippines	•	•	
Poland		•	•
Singapore			•
South Africa			•
Sri Lanka	•		
Surinam		•	•
Syria		•	•
Turkey			
Uganda		•	•
Venezuela			•
Zaire	•		•
Zimbabwe			•

The Cold War

THE STRUGGLE BETWEEN TWO OPPOSING political alliances – one led by the United States, the other by the Soviet Union – overshadowed international affairs after World War II and is known as the Cold War. In the 1930s both these superpowers were only on the margins of a European-dominated world order. The emergence of the Cold War signaled the start of a new world order based on the ideological differences between capitalist and communist states.

In 1946 the British statesman Winston Churchill (1874–1965) warned that an "iron curtain" was descending across Europe as the Soviet Union carved out an exclusive sphere of influence in the eastern part. The threat of a similar communist takeover in Greece and Turkey so alarmed the Western powers that in 1947 United States' president Harry S. Truman (1884–1972) announced his country's intention to oppose communism and support "freedom loving people" everywhere. Usually called the Truman doctrine, this is regarded as the beginning of the Cold War.

Over the decades the Cold War spread to every part of the world. As hostility and suspicion hardened the two sides organized themselves into several major international alliances. The North Atlantic Treaty Organization (NATO, 1949), a defensive alliance of the United States and Canada with 10 Western European countries (later increased to 14), was followed by the South East Asian Treaty Organization (SEATO, 1954) to contain communism in Asia. The communist bloc countered with the Council for Mutual Aid and Assistance (COMECON, 1949) and the Warsaw Pact (1955). Repeated crises, such as the Korean war (1950–53), the Cuban missile crisis (1962) and the Vietnam war (1964–75), were interspersed with periods of detente. For example, tension was reduced after the United States' withdrawal from Vietnam, but Soviet military intervention in Afghanistan in 1979 initiated another phase of severe antagonism.

The United States' policy of intervention in countries thought to be in danger of adopting communism, first expressed by the Truman doctrine, was later justified by the "domino theory". Neighboring countries are equated with a row of standing dominoes so that if one should fall (to communism) so, by a knock-on effect, will all the others. Although the domino theory was discredited after the United States' disastrous experience in Vietnam, it was revived in the 1980s to justify intervention in Central America, which the United States had long regarded as being within its own legitimate sphere of influence. When the Marxist-oriented Sandinista party came to power in Nicaragua in 1979, it raised the specter of communism spreading throughout Central America and reaching the United States–Mexican border.

The end of an era

Mikhail Gorbachev's rise to power in the Soviet Union after 1985, and the introduction of reforms into the communist system, marked the beginning of a series of events that would culminate in the ending of the Cold War less than a decade

later. The political revolutions that saw the dismantling of the communist regimes throughout Eastern Europe in 1989–90 led to the dissolution of the Warsaw Pact and the gradual withdrawal of Soviet troops from Eastern Europe.

The end of the Cold War in Europe was marked by the Conference on Security and Cooperation in Europe (CSCE) held in Paris in November 1990 and attended by all the NATO and former Warsaw Pact countries, at which agreement was reached over the reduction of combat weapons in Europe. Then came the collapse of communist power in the Soviet Union and its breakup into independent republics, and with it the ending of the political world order based on the confrontation between capitalism and communism.

THE NONALIGNED MOVEMENT

Not all states were drawn into the Cold War; many expressed their unwillingness to side with one bloc or the other by becoming part of the nonaligned movement. Nonalignment does not mean neutrality: neutral countries such as Sweden and Switzerland abstain from all conflicts, whereas nonaligned countries abstained only from the Cold War. The movement was first formed in 1961 at a conference of 26 states in Belgrade, Yugoslavia hosted by its president, Josip Broz Tito (1892–1980). As well as Tito, other early leaders of the nonaligned movement were President Gamal Nasser of Egypt (1918–70) and India's Prime Minister Jawaharlal Nehru (1889–1964). By the 1980s the nonaligned movement had over a hundred member states, almost all of them from the Third World.

Three main issues dominated the politics of the nonaligned movement. Protest against the Cold War was the prime issue and defined the group, but later declined in importance. The second characteristic was opposition to colonialism, though as decolonization succeeded this, too, became less important. Protest came to center on economic development and global inequalities. The ending of the Cold War and growing differences of opinion among member states made the future of the nonaligned movement as a distinct political entity increasingly uncertain.

The withdrawal of Soviet troops from Afghanistan in 1989 signified the "new thinking" in Soviet foreign policy, which saw the need to cut down on military commitments and seek reconciliation with the West.

Heralding the start of the Cold War The United States showed its determination to resist communist aggression in Europe by airlifting supplies of food to West Berlin during the Soviet blockade in 1948.

Soviet troops in the streets of Prague in 1968 The use of force to quell reforms in Czechoslovakia was justified by the Brezhnev doctrine upholding the Soviet Union's right to defend socialism wherever it was under threat.

CHRONOLOGY OF THE COLD WAR

Antagonism

1947	US agrees to aid Greece and Turkey repel communism
1948	USSR blockades West Berlin
1949	NATO formed USSR detonates nuclear bomb and ends USA's nuclear monopoly Communists take control of mainland China
1950	Korean war begins
1952	SEATO formed **Dwight D. Eisenhower elected US president**

Alternating antagonism and accommodation

1953	**Death of Soviet leader Joseph Stalin** Armistice brings Korean war to an end
1954	West Germany admitted to NATO and allowed to rearm
1955	Warsaw Pact formed
1956	USSR suppresses Hungarian uprising
1958	**Nikita Khrushchev assumes absolute control in USSR** Fidel Castro leads left-wing revolution in Cuba
1959	Khrushchev visits USA
1960	US U-2 spy plane shot down over USSR **John F. Kennedy elected US president**
1961	Berlin Wall built
1962	Cuban missile crisis brings superpowers to brink of war
1963	Nuclear Test Ban Treaty signed **Lyndon B. Johnson assumes US presidency**
1964	**Khrushchev deposed in USSR**
1965	USA commits troops to Vietnam
1968	Warsaw Pact troops invade Czechoslovakia **Richard Nixon elected US president**

Detente

1969	Nuclear Non-Proliferation Treaty signed
1972	Nixon visits China and Moscow Strategic Arms Limitation talks (SALT I) agreement signed
1973	Vietnam ceasefire agreement signed
1974	**Nixon resigns and Gerald Ford becomes US president**
1975	Saigon falls to North Vietnamese troops
1976	**Jimmy Carter elected US president**
1979	SALT II agreement signed USSR troops move into Afghanistan

Alternating antagonism and accommodation

1980	US Senate halts SALT II debate **Ronald Reagan elected US president**
1981	Intermediate Nuclear Force (INF) treaty talks begin
1982	Strategic Arms Reduction treaty (START) talks begin **Soviet leader Leonid Brezhnev dies**
1983	Reagan announces the Strategic Defence Initiative INF and START talks suspended
1985	**Mikhail Gorbachev assumes leadership of USSR**

End of the Cold War

1986	Gorbachev and Reagan meet in Reykjavik
1987	INF treaty agreed
1988	**George Bush elected US president**
1989	USSR withdraws troops from Afghanistan Berlin Wall comes down Communist rule crumbles in Eastern Europe
1990	Paris Agreement: end of Cold War
1991	**Dissolution of the USSR**

Conflict and the Arms Race

THE 20TH CENTURY MUST DESERVE TO BE remembered as a period of global conflict on a scale never known before. Europeans may celebrate the peace they have enjoyed since 1945, but major wars have proliferated worldwide, claiming an estimated 17 million lives. Some of these have been fought against colonialism, particularly in Africa and Southeast Asia. Some of them have arisen out of border disputes between neighboring states. Wars of this kind have taken place between Somalia and Ethiopia, Cambodia and Vietnam, India and Pakistan, and Iran and Iraq. Sometimes internal civil wars have incurred outside interference,

as happened in Afghanistan in 1979 when the Soviet Union intervened to install a friendly government, or in Angola and Mozambique after independence.

Most of these major conflicts, which have taken place in zones known to military and political analysts as "shatter belts", have had direct Cold War implications. A shatter belt is an area where there is no clear demarcation between the interests of the superpowers, and so a jumble of alliances and highly unstable politics arises. In the last 40 years two main shatter belts have been particularly vulnerable to major conflict: the Middle East and Southeast Asia.

Conflict in the Third World has been almost continual since 1945 in the form of independence wars, disputes over borders and internal struggles for self-determination. Superpower involvement in local disputes was a feature until the 1980s, particularly in the "shatter belts" of the Middle East and Southeast Asia.

STATES AT WAR

On a number of occasions since 1945 disputes over territory and sovereignty have flared up into major confrontation between states.

India/Pakistan	1947–59, 1965
Israel/Arab states	1948–49, 1956, 1967, 1973, 1978, 1982, since 1982
North Korea (China)/South Korea (USA)	1950–53
China/Tibet	1950
North Vietnam (China)/South Vietnam (USA)	1954–75
China/India	1962
Indonesia/Malaysia	1962–66
El Salvador/Honduras	1969
Somalia/Ethiopia (Cuba, USSR)	1977–78
Cambodia/Vietnam	1978–79
China/Vietnam	1979
Tanzania/Uganda	1978–79
Iran/Iraq	1980–88
Argentina/UK	1982
Iraq/Kuwait (US, UK, France, Saudi Arabia)	1991

In the Middle East conflict originally arose from a border dispute following the partition of Palestine into separate Jewish and Arab states in 1948. It has escalated into continuing violent confrontation. The United States is aligned with Israel and the traditional conservative Arab states, while the Soviet Union supported the more radical Arab regimes. The worst casualty has been Lebanon – a once prosperous state now almost destroyed by nearly twenty years of conflict.

Two major wars have been fought in Southeast Asia: the Korean war of 1950–53 and the Vietnam war of 1964–75. The latter – the culmination of the Vietnamese anticolonial struggle – became through the intervention of the superpowers a bloody battleground for the Cold War forces.

The nuclear arms race

The dropping of the atom bomb on Hiroshima by the United States in August 1945 marked the beginning of the nuclear age. The arms race that developed between the United States and the Soviet Union as the Cold War took hold in the

THE COLD WAR IN THE THIRD WORLD

Region	International arms trade (1982–86)			
	USA sales ($US million)		USSR sales ($US million)	
Middle East	Egypt	7,180	Syria	9,180
	Saudi Arabia	4,790	Iraq	7,170
	Israel	2,620	Libya	3,050
	Jordan	760	Afghanistan	1,970
	Kuwait	530		
Indian subcontinent	Pakistan	1,630	India	7,540
Southeast Asia	Thailand	590	Vietnam	970
East Asia	Taiwan	2,080	North Korea	1,260
	South Korea	1,300		

The international buildup in arms Egyptian surface-to-air (SAM) missiles stacked up in the desert near Ismailia on the Suez Canal. The Soviet Union and the United States both supplied weapons to Egypt at various times. In the early years of the Cold War, the arms trade was seen as a way to gain political influence in a region, but with the development of increasingly sophisticated weapons systems and high-technology combat aircraft, the commercial aspects of the trade gained greater prominence. In the mid-1980s total world military spending was about $800 billion, half of which was accounted for by the Soviet Union and the United States. The oil-rich countries of the Middle East and North Africa (Libya) were the next highest spenders.

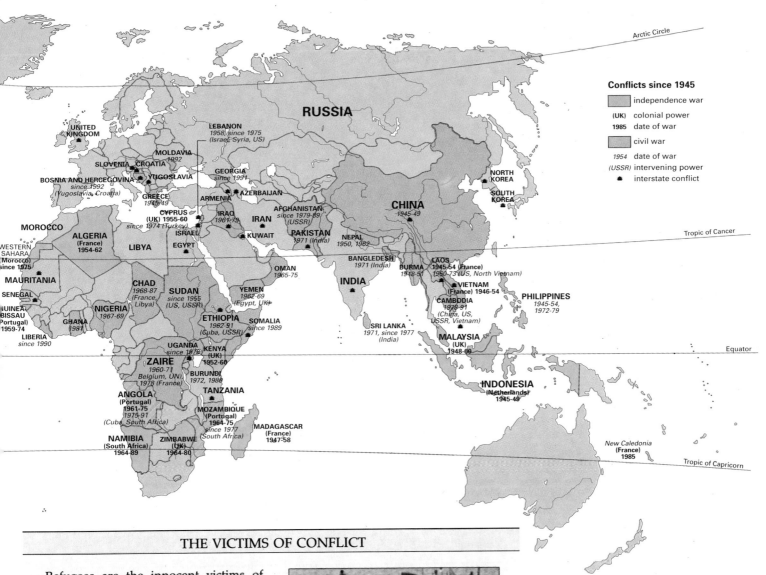

Conflicts since 1945

- independence war
- (UK) colonial power
- 1985 date of war
- civil war
- 1954 date of war
- (USSR) intervening power
- ✦ interstate conflict

THE VICTIMS OF CONFLICT

Refugees are the innocent victims of war and political conflict. Forced out of their own countries, they often have no guarantee of a new place to settle, and are obliged to live in camps that are overcrowded and prone to disease. Only a few are given the chance to rebuild their lives in another country.

The United Nations High Commission for Refugees was established in Geneva in 1951 to deal with the problems of those who had been displaced by the fighting in Europe in World War II. In 1959 it was estimated that there were 1.2 million refugees worldwide; by 1989 the Commission reported that the number had risen to 14 million.

Some refugees flee their country to avoid persecution. Such were the Jews who escaped from Germany in the 1930s, and the thousands who left Chile in the 1970s and 1980s. Communal violence may make others homeless, as in India in 1948, Croatia and Bosnia after 1991, and Somalia in the early 1990s. Others may be expelled by a hostile government, as happened to the Asians living in Uganda in the 1960s. The late 1980s defined a new kind, the economic refugee, who leaves his or her country to seek a viable life elsewhere.

The chief reason for rising refugee numbers is war. Nearly all the major

Boat people Some of the refugees who fled from Vietnam after the fall of the South in 1975.

wars – some 27 in number – in the last two decades have been in the Third World countries. The conflicts in Cambodia, Laos and Vietnam in the 1970s disrupted the lives of millions who fled to neighboring countries or took to the seas in tiny fishing boats. Three million Afghans fled to Pakistan and Iran in the 1980s. Such refugees usually hope to return home as soon as stability returns, but for some this will never be possible. The Palestinians displaced during the Arab–Israeli war of 1947–48 are still housed in refugee camps with no state to return to.

years after World War II thus gained an alarming new aspect.

An arms race develops when, in an atmosphere of tension, one state starts to build up its defenses against another. This prompts a response, and so a rising spiral develops as each side attempts to outdo the other. The most famous arms race in the past was the one that sprang up between the navies of Great Britain and Germany before World War I.

In the Cold War arms race the United States and the Soviet Union produced enough nuclear weapons to exterminate the world's population many times over. The Swedish statesman Olof Palme (1927–86) pointed out that 1,000 years of nursing care for the old could be provided for the price of just one nuclear weapon.

Another consequence of the overproduction of arms fueled by the Cold War is that it made possible the ready supply of weapons to other conflicts. The Iran–Iraq war (1980–88) was the longest and one of the most expensive wars that has taken place this century. During its course some 53 countries sold arms or related supplies to the belligerents, and 28 of these provided them to both sides.

The Politics of Accommodation

Women against the bomb Antinuclear protestors at a US base in Britain. International peace movements claimed to have influenced the new shift in attitudes.

Accommodation in practice A Soviet military observer (on the right) monitors a NATO exercise in West Germany in 1986.

PROLONGED PERIODS OF WAR AND OF confrontation, such as the world experienced in the 40 years after World War II, generate a yearning for stability and peace that is reflected in the development of policies of accommodation. Political accommodation – the recognition of differences between states – emphasizes communication and negotiation, and diplomacy provides the means for direct bargaining between governments.

In their search for stability, diplomats attempt to create international equilibrium through a balance of power. This concept came into being in the 18th century when no single state dominated the European state system. In the Cold War the balance of power has been held by two opposing political blocs, with the United States, the world's dominant economic power, at the center of one of them. The possibility of a balance of power leading to accommodation was largely thwarted by the belligerency of both sides for most of the Cold War period. In the nuclear age the balance of power has frequently resembled a balance of terror.

For the first 25 years of the Cold War there was no real balance of power, as the United States was able to use its economic prowess and military technology to remain the major military power. By the 1970s the Soviet Union had begun to catch up, and it was then that accommodation began to appear on the political agenda in the form of detente.

Detente provided the opportunity for both superpowers to reassert their position at the head of their own political bloc, and agreement between them was often achieved at the expense of other states. It is no coincidence that the major push toward detente in the early 1970s, and again at the end of the 1980s, occurred at times when both powers were feeling the effects of the arms race, which left them economically disadvantaged in relation to other states that did not have to carry their costly burden of defense.

Arms control

Central to the process of detente has been the question of arms control. In 1960 it was agreed to ban nuclear weapons, and indeed any other military activity, from Antarctica, and later treaties extended the

ban to outer space (1966) and the seabed (1970). Specific areas of the world have also been made nuclear-free zones. The first of these covered Central and South America (1967); the South Pacific became nuclear free in 1985, and there are plans to create similar zones for the Nordic countries, the Mediterranean and Africa.

In 1963, a year after the Cuban missile crisis, the Partial Test Ban Treaty, signed in the first instance by Britain, the Soviet Union and the United States, banned atmospheric nuclear testing. It was intended that this ban should later be extended to underground testing, but this has yet to be achieved. More than a hundred countries have since signed the treaty: the notable exceptions are China and France, both of which continued to test nuclear weapons in the atmosphere.

The Nonproliferation Treaty (1968) limited the number of declared nuclear powers to five: Britain, China, France, the Soviet Union and the United States. This did not prevent India being added to the nuclear club in 1974, and several other countries – notably Israel, Pakistan and South Africa – are suspected of having nuclear capability.

The 1970s saw two direct attempts to limit arms through the Strategic Arms Limitation talks (SALT I and II). These became bogged down in discussions over achieving parity of arsenals, and were the victim of a decline in East–West relations: ratification of the SALT II agreement was delayed by the United States' senate after the Soviet invasion of Afghanistan (1979).

In the improved relations that followed Gorbachev's rise to leadership came the Intermediate Nuclear Forces (INF) agreement of 1987. This was the first agreement that led to the actual destruction of nuclear arms, though it accounted for only 4 percent of the world's nuclear arsenal. Events in Eastern Europe in 1989–90 and the dissolution of the Soviet Union in 1991 led to a suspension of nuclear reduction talks, but in June 1992 a large cut in strategic nuclear arms was agreed at the first summit meeting between the United States and Russia.

Nuclear tests France (with China) refused to sign the Test Ban Treaty in 1959, and continued to carry out atmospheric tests in the Pacific. Here a Greenpeace ship monitors a French naval exercise. In 1985 the South Pacific was made a nuclear-free zone.

THE BALANCE OF TERROR IN 1987

Weapon system	Number of warheads	
	USA	USSR
Strategic nuclear forces		
Intercontinental ballistic missiles	2,290	6,900–13,000
Submarine-launched ballistic missiles	6,050	2,400–4100
Bombers	5,343	600–1,200
Antiballistic missiles		100–200
Theater nuclear forces		
Land-based systems	4,137	6,934–11,231
Naval systems	3,021	2,317

In addition, Britain had 536 warheads stockpiled, France 487, and China between 306 and 423.

THE PATH TO NUCLEAR ARMS REDUCTION

1977	USSR deploys SS-20 missiles capable of destroying major targets in Western Europe.
1979	NATO decides to modernize its medium-range nuclear weapons, replacing them with US Pershing-2 and Cruise missiles. Offer of talks to reduce all land-based medium-range missiles in Europe rejected by USSR.
1981	Soviet president Leonid Brezhnev urges freeze on deployment of medium-range missiles in Europe. US president Ronald Reagan responds with "zero-option" offer – NATO will cancel deployment only if USSR dismantles all medium-range missiles. This is rejected. Negotiations on intermediate-range nuclear forces (INF) begin in Geneva. Britain, Italy and West Germany agree to start deploying Pershing-2 and Cruise missiles in 1983.
1983	USSR threatens to put new missiles into Warsaw Pact countries if NATO deployment goes ahead. First Cruise missiles arrive in Britain. USSR suspends talks.
1985	Reagan and new Soviet leader Mikhail Gorbachev meet at Geneva summit: call for interim INF agreement.
1986	Second meeting between Reagan and Gorbachev takes place at Reykjavik. Agree outline plan to abolish INF weapons in Europe, but Gorbachev insists there must be simultaneous agreement on strategic and space (SDI) weapons.
1987	Gorbachev drops this demand but in April INF talks falter as USSR attempts to link agreement to deal over shorter-range (Pershing-1A) missiles in West Germany. *August*: West German chancellor Helmut Kohl pledges to dismantle Pershing-1A once USA and USSR scrap medium-range missiles. *September*: Gorbachev proposes moratorium on the production, testing and deployment of both medium- and shorter-range missiles. *November*: Last-minute hitch as negotiators fail to reach agreement on verification procedures. *December*: INF treaty to eliminate all intermediate-range nuclear weapons is signed by Gorbachev and Reagan in Washington DC.

Keeping the World Peace

TWO WORLD WARS THIS CENTURY SHATTERED belief that "world peace" was a realistic notion. Nevertheless, after both, international organizations were set up expressly to prevent war and promote and maintain world peace: the League of Nations after World War I, and the United Nations after World War II.

The United Nations works through the concept of collective security, which assumes that if a broad consensus can be achieved between states, irresistible pressure can be applied to any state that threatens world peace, either by imposing economic sanctions (as were used against South Africa in the 1980s) or by the application of military force. In the era of the Cold War, however, the inability to achieve consensus between the superpowers usually made the concept of collective security meaningless. Events since its demise suggest that, in a period of increased instability, the United Nations may find a more powerful role on the world stage. This was first seen in the agreement to allow a UN force to intervene to liberate Kuwait from Iraqi occupation in 1991. In October that same year the UN Transitional Authority in Cambodia (UNTAC) sponsored the peacetalks that brought to an end two decades of civil war.

Article 33 of the Charter requires all states to settle their disputes by "negotiation, inquiry, mediation, conciliation, arbitration, judicial settlement ... or any other peaceful means". States are expected to report their grievances to the Security Council, which may call an emergency meeting to set in train one of the settlement procedures. If a solution is still not reached, the dispute will be passed to the General Assembly. The United Nations becomes involved in between 30 and 40 international disputes each year.

Deployment of peacekeeping forces in trouble spots is the most visible way that the United Nations tries to prevent war. The familiar, blue-helmeted peacekeeping forces were formerly drawn from powers that were independent of the superpowers, such as Canada, India, Ireland, the Netherlands and the Nordic countries. Since 1991 they have included countries such as the Ukraine. Typically a ceasefire is arranged and the UN troops, who are under orders to fire only in self-defense, are deployed between the two sides. The role of the peacekeeping forces varies from their being observers of a truce to mediators and investigators. On one occasion they were creators of order: between 1960 and 1964 a UN force of 20,000 troops maintained the territorial

THE TERRORIST THREAT TO PEACE

Terrorism – the systematic use of unpredictable violence such as kidnapping, hijacking, assassination or bombing against governments, individuals or groups of people to attain a political objective – has been practiced throughout history. However, the 20th century has witnessed a growth and change in terrorist methods as technological advances in weaponry and evasion tactics have increased the mobility and effectiveness of terrorists.

The definition of a terrorist act depends on the observer's viewpoint: after all, one politician's "terrorist" may be another's "freedom fighter", and political sabotage and murder be regarded as a legitimate act of protest after all other methods have failed. This argument was used to justify the violent protest of the African National Congress (ANC) in South Africa, and gives legitimacy to violent campaigns against foreign occupation, such as those of the Palestine Liberation Organization (PLO) on the West Bank and of the Irish Republican Army (IRA) in Northern Ireland. Naturally enough,

the South African, Israeli and British governments vigorously dispute these justifications.

Terrorism in the 20th century has developed an international character. Terrorist groups have well-established networks of contacts with groups in other countries, often fighting for very different aims.

There is clear evidence that many governments find the covert support of political violence by disaffected groups a cheap and effective way of achieving certain foreign policy aims. The United States gave support to the Contra rebels, which it termed freedom fighters, in Nicaragua; Iran and Libya have both been accused of aiding terrorist attacks against legitimate governments. There were well-known arms links between Iran and the Nicaraguan Contras and between Libya and the IRA. Although cooperative agreements have been signed between states to combat terrorism, particular interests are more likely to determine the response of individual governments to the fight against terrorism.

integrity of the newly independent Congo (later Zaire) in a situation of political chaos. This was an exceptional use of direct military action.

Peacekeeping forces have been present in all the world's major trouble spots since World War II. They were in Kashmir after the ceasefire between India and Pakistan in 1948, and have been deployed in the Middle East in various roles since 1956. For several periods they separated Israel from Egypt and Syria, and in the 1980s were stationed in Lebanon and

between Iran and Iraq after the ceasefire. They have separated the Turkish- and Greek-Cypriot zones of Cyprus since 1964. In 1992 and 1993 they faced enormous risks in trying to safeguard the passage of supplies to communities in war-torn Bosnia. Although other international organizations such as the Organization of American States (OAS) may become involved in peace efforts from time to time, the United Nations seems likely to remain the principal vehicle for trying to keep the world peace.

Holding the front line for peace The UN and French flags fly outside a war-scarred bunker in Lebanon, where a UN force has been engaged in keeping the peace since 1978. Success has been mixed: Israeli troops swept past it when invading Lebanon in 1982, and the force was withdrawn for a period in 1989 when the situation became too dangerous.

Hijack A terrorist engages in gunpoint negotiation. Cooperative action to combat hijacking, and improved security measures, helped to reduce the number of incidents such as this. But the terrorist threat to air passengers took on a new aspect in 1988 when a bomb placed on board a Pan-Am plane exploded in midflight.

Interstate Cooperation

CONFLICTS MAKE NEWS, AS DO ATTEMPTS at accommodation between hostile powers. International cooperation, on the other hand, rarely makes the headlines. Yet thousands of international organizations are daily promoting cooperation across political boundaries. While wars and treaties capture the attention of the media, these organizations are working toward the achievement of a very different sort of world, one that places common interests above those of individual states in pursuit of an agreed aim or shared purpose.

International mail and telephone services are classic examples of the necessity for cooperation between states in the modern world. They came into being toward the end of the 19th century when increased economic and technological links between states produced the need for regulation across political boundaries. International health organizations were also among the first to appear, recognizing that disease is no respecter of frontiers. By the eve of World War I more than two hundred international organizations had been formed; these ranged from the International Committee of the Red Cross to the "Second International" of the socialist parties.

Today there are 4,500 international bodies, mostly formed since 1945, many of them under the auspices of the United Nations. They are concerned with international politics, economics and trade, education and culture, science, technology and health. While membership of the majority of them is organized through international professional associations, membership of many international organizations, both large and small, is restricted to state governments.

One of the most extensive of these is the Commonwealth, an informal association of nearly fifty independent states, accounting for a quarter of the world's population, of which the only condition of entry is former membership of the British empire. Other major intergovernmental organizations include the Arab League, the Organization of African Unity (OAU) and the Organization of American States (OAS). However, in all of these unity of common purpose is more honored in the word than in the deed.

Economic cooperation

A number of influential cooperative movements formed between states with common economic interests have met with greater success. Important among them is the Organization of Petroleum Exporting Countries (OPEC), which attempts to regulate the price of oil and coordinate policies for world oil production. Its members are drawn mainly from the Middle East, but also come from Africa and Latin America. Not every oil exporting country (for example Norway) is a member.

The European Community (EC), consisting of 12 member states, is one of the world's great trading powers; it has common policies for external customs tariffs, transport and agriculture. In 1986 it was agreed to create a single European market, to come into effect in 1992, that would eliminate existing restrictions on the circulation of capital and labor between member states. The Lomé Convention (1975) established economic cooperation with many African, Caribbean and Pacific states to allow duty-free access for their products into the Community, as well as making provision for development aid to be channeled from the Community to these countries.

Similar regional organizations exist in other parts of the world, and they are sometimes considered the first stage toward a more complete political and economic integration that will eventually render the nation-state redundant. In the vacuum left by the collapse of the communist bloc some experts foresaw a greater role for the EC with a widening of its membership to include Eastern European countries. Others feared that the examples of Yugoslavia and Czechoslovakia pointed to greater disintegration.

INTERNATIONAL SPORT

Sport is often seen as a way of promoting international understanding. This is symbolized by the Olympic Games, where "young people of the world" assemble for two weeks of friendly competition once every four years. The Olympic spirit is supposed to enhance ties of friendship between countries, but even this sporting ideal is hard to achieve. It is ironic that the Olympics, the symbol of internationalism, have become a political pawn. The Cold War affected two successive Games, in 1980 and 1984, as first the United States and then the Soviet Union boycotted them. Antiapartheid boycotts have also become a regular feature of international sport, though sports links were reintroduced in 1992.

International sport has at least as much potential for provoking national rivalry as it does for promoting friendship and understanding. One football match between El Salvador and Honduras in 1969 even sparked a "Football War", resulting in about 2,600 deaths. Sport can reinforce national prejudices and stereotypes, with results being interpreted as national disasters or victories. States often invest in sport improve national prestige, and this sort of atmosphere can lead to excesses of nationalism, which are a sad parody of the spirit of friendly competition.

MAJOR REGIONAL ORGANIZATIONS

	Date established
ALADI (Latin American Integration Association)	1981
Argentina, Bolivia, Brazil, Chile, Colombia, Ecuador, Mexico, Paraguay, Peru, Uruguay, Venezuela	(took over from LAFTA)
Headquarters: Montevideo, Uruguay	
ASEAN (Association of South East Asian States)	1967
Brunei, Indonesia, Malaysia, Philippines, Singapore, Thailand	
Headquarters: Djakarta, Indonesia	
CARICOM (Caribbean Community)	1973
Antigua and Barbuda, Bahamas, Barbados, Belize, Dominica, Grenada, Guyana, Jamaica, Montserrat, St Kitts-Nevis, St Lucia, St Vincent and the Grenadines, Trinidad and Tobago	
Headquarters: Georgetown, Guyana	
Council of Arab Economic Unity	1964
Iraq, Jordan, Kuwait, Libya, Mauritania, PLO, Somalia, Sudan, Syria, UAE, Yemen	
Headquarters: Amman, Jordan	
ECOWAS (Economic Community of West African States)	1975
Benin, Burkina, Cape Verde, Gambia, Ghana, Guinea, Guinea-Bissau, Ivory Coast, Liberia, Mali, Mauritania, Niger, Nigeria, Senegal, Sierra Leone, Togo	
Headquarters: Lagos, Nigeria	
EC (European Community)	1957
Belgium, Britain, Denmark, France, Greece, Ireland, Italy, Luxembourg, Netherlands, Portugal, Spain, Germany	
Headquarters: Brussels, Belgium	
EFTA (European Free Trade Association)	1960
Austria, Finland, Iceland, Norway, Sweden, Switzerland	
Headquarters: Geneva, Switzerland	
SADCC (Southern African Development Coordination Conference)	1980
Angola, Botswana, Lesotho, Malawi, Mozambique, Swaziland, Tanzania, Zambia, Zimbabwe	
Headquarters: Gaborone, Botswana	
SARC (South Asian Regional Cooperation Committee)	1985
Bangladesh, Bhutan, India, Maldives, Nepal, Pakistan, Sri Lanka	
Headquarters: Dhaka, Bangladesh	

Growing cooperation Ceremonial speeches in Brussels herald the entry of Britain, Denmark and Ireland into the European Community in 1973 to join the original six members. Since then Greece, Spain and Portugal have been added, and the Community has grown in influence and importance. It was poised at the beginning of the 1990s for even greater cooperation between its members.

French-speaking heads of state – including Canada's prime minister Brian Mulroney – gather for a summit meeting. France has retained economic and cultural links with most of its former colonies, based on a shared heritage and common language, and continues to arrange for the exchange of teachers and technicians.

Prestige through sport A mass gathering of athletes in East Germany, an Eastern bloc country that set out to achieve international sporting success.

A World without Boundaries

DURING AN APOLLO SPACE FLIGHT ONE OF the astronauts, looking back at Earth, experienced a simple but startling revelation. The Earth is a planet without boundaries. None of our real or imagined frontiers can be seen from space. The future of our planet depends on a deeper appreciation of this simple fact.

The United Nations is the closest we have come to a universal organization. Although some states are still not members, and many of the powerful member states conduct their relations outside the framework of the UN, it is still a major force for promoting world peace and mutual assistance through its many international agencies. During the last few decades it has sponsored a large number of conferences, open to all countries, on concerns such as world food, world population, the rights of women, disarmament, racial discrimination and environmental issues. Particular groups of countries, especially those from the Third World, have also organized conferences to discuss matters of more regional concern. The practical results of these conferences are sometimes disappointing, but they have a vital role in raising world consciousness on matters that are all too often neglected.

The environment

We live in a world with a finely balanced natural environment that we have been exploiting with scant regard for future generations. The riches of our environment are being exhausted. Forests are being cleared, fossil fuels used up, and the vital remaining wildernesses destroyed. Oceans, seas and rivers are being treated as though they were sewers with an infinite capacity, and the burning of fossil fuels is raising the level of carbon dioxide in the atmosphere to unprecedented levels – the major contributory factor to the "greenhouse effect".

Measures to protect the fragile environment of the Antarctic provide a unique

Third World support A well construction project brings water to the people of Mali.

The organization of the UN At the center is the general assembly, consisting of all member states, which supervises and finances the work of the other councils and bodies.

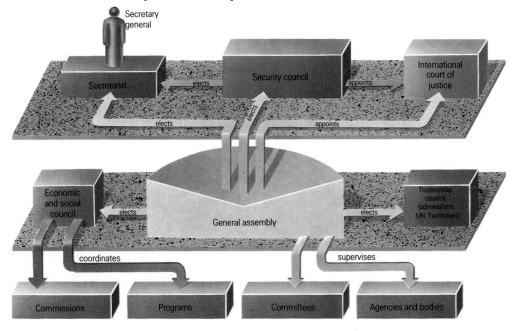

THE SPECIALIZED AGENCIES OF THE UNITED NATIONS

	Established
Food and Agriculture Organization (FAO) Investment in agriculture, emergency food supplies Headquarters: Rome, Italy	1945
General Agreement on Tariffs and Trade (GATT) Reduction of trade barriers, assistance to trade of developing countries Headquarters: Geneva, Switzerland	1948
International Atomic Energy Agency (IAEE) Aids atomic energy programs to contribute to peace and prosperity Headquarters: Vienna, Austria	1957
International Bank for Reconstruction and Development (IBRD) Provides funds to facilitate economic development in poorer member countries; known as World Bank Headquarters: Washington DC, USA	1945
International Civil Aviation Organization (ICAO) Safety and efficiency in air, international facilities and air law Headquarters: Montreal, Canada	1947
International Development Association (IDA) A leading agency administered by the World Bank Headquarters: Washington DC, USA	1970

	Established
International Finance Corporation (IFC) Encourages private enterprise in less developed countries; affiliated to the World Bank Headquarters: Washington DC, USA	1956
International Fund for Agricultural Development (IFAD) Mobilizes additional funds for agricultural development Headquarters: Rome, Italy	1977
International Labor Organization (ILO) Seeks to improve labor relations and raise living standards Headquarters: Geneva, Switzerland	1919
International Maritime Organization (IMO) Safety at sea; pollution control Headquarters: London, UK	1958
International Monetary Fund (IMF) Promotes cooperation, international trade, exchange rate stability; funds countries in need Headquarters: Washington DC, USA	1945

	Established
International Telecommunication Union (ITU) Allocates radio frequencies; promotes low tariffs Headquarters: Geneva, Switzerland	1934
United Nations Educational, Scientific and Cultural Organization (UNESCO) Promotes collaborative action to further the aims of the UN Charter Headquarters: Paris, France	1946
Universal Postal Union (UPU) Promotes the collaboration of international postal services Headquarters: Berne, Switzerland	1875
World Health Organization (WHO) Prevents the spread of disease and works to eradicate them Headquarters: Geneva, Switzerland	1946
World Intellectual Property Organization (WIPO) Protection of copyright in arts, science and industry Headquarters: Geneva, Switzerland	1974
World Meteorological Organization (WMO) Worldwide cooperation in meteorological observation Headquarters: Geneva, Switzerland	1951

One world An early UN poster points the way ahead. The growing threat to the global environment makes more urgent the realization that we all inhabit the same planet, whatever political differences divide us.

example of the kind of collective action that can be taken. The Antarctic Treaty of 1959 guaranteed Antarctica's continuing status as a nonmilitarized zone, and suspended existing territorial claims in the region, including mineral exploitation. Renewal of the treaty in 1988 extended its terms to cover nature conservation. However, many fear that once mineral exploitation of the region becomes economically feasible, these guarantees will be severely tested.

Pressure groups such as Greenpeace, which have been campaigning on environmental issues for many decades, helped to put environmental issues on the political agenda. In the 1980s environmental parties began to achieve notable

success in many countries, and ecological policies became part of many a mainline political party's platform.

More than a hundred countries were represented at the UN conference on the human environment held in Stockholm in 1972. With its theme of "Only One Earth" it made an important contribution to increasing international awareness of environmental issues. Twenty years later, in June 1992, the UN "Earth Summit" in Rio de Janeiro saw international agreement in a number of conservation goals, but many felt that not nearly enough aid was forthcoming from the richer countries for those with the worst environmental pollution problems – Eastern Europe and the former Soviet Union, and the developing

countries of Africa and Asia – to help to shoulder the enormous financial burden of dealing with them. Unsurprisingly, many Third World governments regard environmental degradation as a preoccupation that only the First World can afford to consider.

There are no national solutions to these ecological problems: the warming of the Earth by the "greenhouse effect", the threat to the planet resulting from the thinning of the ozone layer, and the problems of nuclear waste disposal are all part of the global political agenda. Finding a way to ensure international regulation and cooperation on environmental issues is the essential challenge to the world at the end of the 20th century.

REGIONS OF THE WORLD

CANADA AND THE ARCTIC
Canada, Greenland

THE UNITED STATES
United States of America

CENTRAL AMERICA AND THE CARIBBEAN
Antigua and Barbuda, Bahamas, Barbados, Belize, Costa Rica, Cuba, Dominica, Dominican Republic, El Salvador, Grenada, Guatemala, Haiti, Honduras, Jamaica, Mexico, Nicaragua, Panama, St Kitts-Nevis, St Lucia, St Vincent and the Grenadines, Trinidad and Tobago

SOUTH AMERICA
Argentina, Bolivia, Brazil, Chile, Colombia, Ecuador, Guyana, Paraguay, Peru, Uruguay, Surinam, Venezuela

THE NORDIC COUNTRIES
Denmark, Finland, Iceland, Norway, Sweden

THE BRITISH ISLES
Ireland, United Kingdom

FRANCE AND ITS NEIGHBORS
Andorra, France, Monaco

THE LOW COUNTRIES
Belgium, Luxembourg, Netherlands

SPAIN AND PORTUGAL
Portugal, Spain

ITALY AND GREECE
Cyprus, Greece, Italy, Malta, San Marino, Vatican City

CENTRAL EUROPE
Austria, Germany, Liechtenstein, Switzerland

EASTERN EUROPE
Albania, Bosnia and Hercegovina, Bulgaria, Croatia, Czech Republic, Hungary, Macedonia, Poland, Romania, Slovakia, Slovenia, Yugoslavia (Serbia and Montenegro)

NORTHERN EURASIA
Armenia, Azerbaijan, Belorussia, Estonia, Georgia, Kazakhstan, Kirghizia, Latvia, Lithuania, Moldavia, Mongolia, Russia, Tajikistan, Turkmenistan, Ukraine, Uzbekistan

THE MIDDLE EAST
Afghanistan, Bahrain, Iran, Iraq, Israel, Jordan, Kuwait, Lebanon, Oman, Qatar, Saudi Arabia, Syria, Turkey, United Arab Emirates, Yemen

NORTHERN AFRICA
Algeria, Chad, Djibouti, Egypt, Ethiopia, Libya, Mali, Mauritania, Morocco, Niger, Somalia, Sudan, Tunisia

CENTRAL AFRICA
Benin, Burkina, Burundi, Cameroon, Cape Verde, Central African Republic, Congo, Equatorial Guinea, Gabon, Gambia, Ghana, Guinea, Guinea-Bissau, Ivory Coast, Kenya, Liberia, Nigeria, Rwanda, São Tomé and Príncipe, Senegal, Seychelles, Sierra Leone, Tanzania, Togo, Uganda, Zaire

SOUTHERN AFRICA
Angola, Botswana, Comoros, Lesotho, Madagascar, Malawi, Mauritius, Mozambique, Namibia, South Africa, Swaziland, Zambia, Zimbabwe

THE INDIAN SUBCONTINENT
Bangladesh, Bhutan, India, Maldives, Nepal, Pakistan, Sri Lanka

CHINA AND ITS NEIGHBORS
China, Taiwan

SOUTHEAST ASIA
Brunei, Burma, Cambodia, Indonesia, Laos, Malaysia, Philippines, Singapore, Thailand, Vietnam

JAPAN AND KOREA
Japan, North Korea, South Korea

AUSTRALASIA, OCEANIA AND ANTARCTICA
Antarctica, Australia, Fiji, Kiribati, Nauru, New Zealand, Papua New Guinea, Solomon Islands, Tonga, Tuvalu, Vanuatu, Western Samoa

North America

CANADA AND THE ARCTIC

THE UNITED STATES

CENTRAL AMERICA AND THE CARIBBEAN

SOUTH AMERICA

Central and South America

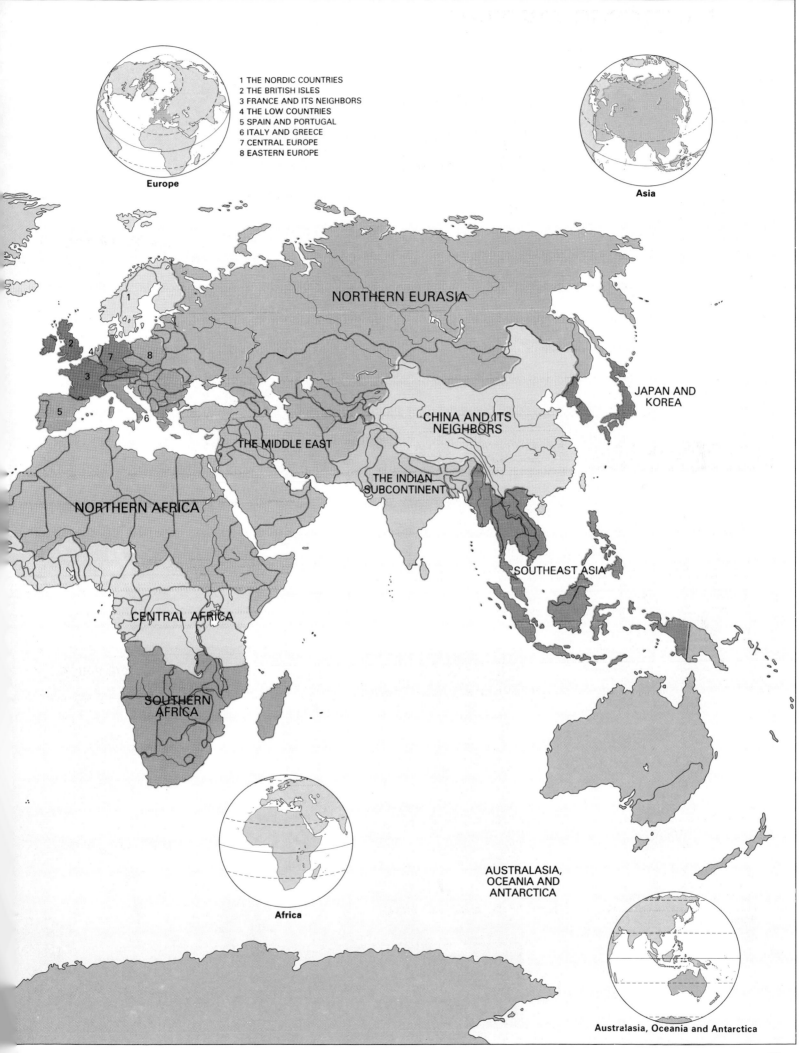

1 THE NORDIC COUNTRIES
2 THE BRITISH ISLES
3 FRANCE AND ITS NEIGHBORS
4 THE LOW COUNTRIES
5 SPAIN AND PORTUGAL
6 ITALY AND GREECE
7 CENTRAL EUROPE
8 EASTERN EUROPE

Europe

Asia

NORTHERN EURASIA

JAPAN AND
KOREA

CHINA AND ITS
NEIGHBORS

THE MIDDLE EAST

THE INDIAN
SUBCONTINENT

NORTHERN AFRICA

SOUTHEAST ASIA

CENTRAL AFRICA

SOUTHERN
AFRICA

Africa

AUSTRALASIA,
OCEANIA AND
ANTARCTICA

Australasia, Oceania and Antarctica

A STABLE DEMOCRACY

WHAT IS A CANADIAN? · A PARLIAMENTARY FEDERAL SYSTEM · THE CANADIAN PERSPECTIVE IN NORTH AMERICA

Two European nations – Britain and France – were the earliest colonizers of Canada. Rivalry between them was great from the 17th century onward. Britain eventually acquired all France's possessions, and in 1791 the territory was divided into English-speaking Upper Canada and French-speaking Lower Canada. As the provinces of Ontario and Quebec they formed the federal Dominion of Canada in 1867, with Nova Scotia and New Brunswick. Westward expansion was rapid, reaching across the continent when British Columbia joined the Dominion in 1871. By 1905 four more provinces had been added; Newfoundland (a British colony since 1583) voted in 1949 to become the tenth. There are also two national territories, the Yukon Territory and the Northwest Territories, administered by the federal government.

COUNTRIES IN THE REGION

Canada

Dependencies of other states   Greenland (Denmark); St Pierre and Miquelon (France)

STYLES OF GOVERNMENT

Monarchy   Canada

Federal state   Canada

Multi-party state   Canada

Two-chamber assembly   Canada

CONFLICTS   (since 1945)

Nationalist movement   Canada (Quebec)

MEMBERSHIP OF INTERNATIONAL ORGANIZATIONS

Colombo Plan   Canada

North Atlantic Treaty Organization (NATO)   Canada

Organization for Economic Cooperation and Development (OECD)   Canada

Note: Greenland remains part of the Danish realm but acquired self-governing status in 1981.

WHAT IS A CANADIAN?

Canada's sheer size always presented something of an obstacle to national integration. Most Canadian citizens live within 320 km (200 mi) of the border with the United States, the world's longest undefended boundary. Regional interests have always tended to override wider national interests; the western provinces in particular have traditionally looked toward the United States not only for economic markets and investment but also for cultural ideas.

Rail links across the continent were built with the political aim of welding the western provinces more effectively into the young state structure, and to counter the northernward thrust of commercial interests from the United States. In the event, however, their construction had the effect of consolidating the political and economic dominance of the central provinces of Quebec and Ontario over the small, less developed maritime provinces in the east and the sparsely populated, though increasingly rich, western prairie provinces.

From the beginning, Canada's federal system of government (modeled on that of its giant neighbor, the United States) took account of the need to guarantee the rights of the French Canadian community within Quebec. Constitutional undertakings were given to protect the French language, the Roman Catholic church and the French tradition of civil law. This enabled a well-developed sense of ethnic community to survive within French Quebec, which expressed itself politically in periodic demands for either a renegotiated position within the confederation or secession from it.

A diverse population

The indigenous Inuit (once known as Eskimo) and Amerindian population probably numbered about 220,000 before the arrival of European settlers. It declined considerably afterward, until improved medical care and living conditions raised levels again in the present century. Growing protest led to increased opportunities for Amerindians – particularly those living on reserves (to whom the franchise was extended only in 1960) – to influence decisions affecting their own lives. The Nunavik district of northern Quebec has been designated as a homeland for the Inuit people.

During the initial period of settlement

Pierre Trudeau's colorful personality made him highly popular; he was elected to office four times between 1969 and 1984. His premiership saw a number of important political changes. Equal status was given to French and to English, and his vigorous defense of federalism defeated Quebec separatism.

Canada, who argued that bilingualism was an eastern preoccupation, and that German or Ukrainian had a greater claim than French to be made the second official language of the prairie provinces.

Efforts were therefore made to place Canada's bilingual and bicultural policies within a multicultural framework that would recognize the permanent and positive contribution of other minority ethnic groups to Canada's national life. Nonetheless, regional and ethnic tensions made themselves felt in the opposition to the Meech Lake constitutional accord proposed by Prime Minister Brian Mulroney, which among other things asserted that Quebec is a distinct society within Canada. The failure of all ten provinces to ratify the clauses of the accord in June 1990 was prompted by the desire of some of them to wrest a number of further concessions for their own ethnic and indigenous minorities.

In 1992, new constitutional proposals put forward by the government to recognize Quebec's distinct status were rejected in a national referendum, amid strengthening demand in Quebec for separatism from the rest of Canada. The debate about national identity, regional competition and the division between federal and provincial authority – carried on in a more or less harmonious fashion – showed little sign of abating.

in Canada until World War I, the population grew from 3 million to just over 7 million. Immigrants, who came mostly from Britain, northwestern Europe and the United States, settled first in eastern Canada, but efforts were later made to encourage settlement of the grasslands of the prairies to the west, particularly by farmers and others from central Europe. After World War II more than 4 million immigrants, many from Africa, Asia and Latin America, entered the country; they settled mainly in the urban centers of central Canada.

Most of the newcomers, who were assimilated rapidly into Canadian life, identified themselves by and large with the Anglo-Saxon, or British, cultural tradition. This heightened dissatisfaction among the French-speaking community, which came to feel that its interests were no longer central to Canadian national concerns and federal policy. As a result of increasing pressure for the separation of Quebec during the 1960s, Canada witnessed one of the most exhaustive debates on national identity ever carried out within a democratic society.

The Official Language Act
Under the premiership of Pierre Elliott Trudeau (Liberal prime minister 1968–79, 1980–84), who was a convinced federalist and opponent of separation though himself a Québécois, the Official Language Act, which made Canada a bilingual and bicultural state, was passed in 1969. The granting of equal status to English and French continued to be vigorously opposed by many people in western

The Canada Act, passed in 1982, established Canada's independence from the British parliament, which until then had the task of ratifying all constitutional changes made by the Canadian parliament. Like all legal and other official documents, it is written in both English and French.

A peaceful dominion The stability of Canada's territorial boundaries means that it has fought no major wars of its own making since the 18th century. Its great size posed problems of integration during the early years of westward expansion; the building of transcontinental railroads strengthened the dominance of Ontario and Quebec. In 1977 Canada extended its maritime limits to 370 km (200 nautical mi) to include resource-rich areas of the Arctic sea floor. These claims are disputed by the United States.

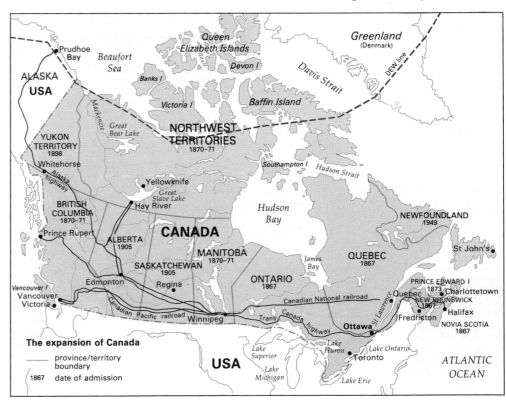

The expansion of Canada

— province/territory boundary
1867 date of admission

A PARLIAMENTARY FEDERAL SYSTEM

Canada's style of parliamentary democracy is based on the British model. The federal parliament in Ottawa, Ontario consists of a lower house (the house of commons), elected from single-member constituencies called ridings, and an appointed upper house (the senate). The head of state is the British monarch, represented by a governor-general who appoints the prime minister and cabinet from the party that has most support in the house of commons. The crown also has a representative in each province in the lieutenant governor. Each of the ten provinces has its own elected single-chamber legislature with a prime minister and cabinet. In the two territories there are locally elected councils that work with a commissioner appointed by the federal government.

Federal and provincial government
The federal government has exclusive or dominant power over external affairs, defense, commerce, finance and criminal justice. The provincial governments are responsible for education, health, social services and civil justice. Federal involvement in provincial matters has been seen to be increasing, especially with regard to the development and management of local resources.

Since World War II relations between federal and provincial government have been strained by a number of issues. These include the debate over Canadian identity and the question of Quebec separatism. A particular problem has been posed by the rapid development of the resource-rich western provinces; for example, Alberta has at times sought greater autonomy for itself, stretching the concept of federalism to its limit. By contrast, the poorer maritime provinces have attempted to secure more concessions and a greater share of federal tax revenues for themselves.

Underlying all these tensions were Canada's persistent efforts to disengage itself from political dependence on Britain. Not until 1982, with the passing of the Canada (or Constitution) Act, did Canada gain total control over its constitution (patriation). Before that date all constitutional changes agreed in the Canadian parliament had to be ratified by

the British parliament before becoming law. The new Act incorporated a charter of rights and freedoms.

Quebec – the key to electoral success
In the past the two main parties – the Liberals and the Conservatives (later the Progressive Conservatives) – received three-quarters of the votes cast in federal elections. The rest went to the third party, the New Democratic Party, and to minor parties. The party system at national level has been described as a "stable two and a half party system".

The Liberals have enjoyed long, unbroken periods in power this century, most notably from 1935 to 1957, and from

A powerful neighbor Canada's former premier, Brian Mulroney with George Bush, then vice president of the United States. Mulroney made the Free Trade Agreement with the United States an election issue in 1988 and won. In 1993 Canada's new liberal prime minister, Jean Chrétien, determined to renegotiate aspects of the Agreement arguing that free trade with Mexico had been obtained at the expense of protectionist measures against Canada.

1963 to 1979. For most of this time the Conservatives – identified as the party of the British interest – failed to win votes in Quebec. It gained fewer than ten seats here in nearly every general election from 1917 to 1958, and it was not unusual for the Liberals to have an absolute majority in Quebec while the rest of Canada favored the other parties.

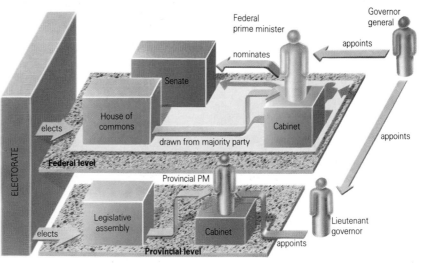

Canada's system of government has grafted the federalism of the neighboring United States onto a parliamentary system modeled on that of the British. The two-chamber federal parliament consists of the senate, whose 104 members are appointed for life, and the 282-member house of commons, elected for a maximum of five years. Legislation must be passed by both houses to become law.

The parliament building in Ottawa was built in Gothic style in homage to the British Houses of Parliament, from which its procedures were borrowed. It stands at the top of Parliament Hill dominating the city of Ottawa in southeastern Ontario, which was chosen as Canada's political and administrative capital by Queen Victoria in 1858.

The 1980s saw a reversal of this traditional arrangement. In 1980 the Progressive Conservatives held only one seat in Quebec, the Liberals 74. In the general election of 1988 the Progressive Conservatives, led by Prime Minister Brian Mulroney, won 63 seats – a truly remarkable turnabout.

The Liberal stranglehold on Quebec provided the national support base for Trudeau's two administrations, from 1968 to 1979 and from 1980 to 1984. These years saw the establishment of a strong federal government, in which the separatist demands of Quebec and regionalist pressure from the western provinces were recognized – and were met head on. Toward the end of his second period in office Trudeau's policies were more and more characterized by their vigorous defense of Canadian economic nationalism in the face of the increasing commercial pressure from the United States.

His successor, Brian Mulroney – like Trudeau, a lawyer from Quebec – turned these policies around. He became prime minister in 1984, when the Conservatives were elected with the largest parliamentary majority in Canadian history. The 1988 general election, fought on the issue of a comprehensive Free Trade Agreement (FTA) with the United States, gave the Progressive Conservatives a 45-seat majority in the house of commons (though only 43 percent of the votes cast).

However, continued economic and ethnic difficulties contributed to a drastic fall in Conservative popularity and in 1993 the Liberals presided over a crushing electoral defeat for the Conservatives who retained only two seats in the house of commons. Jean Chrétien headed the new Liberal government and immediately announced his intention to renegotiate aspects of the Free Trade Agreement in order to secure better terms for Canada, in particular to reduce the likelihood of punitive trade sanctions.

THE PROVINCIAL LEVEL OF GOVERNMENT

Canada's ten provincial governments mirror in structure its federal government. Each lieutenant governor is appointed by the federal cabinet on the advice of the prime minister; they represent the crown by summoning and dissolving the provincial assemblies, and by assenting to provincial legislation. It is an important and potentially powerful office, though in practice no lieutenant governor would act against the advice of the provincial premier or cabinet except in extraordinary circumstances.

The legislative assemblies and their various governmental departments are supported by a burgeoning civil service. This is responsible for providing services at provincial level, and for staffing provincial government enterprises, boards and agencies. With the employees of school boards, municipalities and hospitals they add up to some 75 percent of all Canada's public employees. Provincial government thus touches closely on the lives of very many Canadians. A large number of people, particularly regionalists and other disaffected citizens, have a greater sense of identity with it than they do with federal government, which is frequently perceived as a distant, intrusive element in their lives. Several provincial governments have sought ways of directly involving their citizens in the formulation of policy.

Vast differences exist in the capacity of individual provinces to harmonize national standards of health care and the provision of education and social services. Prince Edward Island, for example, with a total population of only 128,000, can hardly be compared with the two largest provinces: Ontario (9 million) or Quebec (6.5 million).

The provincial government may be formed by a different party from the majority party representing the province in the federal parliament. Some sophisticated electoral behavior may be involved in this. For example, in the period after 1976 voters in Quebec chose the autonomist Parti Québécois at provincial level, but the Liberals at federal level. The third national party, the New Democractic Party, also performs well in provincial elections, and has formed governments in British Columbia (1972–75), Saskatchewan (1944–64) and Manitoba (1969–77, 1981).

THE CANADIAN PERSPECTIVE IN NORTH AMERICA

The single most important factor that affects Canada's external relations is the presence of the economic giant of the United States to the south. Wary of being drawn into its continental influence, Canadians have at the same time been irresistibly attracted by its cultural and material dynamism. Canada and the United States are each other's largest trading partners in goods, services, capital and manpower: three-quarters of Canada's trade is with the United States. But the exchange is clearly unequal, as the United States' economy is ten times larger than that of Canada. To capitalize on this degree of economic dependence the Free Trade Agreement (FTA) was signed with the United States in 1988 (and ratified by the Canadian parliament in 1989).

Defense strategies

Since the end of World War II Canada has been a partner in United States' military defense policies as well as in the economic sphere: it occupies a vital strategic position between the United States and the Soviet Union over the polar ice cap. Under the North American Defense Agreement (NORAD) of 1957, renamed the North American Aerospace Defense Command in 1981, the United States and Canada's air defense forces are integrated in a joint command system with head-

A radar station on the Distant Early Warning (DEW) line, which runs from Alaska across northern Canada was a major component of NORAD's defenses to protect the North American continent against Soviet long-range missile attack.

quarters at Colorado Springs, Colorado. A Canadian officer always serves as the deputy commander.

In 1949 the Canadian government supported the United States' proposal for an alliance of Western European and North American powers – the North Atlantic Treaty Organization (NATO) – that would

Working for peace between nations Since 1945 Canadian forces have served with the United Nations in many trouble spots in an effort to reconcile warring factions, and Canada has been a strong supporter of UN initiatives to aid the Third World.

provide for the collective defense of its member states from the perceived threat of Soviet domination in Eastern Europe. At Canada's insistence, Article 2 of the treaty (sometimes called the "Canadian article") gave a political and economic, as well as a military, dimension to the alliance. However, it has never been put to any significant use.

The discovery of potentially rich oil reserves in the Arctic Ocean reactivated Canada's claims to sovereignty over the icebound Arctic archipelago and all waters within 370 km (200 nautical mi) of the Canadian coastline, the counter interest of its southern neighbor and key defense partner notwithstanding. The United States recognizes Canada's coastal jurisdiction as extending over only a 20 km (12 mi) zone. The growing pollution of these Arctic waters is an enduring concern of the Canadian government.

In the past, Canada's historical links with Britain also helped to shape its external policies. Until World War II Britain and the United States frequently alternated as Canada's first and second most important trading partners. Canada now sends less than 5 percent of its exports to Britain, which accounts for just over 3 percent of its total imports. Japan has replaced Britain as Canada's second largest trading partner. Although most Canadians still regard Britain as an important part of their national heritage, the link has been given less and less political expression since the patriation of the Canadian constitution in 1982. In the same period closer ties have been nurtured with the worldwide association of French-speaking states.

An independent actor

Canada has attempted to distance itself from its image of being merely a northern outpost of the United States. Building on its connections with the Commonwealth and the community of French-speaking states it has sought a role as a middle-ranking power with global presence, prepared to take a responsible role in the resolution of worldwide conflicts. In the frequently uneasy period of decolonization after World War II troops and resources were committed to the joint security missions of the United Nations, and more recently Canada has been a powerful advocate of Third World needs.

Canada's strong commitment to the international community was initially

A base at Thule, on Baffin Bay in northern Greenland, is shared with the United States and forms part of the Arctic line of defense. Greenland is a self-governing province of Denmark, which is a member of NATO.

fostered by Lester Pearson (1897–1972), as foreign minister from 1948 to 1957 and as prime minister from 1963 to 1968. Pierre Trudeau subsequently developed these policies by establishing closer relations with many Third World states. In the early 1970s Canada supported the application of the People's Republic of China for a seat in the UN, and arranged wheat sales at preferential prices to ward off the worst ravages of the Chinese famine.

Successive governments since then have championed the cause of the developing nations of the South in the North–South dialogue by arguing for radical change in international terms of trade. Canada has continued to undertake humanitarian relief programs and to grant asylum to political refugees. It has promoted human rights campaigns to secure individual freedom as well as environmental programs aimed at ensuring global survival.

CANADA'S ARCTIC FRONTIER

In August 1985 the US Coastguard icebreaker *Polar Sea* sailed through the Northwest Passage. Its voyage through the strait, recognized by the United States as an international waterway, was in defiance of Canada's claims to sovereignty and produced a surge of Canadian national pride, leading to a series of measures aimed at strengthening Canada's hold in the Arctic.

The Arctic assumed significance as a bulwark against long-range missile attack across the polar ice cap at the onset of the Cold War period of East–West hostility in the late 1940s and 1950s. Awareness of the area's strategic importance was publicly expressed in the 1947 agreement with the United States to establish five joint weather stations in the Arctic islands, the construction of the Distant Early Warning (DEW) Line and Mid-Canada Line between 1955 and 1957, and in Canada's continuing commitment to its NATO and NORAD obligations.

However, Canadian public opinion was more consciously concerned with the activities of its southern neighbor in the area, and recent recognition of the Arctic's vast natural resource potential translated a rather passive defense policy into a fully fledged assertion of national sovereignty.

The *Polar Sea* incident showed that Canadian claims could not be enforced against challenge by the United States. Implementation of the planned measures to be taken to extend Canada's jurisdiction in the eastern Arctic included the acquisition of a force of between 10 and 12 nuclear-powered submarines to carry out sustained operations beneath the ice. In all, the package would require a growth in defense budgets of 2 percent in real terms for the next 15 years, and many questioned whether these targets could realistically be met.

With the ending of the Cold War the Canadian Arctic ceased to be a superpower buffer zone. This focused attention more sharply on the sovereignty issue, as the Canadian government finds ways to develop the area's resource potential while seeking to safeguard its environment.

Quebec: the French factor

The French colonization of Quebec took place in the first half of the 17th century, when some 10,000 French people came to settle on the shores of the St Lawrence river. After 1763, when the colony of New France was lost to the British, antagonism between the numerically dominant French-speaking and the economically dominant English-speaking populations was high. Quebec has fought hard to maintain itself as a distinct society.

Today about 80 percent of Quebec's population is French-speaking. In Montreal, Canada's leading banking city, where nearly half of Quebec's population lives, the figure drops to 68 percent. Historically most economic activities in Quebec were controlled by English speakers. It was the exclusion of French Canadians from economic and political life that kept alive the cause of Quebec nationalism.

In the early 1960s Quebec's Liberal prime minister, Jean Lesage (1912–80), conceived and directed the "quiet revolution", which successfully strengthened the powers of Quebec's provincial government by attacking the control of the Roman Catholic church hierarchy over education and the professions. Quebec was revolutionized and modernized. Social attitudes changed markedly, and a new generation of teachers, scientists, businessmen and accountants chose to use the French language, rather than English, in schools and universities, in the workplace and elsewhere.

Revived nationalist aspirations in Quebec, the cultural cradle of the French tradition in Canada, seemed to beckon in two directions. There were those like Lesage, who favored gradual social reform and language legislation – a policy summarized by the term *rattrapage*, to catch up – while remaining within the federal system. Others, led by René Lévesque, a former cabinet minister in Lesage's government, came to seek separate sovereign status for Quebec.

In 1967 Lévesque split with the Liberals to found the separatist Parti Québécois (PQ). As Lesage's revolution seemed to falter, the popularity of the PQ increased. It won power in 1976 with a program to give Quebec self-governing sovereign status while maintaining close economic ties with Canada (Sovereignty–Association). Although rejected by 60 percent of voters in a referendum in May 1980, the PQ was re-elected in 1981. Three bills

STOP
ARRÊT

passed by the Liberal government in the 1970s had strengthened French language rights. The PQ took these cultural policies even further, but alienation of the English-speaking minority and of many New Canadians, coupled with the threat of economic recession, led to their defeat by the Liberals in 1985.

In the late 1980s attention centered on the Meech Lake accord, which proposed assigning a special constitutional position to Quebec as a "distinct society" within Canada. Further extension of French language rights included a measure that all public signs on the outside of buildings should be in French only. This provoked a strong response from English-speakers, who argued that their right to their own language was being eroded within Quebec. Controversy over the accord brought the threat of Quebec's secession into sharp focus again. In 1992 a nationwide referendum to endorse Quebec's special status within Canada was rejected by a majority of Canadians. Those in Quebec voting against sought a greater degree of separation; many of those outside, particularly in the western and maritime provinces, wanted the same recognition for themselves.

Quebec's success

Quebec's language policies and its attempts to differentiate itself from other parts of Canada are often interpreted as cultural perversity and intransigence that

fly against current trends, but Quebec today is one of the most progressive and modern of Canadian provinces. It has an active energy and resource development program, based on its hydroelectricity, a fine record on social welfare, pension rights, medical provision and education, and a confidence in self-directed social change that others might envy.

Quebec has almost come to act as a sovereign state on the international stage. It has championed the rights of French speakers throughout the world, and has materially furthered both French culture and Canadian economic development. By challenging the Anglo-conformist basis of Canadian statehood, Quebec strengthened the bilingual and bicultural character of Canadian society, moving Canada farther away from its British past as well as from the assimilationist trends emanating from the United States.

Reevaluation of the cultural bases of Canadian identity also helped to redefine the role of its citizens of non-British and non-French ethnic origin, particularly Native Indians, Canada's policy of multiculturalism within a bicultural context offers a promising means to achieve state harmony without destroying group cultural identity. The majority of Canadians may not have realized it at the time, but Quebec's challenge in questioning the very foundation of Canadian statehood was the catalyst that produced this more open and honest approach.

WE THE PEOPLE...

The United States of America came into being when thirteen colonies on the eastern seaboard of the North American continent revolted against British rule (the War of Independence, 1775–83). The new nation won control of the land between the Atlantic and the Mississippi river north of Florida. This was more than doubled when Louisiana was purchased from the French in 1803, and Florida won from Spain in 1819. Westward expansion across the Rocky Mountains soon followed. Texas joined in 1845, and other territories in the southwest – then belonging to Mexico, but once Spanish – including California, New Mexico and Utah, were acquired by 1851. The 49th parallel of latitude was fixed as the boundary with Canada in 1848, Alaska was purchased from Russia in 1867 and Hawaii, in the Pacific Ocean, annexed in 1898.

COUNTRIES IN THE REGION

United States of America

Territories outside the region American Samoa, Guam, Johnston atoll, Midway Islands, Northern Marianas, Puerto Rico, US Trust Territory of the Pacific Islands (Marshall Islands, Micronesia, Palau), US Virgin Islands, Wake Island (USA)

STYLES OF GOVERNMENT

Republic USA

Federal state USA

Multi-party state USA

Two-chamber assembly USA

CONFLICTS (since 1945)

Overseas intervention USA/Cuba 1961; USA/Dominican Republic 1965–66; USA/Nicaragua 1981–90; USA/Grenada 1983; USA/Libya 1986; USA/Panama 1989

Interstate conflicts USA/North Korea 1950–53; USA/North Vietnam 1965–73; USA (heading UN force)/Iraq 1991

MEMBERSHIP OF INTERNATIONAL ORGANIZATIONS

Colombo Plan USA

North Atlantic Treaty Organization (NATO) USA

Organization of American States (OAS) USA

Organization for Economic Cooperation and Development (OECD) USA

Note: The USA is composed of 50 states, including Alaska and Hawaii.

"FROM SEA TO SHINING SEA"

No sooner did the United States' territorial sovereignty extend right across the North American continent than internal conflict threatened to split the new nation. The southern – or Confederate – states (Alabama, Florida, Georgia, Louisiana, Mississippi, South Carolina and Texas, joined later by Arkansas, Tennessee, North Carolina and Virginia) resisted the attempts of the northern (Federal) states to limit their autonomy. In particular, the southern states refused to abolish the use of black slaves, originally brought from Africa to work on their sugar and cotton plantations, on which their agrarian economy was based. Instead they claimed the right to secede from the United States.

Northern victory in the resultant Civil War (1861–65) strengthened the economic domination of its rapidly industrializing cities, and speeded up the construction of the roads and railroads that had helped to make the United States the world's leading industrial power by the end of the century. The war left the South impoverished and embittered. More than a century later it remains politically, economically and socially distinct

Following the war, the western half of the country was settled rapidly, and by 1912 all 48 states on the continent south of the 49th parallel had been incorporated into the federal structure of the United States. (Alaska and Hawaii became full states in 1959.) In the process of settling the west large number of indigenous Americans – or Indians – were displaced or exterminated. By 1900 most of them were restricted to reservations. Today approximately 2 million live on reservation lands or in the cities. They are the least integrated group in society. While they now enjoy full legal and political rights, many live in deep poverty.

Black equality

Although slavery was legally abolished throughout the United States in 1865, blacks in the South did not enjoy full political rights for over a century, and racial discrimination was openly practiced. In the late 1950s and early 1960s a black civil rights movement developed, led by Martin Luther King, Jr (1929–68), a passionate advocate of nonviolence. The campaign for integration and equal rights

with whites culminated in a massive march on Washington DC, the federal center of government, in 1963.

The passing of the Civil Rights Act of 1964 and the Voting Rights Act of 1965, along with other laws, guaranteed full political rights to blacks. However, many of the 30 million blacks in the United States today suffer economic discrimination. Very many live in rundown and overcrowded inner city areas, and it is in governing the major cities that black political influence has been most strongly felt. Many cities, including Atlanta, Baltimore, Chicago, Detroit, Los Angeles, New York, Philadelphia and Washington have had, or have, black mayors.

A "melting pot" of nations?

The United States' population is among the most diverse in the world, yet the country has not had experience of the ethnically based political fragmentation found elsewhere in the world. Religious and ethnic divisions are not principally territorial, and each wave of immigrants has been absorbed relatively easily.

The early Protestant settlers from Britain and other northeast European countries brought with them values of political dissent later embodied in the constitution in the form of individual liberty and self-government. By the 19th century a major political divide had formed between the American-born Protestant skilled working class and later Roman Catholic arrivals from Ireland and Germany. It was intensified by the arrival of a further wave of immigrants, also Roman Catholic, who came from southern and eastern Europe at the end of the century. Anticatholicism subsequently became a major force in shaping political loyalties and the development of trades unions. When strict immigration controls were introduced in 1924 the national quota system devised was biased in favor of the older nations (Britain, Germany and Ireland) against the newer ones (Italy, Hungary, Poland and Russia).

The split in the working class hindered the growth of a socialist party comparable to those in Europe and Australia. It lay behind the collapse of a number of radical and labor-based movements in the late 19th and early 20th centuries, ensuring the dominance of a two-party system. Communism, largely confined to German Jews, strengthened the identification of radical labor politics with immigrants and

The Capitol in Washington DC is the country's political heart. Its cornerstone was laid by George Washington and the dome completed in 1863 at the urging of Abraham Lincoln as a sign of national unity.

THE WORLD'S FIRST WRITTEN CONSTITUTION

During the War of Independence the goal that united most Americans was simply to overthrow the British government. Once the fighting was over, the citizens of the new state were able to debate how its government was to be organized. Of particular concern was the extent of power the central government should have toward the individual states of the Union. It was probably the first time in history that the nature and function of government were so thoroughly argued and discussed.

A new form of government

In the summer of 1787 the leaders of the various states met in Philadelphia, Pennsylvania in order to revise the Articles of Confederation. This *ad hoc* form of government had been drawn up by the 13 colonies during the war, but was by now clearly outmoded. The delegates to the Philadelphia convention – men such as Benjamin Franklin (1706–90) and George Washington (1732–99) – came to be known as the framers of the constitution. What they forged among themselves was an entirely new structure of government. The constitution they drew up described the federal system of government that is the particular characteristic of the United States, in which certain powers are delegated to the national government and all other powers fall to the states. It was the first constitution ever to be written

made then vulnerable to charges of being subversive or antiAmerican.

More recently, the black civil rights struggle and the campaigns by Latinos (Hispanics) and others for such things as bilingual education and political representation have raised controversial issues of "group rights". A former director of the Central Intelligence Agency (CIA) has even stated that Latino separatism, or as he called it, "the threat of a Spanish-speaking Quebec", poses the greatest challenge to national security. Tensions continue, but the nation's political culture has proved generally adept at accommodating them.

The growth of a nation The 13 original states of the Union on the Atlantic seaboard had risen to 48 by 1912 (Alaska and Hawaii were added later) as the population flowed west in successive waves of migration; the "frontier" had ceased to exist by the end of the century. The clash of political and economic interests between the industrializing north and the 11 agrarian Confederate states of the southeast, which gave rise to the Civil War, threatened to destroy the integrity of the Union and left a lasting political division.

The United States

— state boundary

▨ Confederate state 1861

ALA	ALABAMA	MD	MARYLAND
CONN	CONNECTICUT	MISS	MISSISSIPPI
DC	DISTRICT OF COLUMBIA	NH	NEW HAMPSHIRE
		NJ	NEW JERSEY
DEL	DELAWARE	PA	PENNSYLVANIA
IND	INDIANA	RI	RHODE ISLAND
LA	LOUISIANA	VT	VERMONT
MASS	MASSACHUSETTS	W VA	WEST VIRGINIA

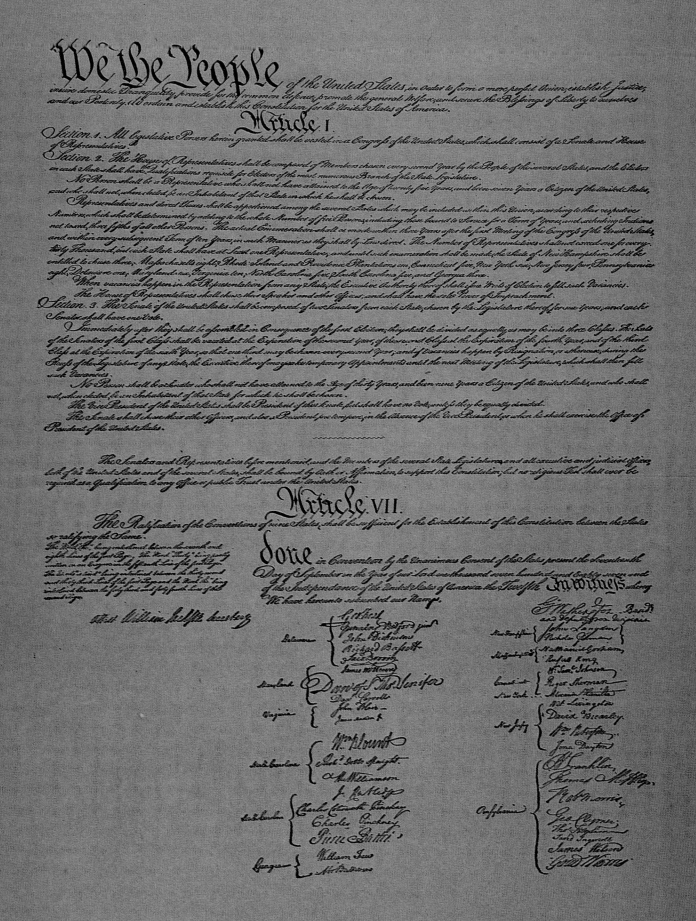

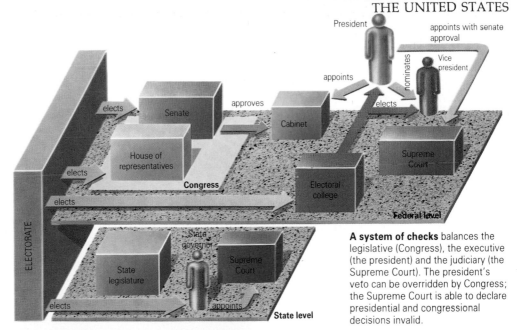

A system of checks balances the legislative (Congress), the executive (the president) and the judiciary (the Supreme Court). The president's veto can be overridden by Congress; the Supreme Court is able to declare presidential and congressional decisions invalid.

down, and came to serve as a model for many other countries.

The original constitution allowed individual states considerable room for self-government. The federal, or central, government was concerned with matters of defense, foreign affairs and the co-ordination of business between states. Legislation in other spheres was left to the states, each of which had its own constitution, elected governor, supreme court and local taxation powers. During this century the federal government has tended to become more involved in state affairs, and is the principal agency for raising and spending revenue.

A system of checks and balances

At the center of the constitution are a number of checks and balances between the three major branches of federal government – the legislative, executive and judicial branches. Legislative power is vested in Congress, which consists of the 100-seat senate and the 435-seat house of representatives. Two members from each state are elected to the Senate for a six-year term. The senate's approval is required for key presidential appointments to federal office, and to ratify treaties. Seats in the house of representatives are apportioned to each state on the basis of population. Members are elected for two-year terms.

In order to become law, a bill must be passed by majority vote in both houses. Legislation may originate in either, except for financial bills, which must be raised in the house of representatives. A bill becomes law when it is signed by the president, who may also choose to veto it; the veto may be overridden by a two-thirds vote in both houses.

Executive power is held by the president who, with the vice president, is elected for a four-year term; no person may serve more than two terms. The president is not elected by direct vote, but by an electoral college, to which each state is assigned votes on the basis of its representation in Congress. He or she serves as head of state, the armed forces and the civil service, and appoints the heads of the executive departments of

We the people... The opening words of the American constitution proclaim the revolutionary principle of the popular will. The constitution embodies a set of compromises that divided power between the states and a strong central government. It has proved a flexible and enduring instrument of government.

A vice president takes over With the former president's widow beside him, Lyndon B. Johnson takes the oath after Kennedy's assassination.

government, who collectively form the cabinet. However, the president's executive policies need the approval of the senate. Each year, often in January, the president addresses both houses in the "state of the Union" speech.

The vice president's single formal duty is to preside over the senate; he votes only to resolve a tie. He takes over the presidency in the event of the president's resignation or death in office. This happened in 1963 when Lyndon B. Johnson (1908–73) assumed office after the assassination of President John F. Kennedy (1917–63), and in 1974 when Gerald Ford became president after the resignation of President Richard Nixon.

The primary judicial power of the United States is the Supreme Court. This consists of the chief justice and nine associate justices appointed by the president for life with the majority consent of the senate. Presidents have been accused of packing the Supreme Court with political appointees. The Supreme Court hears cases involving disputes between states as well as those involving federal law arising out of the lower courts. The Supreme Court determines and interprets the constitutional basis of laws passed by Congress.

AMENDING THE CONSTITUTION

Written into the constitution are provisions that allow it to be amended. Amendments may be proposed by a two-thirds vote of both houses and ratified by three-quarters of the state legislatures – the method used for all 26 amendments adopted so far. A constitutional convention to consider amendments may be called if two-thirds of the states wish it.

The first 10 amendments are known collectively as the Bill of Rights and were adopted in 1791. They explicitly guarantee freedom of speech, religion, and the press and the right to a speedy and fair trial. The Tenth Amendment reserves all governmental powers not explicitly mentioned in the constitution to the states or to the people.

Three amendments – the Thirteenth, Fourteenth and Fifteenth – were adopted after the Civil War. These

extended political rights to black Americans by abolishing slavery, establishing the citizenship rights of native-born and naturalized Americans and recognizing the right of blacks to vote. The Fourteenth Amendment guarantees "equal protection under the laws" to all Americans.

Later amendments have continued to enlarge the political rights of ordinary Americans and to clarify archaic and confusing clauses in the original constitution. The Nineteenth Amendment (passed in 1919) extended the franchise to women in federal elections (some states had done so earlier). The Twentieth streamlined the period of transition between administrations after a presidential election, and the Twenty-fifth clarified the succession process following the death or resignation of a president in mid-term.

NO OTHER SYSTEM LIKE IT

Government at federal and at state level has always been contested between two major parties in the United States. The parties that dominate politics today, the Democrats and the Republicans, had taken on their present form by the middle of the 19th century. The Democrats, who were strongest in the south, stood for individual states' rights and for limited interference from central government; the Republican Party developed out of the antislavery movement in the north during the 1850s. Its status as the second major party was secured by Abraham Lincoln in the presidential election of 1860 that precipitated the Civil War.

Since World War II the Republicans have won 7 of the 13 presidential elections that have taken place while the Democrats dominated at local and congressional level. This has often meant that the ruling administration had to work with minority support in Congress. The victory of Democratic candidate Bill Clinton in the 1992 presidential elections interrupted 12 years of Republican rule. With the Democrats winning 259 seats in the House of Representatives and 56 seats in the Senate, Clinton embarked on an ambitious program of economic and healthcare reform but discovered that a Democratic majority was not necessarily enough to overcome Republican opposition.

Sectional interests

Party support has always owed a great deal to regional interests. From the beginning regional differences in the early settlement of the 13 colonies divided national politics. The plantation economy of the southern states meant that most of the land was owned by a very few, and a conservative political culture developed in support of landowners. The colonists of New England, self-sufficient and religiously oriented, were more moralistic in outlook: the role of government was to maintain the common good; and the Middle Atlantic states, at the economic core of the nation, stressed the benefits of political and economic competition.

The Civil War polarized the country. Until the mid-20th century the defeated Confederate states of the south voted solidly for the Democrats. The industrialized northeast was Republican. The western states tended toward Republican,

An enthusiastic supporter wearing a mask of Jimmy Carter leaves his fellow delegates in no doubt about his choice of presidential candidate. The national conventions are celebrated for their boisterous antics, but President Truman remarked, "I do not know a better way of choosing a presidential nominee."

His party's choice Bill Clinton enjoys the enthusiasm of the crowd after winning nomination as the Democrats' choice for the presidency at the national convention in New York in June 1992. Clinton was always a frontrunner in the race for nomination, which had started more than three years earlier at the state primaries and party caucuses.

but were highly volatile. They distrusted the influence of the northeastern Republicans, who were closely identified with big business interests.

An upsurge of Republican strength in the south after World War II began to break down the traditional sectional pattern of American politics. The development can be traced to white opposition to Democrat-sponsored civil rights legislation, as well as to the migration of northern-born Republicans to the south. By the 1960s the Democrats were becoming increasingly strong in the northeast, while Republicans began to dominate at elections in the west. The Republican domination of the south and west, strong throughout the 1970s and 1980s, was dented in 1992 when the Democrats regained the west, including California. They increased their strength in the northeast, while the Republicans were solid in the midwest and in parts of the south.

Substantial differences exist between the electoral behavior of the cities and suburbs. In the inner cities blacks and other ethnic minority groups vote Democrat; the Republicans are strongest in suburban areas and among more affluent voters.

Volatile states

While most areas tend to shift politically from one election to the next in close correspondence to national trends, certain states in the south, such as Florida and Texas, are highly volatile in their electoral behavior. This is partly due to the enfranchisement of large numbers of black voters, who tend to vote overwhelmingly in favor of the Democrats, and to the large-scale defection of white voters from the Democratic Party in presidential elections. These states have given their electoral votes to the winner of most presidential elections since the 1920s.

Both parties tend to be moderate in their programs, and there is often very

THE ROLE OF THE PARTY CONVENTIONS

The two major parties hold national conventions every four years as the final stage in the process of selecting a presidential candidate. This may have started three and a half years earlier. As well as confirming the choice of candidate, the national conventions decide party programs and unify strategy for the forthcoming presidential campaign. The chosen candidate will announce his running mate for the vice presidency.

The party conventions have become outsized extravaganzas, attracting huge crowds. Television coverage is carried right across the world, and the raucous antics of many delegates have turned the events into media circuses. In 1988 the Democratic convention in Atlanta, Georgia, was attended by 4,212 voting delegates as well as several thousand party functionaries. They were joined by more than 13,500 journalists who overflowed the 15,500-seat stadium, so a kind of "musical chairs" took place throughout the week-long proceedings.

As late as the early 1960s about two-thirds of the delegates to both national conventions were still chosen by party regulars in state-level conventions. National television screening of the turmoil at the 1968 Democratic national convention in Chicago, when Hubert Humphrey was selected, helped to change this by provoking loud rank and file calls for reform.

The effect has been to prolong the nomination process, as the majority of delegates are now chosen at state primaries and at local party caucuses where the electors themselves select who shall run for party office. These are in many cases held way ahead of the conventions, so candidates for the presidency start to whip up popular support even earlier, and campaigns, carried out under the glare of television publicity, become ever more expensive and protracted.

little to distinguish them. Each of them has a conservative and a more liberal wing. Party organization is strongest at state and local level. Their supporters come vociferously into prominence at the four-yearly national party conventions, which are held in the late spring or summer before the forthcoming fall general election, to select each party's presidential and vice presidential candidates. For the rest of the time party organization is rudimentary.

Party members seldom vote as a block in Congress. Party identification will influence the way representatives vote on major issues, but local interests will often decide how votes are cast on other matters. Many representatives work actively to divert federal funds for development and other government schemes into their districts, and the degree of success they enjoy at these so-called "pork barrel" politics influences whether or not they will be reelected.

Electing the President

The process of electing a candidate to the powerful position of president of the United States is extraordinarily complicated. He or she is elected not by direct popular vote but by an electoral college; every state is entitled to one vote for each of its senators and representatives in Congress. The selection of the electoral colleges takes place every four years on the first Tuesday in November, when the citizens of every state vote for the candidate they wish their electors too cast their votes for in the actual presidential election six weeks later, on the first Monday after the second Wednesday in December.

The vote list from each state's electoral college is signed, certified, addressed and delivered to the president of the Senate (the sitting vice president of the United States) in Washington DC. He counts the lists in front of all the members of both houses. A majority of 270 votes out of 538 is required for victory. It is theoretically possible for a president to win a majority in the electoral college without securing a plurality of popular votes, though this has not happened since 1888.

Campaign strategies

Since the voting in the presidential elections in November is on a "general ticket" vote – that is, rather than voting in single-member districts, the winner in each state takes all the electoral college votes – the larger states have enormous influence in deciding the choice of president. For example, California – the most populous state – has 54 electoral votes,

"The people have spoken" George Bush acknowledges the cheers of his supporters after winning 54 percent of the popular vote – securing 426 votes in the electoral college – on 8 November 1988.

Debating the issues The three contenders in the 1992 campaign – Clinton, Perot and Bush – meet to debate live on television, a confrontation that is now a standard part of the election process.

A tickertape welcome greets Kennedy during his triumphant 1960 campaign. The first Roman Catholic and the youngest person ever to be elected president, his assassination on a vote-winning visit to Dallas in 1963 brought an end to open-top car campaigns.

and Alaska, Delaware, North and South Dakota, Vermont and Wyoming only three each. This naturally affects campaign strategy: in the last two months of the 1988 election campaign the two candidates for the presidency – Republican George Bush and Democrat Michael Dukakis – paid more than 20 visits each to California.

In 1992, however, California showed early support for the Democrat candidate, Bill Clinton, and the campaign was fought most fiercely in states such as New Jersey, Ohio and Michigan. Clinton's strategy saw a return to the oldstyle whistlestop tour from state to state. In the closeness of the contest, George Bush was forced to adopt similar tactics, visiting even the smallest states several times.

The size of the American electorate and the growing pace and length of presidential campaigns means that television has become of paramount importance. Contenders for nomination vie for funding, media attention and for popular support during the long process to find the two major party candidates. After the nominating conventions, the direct outlays of the general election campaign committees, are paid for out of federal funds.

Television dominates the general election campaigns from this stage on, and appearances and statements by both candidates are carefully timed and staged to gain maximum impact on national evening news broadcasts. In recent campaigns paid advertising in the run-up to the fall election has become noted for the degree of personalized, often vicious attacks that each side makes on the other candidate's character, his background and elements of campaign strategy.

A particular aspect of American presidential elections is that any individual can stand so long as enough voters in a state petition to have his or her name added to the ballot paper. (The number varies from state to state, but is usually about 200,000.) Most elections field a handful of candidates keen to air a cause or grievance, but in recent times only George Wallace in 1968 and Ross Perot in 1992 have achieved sufficient nationwide support to affect the size of the vote cast for the main party candidates.

STATE AND LOCAL GOVERNMENT

The United States has a strong tradition of state and local government. The structure of state government mirrors that of the federal government – each of the 50 states has an elected governor and state legislature, and a judiciary, which may be appointed or elected. The states are responsible for certain aspects of public education, health and welfare, and for penal correction; they provide numerous other public services such as highways and public safety.

A multiplicity of local governments

Beneath this layer the states are divided into counties – as many as 3,000 throughout the United States. County authority varies from state to state. It is strongest in the south and the middle west, and virtually non-existent in New England, where the town is the primary unit of local government. There are an additional 75,000 public administration bodies. These include city, town and village governments, school districts and other administrative districts for flood control, fire and police protection and transportation. Each of them has the authority to levy taxes to provide the specific services for which it is responsible. The extent of their powers is determined by the state government – unlike those of the states themselves, which are constitutionally guaranteed under the Tenth Amendment.

The federal government has made substantial inroads into some areas of public administration this century, but state and local government remains relatively free of central control. Numerous surveys confirm that Americans value their tradition of locally controlled government, with its encouragement of popular participation in decision making.

Not surprisingly, the three levels of government frequently come into conflict over some areas of administration such as education. Schools were originally controlled through school districts, which were responsible to the state legislatures. In the 1940s, however, federal government began to involve itself more and more in the running of the schools, by providing grants to local authorities so long as they adopted federally approved policies. This arose partly from recognition that some states were not extending to blacks and other groups the equality of education guaranteed in the constitution; it was feared, at a time of growing fear of Soviet military domination, that the United States was falling behind in mathematics and science teaching.

Governing the cities

In some areas of the United States fragmentation of local government is a major problem. Large cities such as Chicago, Los Angeles or New York may contain upward of a hundred suburban municipalities, each with its own area of jurisdiction. This situation is exacerbated by the substantial differences in wealth and income between the central cities and their suburbs. An increasingly high proportion of the poor, elderly and other deprived groups lives in the inner cities, while the affluent middle and upper classes populate the suburbs. This means that the inner cities are faced with growing demands for public services, despite having dwindling taxes to pay for them, while the suburbs have lower demands but greater resources.

Some efforts have been made in a few areas to alleviate the effects of this fragmentation. In Minnesota, for example, local governments in the Minneapolis–St Paul metropolitan area are required to place 40 percent of their locally raised taxes in a common fund. This is subsequently distributed on the basis of need among the numerous municipalities in the area. A number of other states have adopted policies that encourage cooperation between local governments in resolving problems of regional transportation, air and water quality and environmental protection.

Black and Latino politicians have achieved their greatest successes in these major metropolitan areas. They have been less successful in white-dominated constituencies at all levels of government, though David Dinkins, elected as the first black mayor of New York, and Douglas Wilder of Virginia as the first black

GERRYMANDERING

Gerrymandering is the deliberate drawing of electoral districts or constituencies to secure advantage for one party over another and to distort the outcome of an election. In 1812 Governor Elbridge Gerry of Massachusetts approved the redrawing of electoral districts in one county of the state to benefit his party in the forthcoming election. The newly created districts seemed to a contemporary cartoonist to resemble a salamander, and so it was that a new word, gerrymander, entered the political vocabulary.

The drawing of the district boundaries in congressional and state elections has always been a contentious issue in the United States. In 1962 the Supreme Court ruled that districts should be equally apportioned, compact and contiguous. This reflected the realization that, as previously delineated, electoral districts in heavily populated urban areas, by comparison with rural ones, were under represented in Congress and state legislatures. Black, Latino and other urban minorities were consequently likely to receive unfair treatment in most legislative decisions.

The ruling unleashed a flood of political and legal battles over racial gerrymandering. Electoral abuses such as at-large elections (without single-member districts), which allowed white majorities to block minority representa-

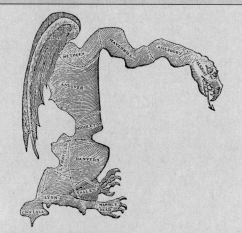

The original gerrymander The cartoonist has added a head, wings, claws and tail.

tion, were exposed. There was also controversy over whether minority populations were under-counted in the United States' census.

Districts can be fairly easily matched in size (though this does not prevent variations in the criteria applied by different courts to adjudge equality), but sophisticated counting techniques may still be used to create districts that favor one party or group over another. In the late 1980s the Supreme Court ruled that partisan as well as racial gerrymandering is a justiciable offense. This decision opened the door to a new round of lawsuits involving charges of electoral bias.

Black students are escorted by National Guardsmen into school in Little Rock, Alabama in 1957. The enforcement of centrally made laws to eradicate racial segregation in schools and bring an end to "whites only" areas brought the federal government into conflict with several southern state governments.

The mayor of Chicago, Harold Washington, addressing a meeting. Municipal governments are of many kinds. Some major cities, including Boston, Chicago, New York and Seattle, are governed by the mayor and city council: the council is nominally responsible for policy, which the mayor enforces. In practice, however, the mayor often controls the council and the position is consequently a very powerful one.

governor anywhere in the United States, achieved notable victories in 1989.

The relative autonomy of state and local governments makes local issue politics particularly important. Party member candidates for state and local office are not expected to advocate the policy positions espoused by the leadership of the national party, party platforms vary a great deal between regions. Though the Democratic Party is generally more liberal than the Republican Party, the Democrats in the nonmetropolitan areas of the South are decidedly conservative, and local candidates there will often disavow the liberal positions on issues adopted by the national party.

LEADING THE WESTERN WORLD

At the end of World War II the United States emerged as the major military and political power in the Western world. From this position of strength it pursued in the next two decades an internationalist foreign policy under successive presidents, which was strongly supported by the leaders of both political parties in Congress. This was carried out in the atmosphere of growing suspicion of and hostility toward the Soviet-dominated bloc of communist states that came to be known as the "Cold War".

Drawing the line against communism
The key figure behind the United States' response to the threat of Soviet domination was Harry Truman (1884–1972), president from 1945 to 1953. His postwar policies included the granting of massive financial aid to the war-shattered economies of Europe and Japan. Under the Marshall Plan – a program in large part prompted by humanitarian concern but also very heavily influenced by long-term strategic considerations – some $12 billion was spent on European reconstruction between 1947 and 1950.

Harry Truman was instrumental in establishing the North Atlantic Treaty Organization (NATO) in 1949, which provided for the collective defense of its members in Europe and North America. The "Truman Doctrine", pledging United States' support for "free peoples who are resisting attempted subjugation by armed minorities or by outside pressures", was first invoked to combat communist guerrillas in Greece and Turkey; it was later used to justify United States' intervention in Asia, particularly Korea from 1950 to 1953, following the assumption of power by the Communist Party on mainland China in 1949.

By 1952, when the first hydrogen bombs were tested on both sides, the new superpowers of East and West were engaged in a mounting nuclear arms race. The military expenditure of the United States stayed above 10 percent of GNP throughout the decade – an unprecedentedly high level in peacetime.

Competition for influence spread in the late 1950s and 1960s as the United States and the Soviet Union vied to outdo each other in lending military and other aid to

the countries achieving independence in Africa, Asia and the Caribbean. The overthrow of the corrupt Batista dictatorship in Cuba by Fidel Castro in 1959 brought the Cold War to the United States' own backyard. The new regime fostered increasingly close relations with the Soviet Union, and the installation of Soviet missiles on Cuba provoked a major confrontation between the two super

The Cuban missile crisis, 1962 A US reconnaissance photograph shows a Soviet ship returning home with eight missiles on board. Soviet leader Nikita Khrushchev called it "the first direct nuclear confrontation, unlike any in the history of our planet".

Operation Desert Storm (*right*) American soldiers set up the Stars and Stripes in Iraq during the Gulf war in 1991. In the first major conflict of the post-Cold War era, the Soviet Union backed the United Nations' decision to send a US-led international force to eject Iraqi troops from Kuwait.

CONTAINMENT BY TREATY

At the center of the United States' Cold War strategies in the 1950s was the creation of a number of regional treaties for collective defense to contain the threat of Soviet military domination worldwide. The first of these, NATO, brought together in 1949 the United States with Canada and 10 European states. The number later increased to 14, and included West Germany, though France withdrew from the integrated military command in 1966.

The analogous South East Asia Treaty Organization (SEATO) was established in 1954, and in 1959 the United States became an associate member of the Central Treaty Organization (CENTO) set up to counter Soviet influence in the oil-producing areas of the Middle East. Neither was as enduring as NATO, and did not survive into the 1980s.

The NATO agreement stated that an attack against one or more of its members in Europe and North America should be considered an attack against

them all, and that members should collectively and individually develop their capacity to resist attack. As a consequence, during the first 20 years of NATO's existence the United States contributed more than $25 billion in military and defense aid to its European allies.

Many aspects of the treaty were being called into question by the 1980s. The modernization of NATO's intermediate range nuclear weapons was fiercely opposed by pacifist groups in Europe. At the same time, European NATO governments were concerned that bilateral agreements to reduce nuclear arsenals were being reached with the Soviet Union without insufficient consultation of NATO's interests. But these debates were soon thrown into the shade by the ending of the Cold War and the dissolution of the Warsaw Pact in July 1991. The removal of a credible enemy in Europe called into question the very future of NATO itself.

powers. Crisis was only averted when the Soviet Union backed down in the face of President Kennedy's demands that they should be removed.

In the 1960s United States' policy turned to Southeast Asia, where support had been given for several years to the government of South Vietnam to combat communist guerrilla activity directed from North Vietnam. The supply of military aid and advisors increased, until by 1964 the United States had become embroiled in an escalating war. Opposition to the fighting grew at home as casualties mounted and military success seemed impossible. The systematic withdrawal of troops, begun by President Nixon, was completed by 1973, bringing to an end the longest war in the history of the United States.

New directions

Nixon dramatically reshaped United States' foreign policy when he visited the People's Republic of China in March 1972, thus ending years of tension between the two nations. Relations with the Soviet Union were also improved, though strains still remained. By the end of the decade military spending had fallen to less than a quarter of total federal expenditures, just above 5 percent of GNP; and though President Ronald Reagan entered office in 1981 determined to raise military preparedness – his provocative rhetoric included calling the Soviet Union an "evil empire" – it did not exceed 30 percent at any time during his presidency. Many of his policies were rebuffed by Democratic majorities in Congress, and opposition was voiced to his costly Strategic Defense Initiative ("Star Wars") plans.

By the late 1980s, the pendulum had swung toward detente with the Soviet Union: the signing of the Intermediate Nuclear Forces (INF) treaty in 1987 led to the destruction of over 2,500 missiles on each side. The ending of the Cold War in the 1990s followed by shifting world politics prompted the United States to reassess its international role. Victory in Kuwait demonstrated superiority, but at a huge cost. The Clinton administration made the domestic economy a priority, initiating defence cuts and insisting on clear guidelines before committing itself to United Nations initiatives. Nevertheless, the United States has continued peacekeeping activities in troublespots, including Bosnia, Haiti, Israel and Palestine.

BACKYARD POLICIES

In 1823 United States' President James Monroe (1758–1831), in what later came to be known as the Monroe Doctrine, declared that interference by a European power in any of the newly emerging independent states of Central and South America would be considered an unfriendly act toward the United States itself. Originally intended to warn off European countries from any attempt to help Spain regain control of its disintegrating American empire, the doctrine was invoked on several occasions during the 19th century (most notably in persuading Britain to cede the Mosquito Coast to Nicaragua in 1860), and was later extended to mean that any vital interest of the United States or its citizens throughout the American continent could be protected by military action.

An imperialist power

Toward the end of the 19th century the increasingly aggressive foreign policies of the United States led it to enlarge its imperialist stance in Latin America and the Pacific. Following its victory in the Spanish–American war of 1898 (fought in support of Cuban independence from Spanish rule), the United States acquired Puerto Rico in the Caribbean and Guam and the Philippines in the Pacific. American Samoa and Hawaii were added at about the same time, and the American Virgin Islands were purchased from Denmark in 1917. A number of islands in the Pacific, including the Caroline, Mariana and Marshall Islands, were taken under United States' protection on their liberation from Japan at the end of World War II. Today all these territories (except the Philippines, given their independence in 1946) remain under United States' administration, with varying degrees of self-government.

The Monroe Doctrine was often used to justify the United States' assumption of an active guardianship role in Central America, which it came to regard as its backyard. In 1903, when Panama gained its independence from Colombia with the military support of the United States, it granted the northern power, in return for guaranteed protection and an annuity, the right to build the Panama Canal as well as control in perpetuity of the Panama Canal Zone, a strip extending

An insecure border Mexicans head across the Rio Grande in a dinghy, seeking a new life in the United States. Hundreds of thousands enter illegally every year, despite the efforts of immigration officials to stop them.

Intervention in Panama Traditional backyard policy seemed to reassert itself in 1989 when President Bush ordered troops into Panama to seize General Noriega, formerly an ally but now considered an undesirable presence, in order to face drug-trafficking charges in the United States. The action aroused protest from the neighboring states of Central and South America as well as the Soviet Union.

United States' forces on the Caribbean island of Grenada, which they invaded in 1983 to overthrow a Marxist-oriented government that had seized power the previous year. The United States produced reconnaissance pictures to back their claim that a new international airport on the island, being built with Cuban aid, was capable of being used for military operations by Cuba and the Soviet Union.

5 km (3 mi) on either side of the canal. During the early years of the 20th century the United States dispatched warships or troops to Cuba, the Dominican Republic, Guatemala, Haiti, Honduras, Mexico and Nicaragua, confirming its dominant role in the region.

Carrot and stick policies

In 1933 President Franklin D. Roosevelt (1882–1945) declared his opposition to armed intervention: the foreign policy of the United States in the future would be based on the principle of the "good neighbor". But this did not prevent the United States reverting to the stick to enforce its policies as the Cold War set in after World War II. Military support was given to the conservative revolution that ousted the socialist government of Jacobo Arbenz Guzman in Guatemala in 1954. Following Fidel Castro's socialist revolution in Cuba (1959), Cold War tensions intensified in the region, and military spending to avert left-wing activity rose. Backing was given to the Bay of Pigs invasion of Cuba in 1961, and direct action was taken against the threat of Marxist governments in the Dominican Republic in 1965 and in Grenada in 1983.

Under the presidency of Democrat Jimmy Carter (1977–81) the carrot was reintroduced. Aid was given in return for human rights guarantees from the recipient states. A treaty signed in 1978 arranged for the gradual transfer of United States' ownership of the Canal Zone to Panama; the management and defense of the canal itself would remain with the United States until 2000.

During the Reagan years (1981–89) traditional attitudes reasserted themselves. President Reagan's policy statement of 1982 (his Caribbean Basin Initiative) isolated the United States' socialist "enemies" in the region and encouraged the capitalist development of its "friends", notably Costa Rica, the Dominican Republic and Jamaica. United States bases were maintained in Honduras to assist the right-wing Contra guerrillas fighting the socialist Sandinista government in Nicaragua. When Congress refused to allow continuation of aid to the Contras, other channels, mostly illegal, were found to direct the funds.

In the 1990s, President Clinton's assertion of special interest in Haiti suggested that he intended to pursue an active foreign policy in Central America. With United Nations embargoes failing to bring down the military leaders who deposed Haiti's elected president, Jean-Bertrand Aristide, in 1991, United States military intervention seemed a strong possibility. Congress feared that Haiti's undemocratic regime would create a refugee problem, and encourage illegal drug trading in its own backyard.

BILLY JACK ROBERTS · WALTER T SMITH · GREGORY SQUARE · EDWARD J VAILLANCOURT
MARK E WILLIAMS · EUGENE CARTER · MICHAEL J BLOTTENBERGER · TERRANCE L BOWELL
SAMMY W GOLDMAN · JIMMIE LEE JOHNSON · MICHAEL E KELLY · HERBERT W KLUG
EVERETT C PUGH · JAMES J L PUMPHREY · ROGER J ROWELL · DONALD R SAMUELS
JOHN J BENNETT · GUY RAY BRISTOL · JAMES J BURGOYNE · SYLVESTER COLLINS
VICTOR DEL GRECO Jr · MILTON GOLDSMITH · RICHARD A HOSTIKKA · MARTIN J KEOGH
THOMAS G LOVLEY · LA VOY D McVEY · WILLIAM J PARTINGTON · DANNY JOE QUAITE
ERVIN L TUTTLE · DANIEL L WALLACE · HARLEY D WHITNEY · JOSEPH W WIRTH
LARRY D ARONHALT · PAUL D BRUMMET · JAMES E BURTON Jr · KENNETH S DEE
DANNY D HANKS · JAMES D JOHNSON · KENNETH H KANAMAN · GEORGE G KARR
BRADLEY T MACE · STEPHEN B MURRAY · GEORGE T OLSEN · RICHARD V PAWLAK
JARED B WALDING · RUDOLFO R ALURA · REGALADO M D APEROCHO · HARMON J BOVE Jr
JAMES M STANLEY · DENNIS M DAUGHERTY · THOMAS DRUM · BARRY C HIATT
JAMES A PARKER · JOHN J PARKER · EDMUND A PETRECHKO Jr · TIMOTHY E QUIGLEY
JUSTIN R SPISTO · JOHN MAREZ CHAVARRIA · JOE N TUCKER · BRIAN L WILSON
RICHARD D ATOR · HECTOR L CARABALLO · RONALD D CAREY · JOHN J CRAVEY · MIGUEL RAMOS
MARIO GONSALEZ · JAMES L HOWELL · WILLIE JOHNSON · BRUCE D KLINGAMAN
ELMER R PERRY · DENNIS H RILEY · JOHNNY F ROACH · ROBERT P ROSENBACH
WALTER L BECKWITH Jr · THOMAS C CHANEY · MICHAEL J CONNELL · OREN L CROOK
WALTER B FOOTE · BERT A HAMILTON Jr · IRVING G PARK · JAMES D PRESSON
OSSIE RICHARDSON · JERRY W RUSSELL · ROBERT D SHIPMAN · WILLIE D STEPHENS
ROBERT L WEBSTER · CHARLES WHATLEY · JEFFREY L WILSON · KENT M ZERR
GLENN F AUSTIN · JAMES R BOWERS Jr · THOMAS M BROCKMEIER · CLIFFORD M DAVIS Jr
ALBERT H GATES Jr · DENNIS R HEERMAN · PAUL J JAMES · JIMMIE JACK JONES Jr
JAY W LADNER · STEPHEN L LANE · MAX LIEBERMAN · GLENN A LOVETT
JAMES R MUTH · MARVIN L RINGOEN · THEODORE M ROPCHOCK · ANDREW W SMITH
THOMAS J WHITLOW Jr · DONALD R WISHON · CHARLES E BROOKS · THOMAS L BUDDI
JOSEPH W DEVLIN · JAMES E DODD · DAVID K ERENSTOFT · WIM GOEREE
DOUGLAS R JONES · JACOB L KINSER · JOHN E LOCKHORST Jr
JOSEPH J POSTIGLIONE · WILLIAM R ROGNE · PATRICK TRACY
RICHARD A BRUECK · PHILLIP L CLARK · LARRY W COTTEN
MICHAEL J McCARRON · REX L PARCELS Jr
LEONARD J SCHOEPPNER · RICHARD L STILL
THEODORE L CADENHEAD
DONALD R DONALDSON · LARRY W DUKE
CHARLES D HICKS Jr · DONALD W HOLMAN
FLOYD MOYE · ROBERT D McFALL
DOMINIC N SCATUORCHIO Jr
THOMAS E SEVERSON
LARRY A STRUSS
JAMES W CLEMENT
OLLY G BELL
HACKETT Jr
LEESER
PRUETT
BROWN Jr
KREBS
POLASKI
WASHINGTON
DAVID L HEMKE
MENDIOLA
RICHARDSON
ALD D WEST
CRAIG
GREEN
RIDDLE
TRUSSELL
BERRY
GUILLEN
LEWIS
PELUSO Jr
CLARK
JOHNSON
STOCKSTILL Jr
BURNLEY
DYER
NEUDAHL
ALANIZ
KIMBLE
PATE
STEPHENSON
LEE DRURY
JOHNSON
SHIELDS
WALTON
BAIRD
L CATINO
GREEN
LOCKETT
McCORMICK
SCHUETTE

The war in Vietnam

The Vietnam war not only eroded the assumptions of the United States postwar foreign and military policy, but also weakened the national consensus upon which it depended. Its involvement began with the financing of France's efforts to hold on to its colony of Vietnam by defeating the Vietminh, a coalition of nationalists and communists. Following France's defeat and the country's partition in 1954, President Dwight D. Eisenhower (1890–1969) evoked the threat of a communist takeover as a reason to support the South Vietnam regime against the North. He likened the region to a set of dominoes stood on end; once one fell to communism, the others would inevitably follow – hence the "domino theory".

At first the United States provided weapons and advisers to the South in its fight against internal guerrillas (the Vietcong), who were backed by the North. This initial period of counterinsurgency (1961–65) was followed by the commitment of ground forces backed up by aerial bombing, until in 1968 there were over 500,000 US troops in the country. President Nixon promised to end the war by combining mass bombing and the substitution of Vietnamese for American troops with secret peace negotiations (1968–73). These developments permitted the United States to withdraw, but failed to prevent the eventual fall of the South to the North Vietnamese in 1975. The war cost the lives of 46,000 United States'

citizens. A further 270,000 were wounded. The total expenditure was $300 billion.

Not only did its intervention fail to prevent a communist takeover but, by extending the war into neighboring countries, the United States ensured that Cambodia would fall to the communist Khmer Rouge (1975) and that Laos would be destabilized. Its own action made the domino theory true. The world's most powerful and technologically advanced military could not defeat a peasant army that enjoyed popular support, high morale and a genius for logistics. The North's success during the Tet offensive (1968) showed how the United States underestimated the other side's ability to mobilize forces and to supply them along the Ho Chi Minh Trail, a series of routes that ran through the mountains of Vietnam, Laos and Cambodia. The United States refrained from total war. It largely avoided attacking civilian targets in the North, choosing to judge success in terms of the daily "bodycount" rather than by capturing territory.

An unpopular war

Perhaps the fundamental reason for failure was mounting opposition at home to the war. Presidents Kennedy and Johnson had sought to play down the war, assuming victory could be gained quickly and cheaply. Their calculated deception of the public was undermined by television pictures of the Tet offensive and the

reporting of United States' atrocities such as the massacre of 300 civilians at My Lai in 1968. Opposition came from many quarters. College students condemned both the war and the draft; 750,000 of them marched on Washington DC in 1971. Black nationalists saw the war as an example of United States' racism and imperialism, while the white working class bore a disproportionate amount of the casualties. Congress was alarmed at its lack of control over an expensive war that had never been officially declared.

These deep divisions eroded the American public's support for the war, and enabled Congress to force Nixon to end it. They also meant that returning soldiers were not treated as heroes. Coming back to an economy in recession, the veterans suffered high rates of unemployment, mental illness, drug dependency, suicide and rejection. Only in the 1980s did these wounds begin to heal, a process aided by the making of a number of films that examined the war and its consequences for those involved.

The effect on United States foreign policy was to introduce a new period of isolationism under President Carter and a reluctance to intervene militarily. The 1972 War Powers Act enlarged congressional control over the use of US troops. This encouraged the Reagan administration to pursue military policy by more covert means, a practice that was to lead to the Irangate scandal of the 1980s.

Honoring the fallen A veteran kneels in front of the memorial listing the names of the 46,000 United States' citizens killed in Vietnam. The dedication of the memorial in Washington DC was part of the process that in the 1980s started to heal the wounds left by the war. Thousands of veterans who survived the fighting have suffered crippling mental illness, and suicide rates among them are very high. Americans have only recently begun fully to understand the effect of the war on the minds of a whole generation.

A "search and destroy" mission gets underway in South Vietnam. In spite of their superior firepower, technology and weight of numbers, the United States' forces were unable to win the war. They were pitched against a well-organized guerrilla army, who believed in the cause they were fighting for, while their own men increasingly did not. The American public, frustrated at the cost and length of the war, was reluctant to treat returning soldiers as heroes, and the war became more and more unpopular as televised pictures of the Tet offensive showed scenes of ferocious fighting in Vietnamese cities.

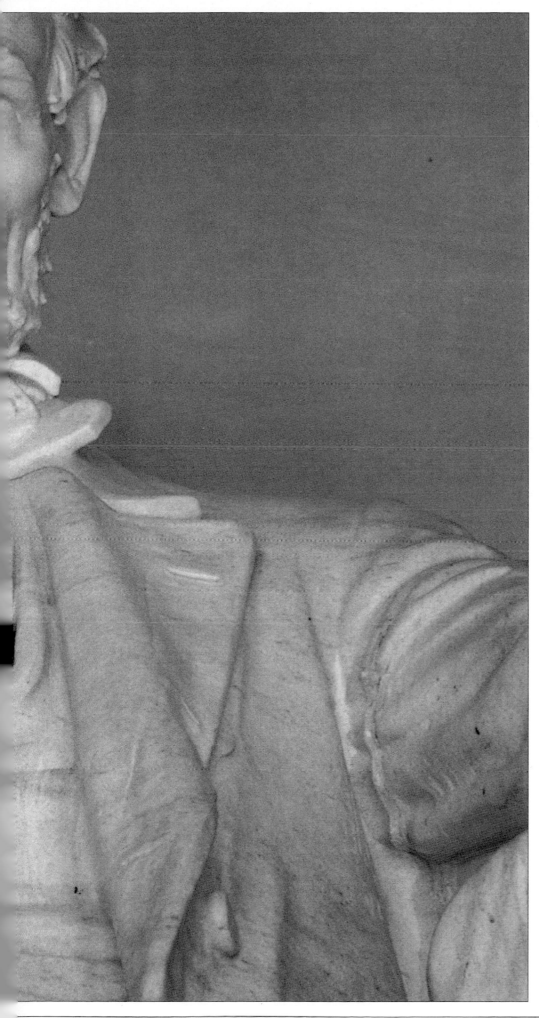

"I have a dream"

On 28 August 1963 more than 200,000 people gathered in the shadow of the Lincoln Memorial in Washington DC to demand equal justice for all people under the law – a right guaranteed by the Thirteenth, Fourteenth and Fifteenth Amendments of the constitution. They were inspired by the prophetic quality of Martin Luther King's famous "I have a dream" speech. The gathering was the emotional high point of the civil rights protest movement, which had been growing since the 1940s.

Racial discrimination was openly practiced in many states and communities, and the protest movement centered on three particular issues: segregated education, voting rights and employment. Political tactics ranged from judicial appeals to the Supreme Court to mass freedom marches through the American South.

Success came in three stages: in the 1950s the Supreme Court outlawed segregated education (the 1954 Brown versus Board of Education decision), in the 1960s voting rights were won in practice as well as theory (the 1965 Civil Rights Act), and since the 1970s Affirmative Action (or reverse discrimination) with racial quotas has reserved access to employment for ethnic minorities.

Afro-Americans achieved their highest political profile during the protests of the late 1960s, when the civil rights movement coincided with the antiVietnam war protests. The assassination of Martin Luther King in Memphis, Tennessee in 1968 sparked off riots throughout the country. Today he is officially commemorated by Martin Luther King Day. In 1988 one of his followers, Jesse Jackson, became the first Afro-American to offer a serious challenge in the presidential election.

The Lincoln Memorial in Washington DC was the setting for the massive civil rights demonstration in August 1963 addressed by Martin Luther King.

THE POLITICS OF INSTABILITY

TWO COLONIAL TRADITONS · DEGREES OF DEMOCRACY · THE GEOPOLITICS OF CONFLICT

The 33 states and dependencies of the Central American region were among the oldest European colonies in the world. Spanish settlement of the island of Hispaniola began as early as 1493; the rest of the region was later colonized by a number of European powers. Mexico and the states of the Central American isthmus (apart from Belize, formerly British Honduras), with Cuba and the Dominican Republic formed part of the vast Spanish American empire. The tiny island colonies of the Caribbean basin changed hands many times as contending European states sought to exploit their rich natural resources. Independence came to the Spanish colonies and to the French colony of Haiti through armed struggle in the 19th century; decolonization in the Caribbean has taken place peacefully since the 1960s.

COUNTRIES IN THE REGION

Antigua and Barbuda, Bahamas, Barbados, Belize, Costa Rica, Cuba, Dominica, Dominican Republic, El Salvador, Grenada, Guatemala, Haiti, Honduras, Jamaica, Mexico, Nicaragua, Panama, St Kitts-Nevis, St Lucia, St Vincent and the Grenadines, Trinidad and Tobago

Dependencies of other states Anguilla, Bermuda, British Virgin Islands, Cayman Islands (UK); Aruba, Netherlands Antilles (Netherlands); Guadeloupe, Martinique (France); Puerto Rico, US Virgin Islands (USA)

STYLES OF GOVERNMENT

Republics Costa Rica, Cuba, Dominica, Dominican Republic, El Salvador, Guatemala, Haiti, Honduras, Mexico, Nicaragua, Panama, Trinidad and Tobago

Monarchies All other countries of the region

Multi-party states All countries except Cuba, Haiti

One-party states Cuba, Haiti

Military influence Guatemala, Haiti, Honduras

CONFLICTS (since 1945)

Coups Dominican Republic 1961, 1963; El Salvador 1948, 1960–61, 1979; Grenada 1979, 1983; Guatemala 1954, 1957, 1963, 1982, 1983; Haiti 1950, 1956, 1986, 1991; Honduras 1963; Panama 1968, 1988

Revolutions Cuba 1959; Nicaragua 1979

Civil wars Costa Rica 1948, 1955; Dominican Republic 1965–66 (US involvement); El Salvador 1979–92; Guatemala 1967–85; Nicaragua 1962– (US involvement after 1981); Panama 1958–59

Interstate conflicts Nicaragua/Honduras 1957–60; Cuba/USA 1961; El Salvador/Honduras 1969; Grenada/USA 1983

MEMBERSHIP OF INTERNATIONAL ORGANIZATIONS

Caribbean Community (CARICOM) Antigua and Barbuda, Bahamas, Barbados, Belize, Dominica, Dominican Republic, Grenada, Jamaica, St Kitts-Nevis, St Lucia, St Vincent and the Grenadines, Trinidad and Tobago

Organization of American States (OAS) All countries of the region

---

TWO COLONIAL TRADITIONS

The struggle for independence in Mexico, begun in 1810, was won in 1821. A short-lived Mexican empire was set up, which included the territories of California, New Mexico, Texas and Utah in North America and extended southward as far as the present border of Panama. In 1823 Central America broke away to form a federation. El Salvador was the first of the isthmus states to declare its independence in 1838, and the other remote settlement clusters of Central America became the cores of the independent states of Costa Rica, Honduras, Guatemala and Nicaragua. In 1836 Texas rebelled against the Mexican dictator Santa Anna (1795–1876). Its annexation by the United States in 1845 led to the US–Mexican War, which ended in 1848 with the sale to the United States for $15 million of all territory formerly held by Mexico north of the Rio Grande.

The creation of Panama

Panama, which had been part of the South American viceroyalty of New Granada under Spanish rule, became part of the newly independent state of Colombia. The United States had long sought to construct a canal across the isthmus to secure a sea route between its west and east coasts, and in 1903 underwrote a revolution that secured Panama's independence under United States protection. Work on the Panama Canal began in 1906, and the United States was granted in perpetuity the use, occupation and control of a zone 5 km (3 mi) wide on either side. Panama's protectorate status was ended in 1939; the Canal Zone, which became a focus of anti-US feeling in the 1960s, was formally transferred to Panamanian sovereignty on 1 October 1979. Final transfer of the ownership of the canal itself will take place in 2000.

Belize, formerly British Honduras, the only state on the Central American isthmus not colonized by Spain, achieved full internal self-government in 1964. A long-standing territorial dispute with neighboring Guatemala delayed the granting of full independence until 1981, and after that Britain still maintained a military defensive presence there.

Spain was the colonizing power of the larger islands of the Caribbean: Cuba, Hispaniola and Puerto Rico (the Greater

Cuba's revolution, led by Fidel Castro – here cementing friendship with Soviet leader Nikita Khrushchev – affected much of the region.

Armed soldiers go unremarked in El Salvador, scene of some of the worst excesses of political violence in the region. In the early 1980s more than 20,000 civilians were killed as army "death squads" terrorized the population in its campaign against left-wing guerrillas.

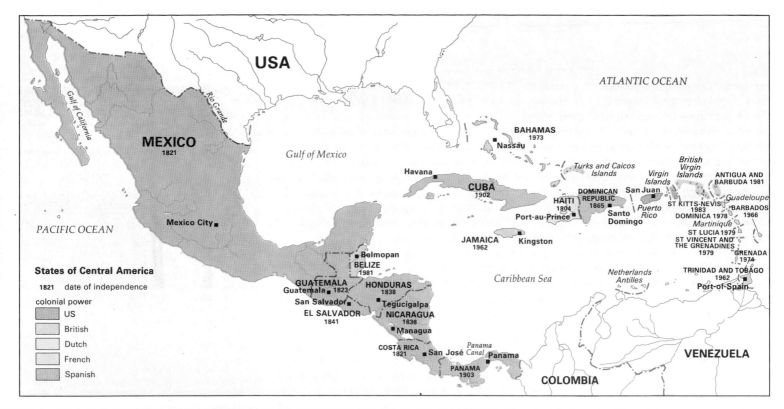

States of Central America

1821 date of independence

colonial power

- US
- British
- Dutch
- French
- Spanish

Vanished empires Most of the states in the region belonged to the Spanish empire, which overran and destroyed the indigenous Indian cultures in the 16th century. The smaller islands, colonized by several European nations, have generally been more stable.

Antilles). Haiti, occupying the western one-third of Hispaniola, was a French colony. Its independence was achieved with great violence in 1804 (when it became the first independent state in the region and the first black republic in the world). The Spanish colony on Hispaniola, the Dominican Republic, was also ruled by France from 1795 to 1808, and by Haiti from 1822 to 1844, when it was declared a republic. It was briefly reannexed to Spain from 1861 to 1865; it had a troubled history of weak governments, and was occupied by the United States from 1916 to 1924. Haiti was similarly occupied from 1915 to 1934.

Following the Spanish–American War of 1898 Puerto Rico was ceded to the United States, and in 1952 became a Commonwealth voluntarily associated with the United States. Cuba also gained its independence from Spain in 1898, with military assistance from the United States. It became fully independent in 1902, though the United States retained naval bases there and reserved the right to intervene in Cuba's domestic affairs. This it did several times before relinquishing the right in 1934. In 1959 Cuban nationalists, led by Fidel Castro, overthrew the right-wing Batista dictatorship (1935–59). In 1961 Castro declared a Marxist regime in Cuba, following the unsuccessful "Bay of Pigs" invasion attempt made by Cuban exiles with the support of the United States. The Cuban government became a source of support for left-wing groups in the Caribbean and in South America as well as Africa.

An archipelago of island states

The smaller islands of the Caribbean were colonized mainly by Britain, France and the Netherlands, though Denmark and Sweden have both had colonial interests there. The United States purchased the Danish-owned Virgin Islands in 1917.

British plans for decolonization in its Caribbean possessions after World War II involved the creation of a single federal state – the Federation of the West Indies – which it was hoped would incorporate all its colonies. This failed to provide a workable solution to self-government in the area, however, and following Jamaica's defection from the Federation in 1961, the British territories became independent in two waves: Jamaica and Trinidad, and then Barbados, in the 1960s; the remaining (smaller) islands in the 1970s and 1980s.

Britain retains one colony (Bermuda) and also five Caribbean dependencies. Martinique and Guadeloupe are French overseas departments; the Netherlands Antilles form an autonomous part of the kingdom of the Netherlands; the westernmost island, Aruba, achieved separate self-governing status in 1986.

DEGREES OF DEMOCRACY

The two distinct patterns of colonial rule, with their differing experiences of decolonization, have resulted in contrasting styles of government. The former Spanish-ruled states, and Haiti, have been characterized by long periods of authoritarian military or civilian dictatorship.

The military factor

In Haiti a long history of military dictatorship was modified in 1956 by the installation – first by election and then by presidential fiat – of François Duvalier ("Papa Doc", 1907–71). In the early 1960s he established a perverted version of black power, using the folk religion, *vodun*, and his private army, the *tontons macoutes*, as a means of social control. Following a coup to remove his son Jean Claude in 1986, Haiti returned to military rule. Elections were restored in 1990 but the winner, Jean Bertrand Aristide, was overthrown by the army in 1991.

Cuba is the only left-wing authoritarian regime in the region. During the 1970s it began to adopt a political system based on that of the Soviet Union. In 1975 the first congress of the Cuban Communist Party was held, and a new socialist constitution was approved by referendum the following year, when Fidel Castro was elected head of state.

Nearly all the Central American main-

THE PRICE OF MEXICO'S POLITICAL STABILITY

As its name implies, the dominant political party in Mexico, the Institutional Revolutionary Party (PRI), considers itself the institutional guardian of the Mexican Revolution (1911–17), which swept away the authoritarian regime of Porfirio Diaz (1830–1915), who had ruled Mexico for 27 uninterrupted years. Unlike most of its Central and South American neighbors, Mexico has experienced political stability since 1920. The price paid has been single-party government, with democracy until recently reduced to appearances only. Since World War II the PRI has attempted to generate growth through industrialization and the development of the oil industry (nationalized in 1938), while spending substantial amounts on welfare, supplemented by a fluctuating program of land reform. This policy has generated a rapid process of urbanization and led to the development of a substantial middle and working class centered on Mexico City (with a population of 19 million) and the other large towns.

Paradoxically, in recent years disaffection from the PRI has been greatest among urban Mexicans. At the 1988 election Salinas de Gortari, the PRI candidate, polled barely 50 percent of the votes cast (with almost half the registered electorate abstaining). The massive foreign debt, economic recession since the oil boom of 1978–81, and allegations of pervasive government corruption can be laid at no other door than that of the PRI. Opposition candidates of the right, the National Action Party (PAN), and the left (Cuauhtemoc Cardenas, fighting the election as part of the National Democratic Front (NDF) coalition) made spectacular inroads into PRI dominance, which promised a transition to genuine democracy.

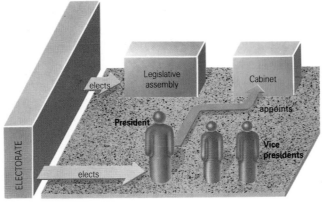

A stable democracy Costa Rica has a unique record of ininterrupted government on the Central American isthmus. Its constitution, dating from 1949, forbids the establishment or maintenance of an army. Elections are held every four years to choose a president and two vice presidents. The candidate who receives the largest vote, provided it is over 40 percent, is elected; if none does so a second election is held. There is a single-chamber assembly of 57 members, also elected for a four-year term. Voting is compulsory.

land states are potentially unstable. Small but powerful, mainly European elites, backed by the army (often with United States support), still seek to maintain themselves in power against a large, impoverished lower class, which is often ethnically distinct, demanding a share in government. In several states (El Salvador, Guatemala, Nicaragua) the long-term operation of guerrilla groups has killed many thousands of people, and human rights go largely unrecognized.

Costa Rica, which has the distinction of being the most stable democracy in the isthmus, is the only state that does not have a standing army. Since 1948 it has had uninterrupted democratic government: the president is elected for a four-year term by compulsory adult suffrage. Oscar Arias Sanchez, elected in 1986, won the Nobel Peace Prize in 1987.

In 1966, following United States' intervention, democracy was established in the Dominican Republic after the long dictatorship of Rafael Trujillo from 1930 to 1961. Elections in 1978 brought the first peaceful transfer of power for an opposition party, but the Christian Democrats remain the dominant party. Mexico has the longest experience of stable government; it holds elections at six-yearly intervals, but the institutional Revolutionary Party (PRI) has been in power since 1929. In El Salvador the 1989 elections marked the first transition of power in that country from one elected government to another. However, the winning right-wing Arena party had previously been linked with anti-leftist death squads; the Farabundo Marti National Liberation Front (FMLN) boycotted the elections, and the guerrilla war that had lasted since 1972 continued.

Despite the apparatus of democratic elections the army remained the final arbiter of power in both Guatemala and Honduras; in Panama the legally elected president was ousted in 1988 by General Manuel Noriega, allegedly a narcotics baron. In early 1989 Noriega declared the Panamanian elections void after observers backed the opposition's claim of fraud. When he refused to give way, the United States sent in a military force to eject him.

In Nicaragua the right-wing Somoza dictatorship (1934–79) was violently overthrown by the Sandinista National Liberation Front (FSLN). Despite attempts by the United States to undermine the new socialist regime, the Sandinistas won the 1984 elections in a multi-party contest, and Daniel Ortega, the FSLN leader, assumed the presidency. Six years later, in democratic elections, an alliance of opposition parties toppled the Sandinistas from power.

Westminster constitutions
In the former British colonies parliamentary democracy, based on the model of Westminster, has taken deeper root, though the government of Grenada was subject to a left-wing revolution in 1979. Following a further Marxist coup in 1983, an invasion headed by the United States, which feared the creation of a Cuban base on the island, led to the restoration of its former consititution.

In Jamaica the People's National Party (PNP), led by Michael Manley, formed the government between 1972 and 1980. Then the spiraling economic crisis and the rejection of Manley's socialist programs brought its electoral defeat by Edward Seaga's right-wing Jamaica Labor Party (JLP), following a violent campaign in which more than 600 people were killed. The PNP refused to contest Seaga's snap election of 1983 on the grounds that the electoral roll was out of date, but won a landslide victory in 1989 under peaceful circumstances very different from those that accompanied the 1980 elections.

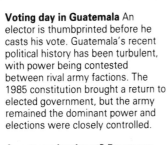

Voting day in Guatemala An elector is thumbprinted before he casts his vote. Guatemala's recent political history has been turbulent, with power being contested between rival army factions. The 1985 constitution brought a return to elected government, but the army remained the dominant power and elections were closely controlled.

A party under threat? Enormous crowds are attracted to a rally of the PRI, Mexico's ruling party since 1929, founded to safeguard the ideals of the Mexican revolution. Massive foreign debts and charges of corruption eroded its former popularity and at the 1988 election a number of opposition parties formed an electoral pact to reduce its majority.

THE GEOPOLITICS OF CONFLICT

The presence since 1959 of a socialist state – Cuba – in the Caribbean, committed to assisting left-wing revolutionary movements in neighboring (and more distant) states, has had widespread repercussions in the region. The United States has long exercised influence there. Puerto Rico, as a Commonwealth of the United States, enjoys a quasi colonial status; the Dominican Republic and Haiti also lie within its sphere of influence. From 1972 to 1991 Cuba was a member of the Eastern bloc's trading organization, COMECON.

Regional cooperation
Factors of geography and historical evolution have determined that with a few exceptions the states of the region fall traditionally into two major groupings. Most of them belong either to the Central American Common Market (CACOM) or to the Caribbean Community (CARICOM – formerly the Caribbean Free Trade Area, CARIFTA). CACOM includes Costa Rica, El Salvador, Guatemala, Honduras and Nicaragua, but not Mexico or Panama. It developed rapidly after 1951, creating important regional manufacturing strategies until Honduras withdrew in 1970, following the "soccer war" – sparked off by a football match – with El Salvador. The decline in trade that took place after 1980 was due in part to economic factors (exacerbated by the heavy external debts of its member states) and in part to the damaging impact of civil wars in El Salvador and Nicaragua.

CARIFTA was created in 1968 among former British colonies, including mainland Belize and Guyana, as a counterbalance to the threat of West Indian fragmentation following decolonization. It was established as CARICOM in 1973 with the aim of coordinating both economic and foreign policy in the Caribbean region. However, many of the tensions that contributed to the breakup of the Federation of the West Indies (1958–62) – such as competitive development strategies, inter-island jealousies and distance between member states – have resurfaced in CARICOM.

The less developed countries of the Lesser Antilles, which want closer economic integration than do the larger members (Jamaica and Guyana), have

US marines in Grenada The threat of a Marxist government provoked the United States to invade the island in 1983 with a force that included 12 ships and as many as 6,000 marines and rangers.

General Manuel Noriega subverted the democratic process in Panama until US troops removed him from power in 1989 – just one example of US intervention in Central America.

Cuban troops were recalled from Angola as part of the agreement for Namibia's independence in 1989. They had been backing the government forces.

complained persistently that they have been exploited by the others. They have sought coordinated diplomatic representation overseas, and in 1983 formed the Organization of Eastern Caribbean States (OECS). It was on the basis of the defense pact involved in this agreement that the United States was invited by the OECS to take part in the 1983 invasion of Grenada, on the grounds that Cuba was planning to build an airport there to use as a military base. In fact the airport was needed to expand the island's tourist economy, and had been planned while Grenada was still a British colony.

By 1981, when Ronald Reagan became president of the United States, the US administration perceived the Caribbean and Central America as penetrated by Marxist activity from Cuba (backed by the Soviet Union). This had supported the Sandinista revolution in Nicaragua and the left-wing FMLN guerrillas in El Salvador. Through its support for the anti-Sandinista Contra guerrillas, United States' efforts in Central America were therefore directed toward toppling the government.

Frustrated plans for peace

The only counterweight to this policy in the mid-1980s was given by the Contadora Group (named after the island where they first met), which included Mexico, Panama, and two South American states, Colombia and Venezuela. Contadora's 21-point peace plan, which proposed the partial demilitarization of Central America to stop the civil wars in El Salvador and Nicaragua was rejected by the United States. The Arias plan, named for the president of Costa Rica who devised it and signed by five Central American presidents in August 1987, proposed the declaration of a ceasefire and the holding of free elections, but collapsed because of the United States' refusal to halt aid to the Contras.

It was the reduction of aid from the Soviet Union as a consequence of its own economic crisis and its desire to withdraw from its commitments to socialist movements and regimes around the world that indirectly brought the civil war to an end. In the face of mounting economic pressure, the Sandinista government was defeated in national elections in 1990, leading to the demobilization of the army and the disbanding of the Contras.

CUBA'S ROLE IN THE WORLD

Until the revolution of 1959 the dominant influence in Cuba had been the United States. Its hostility toward the new socialist regime, and particularly its part in the Bay of Pigs invasion, contributed to Fidel Castro's declaration of a Marxist–Leninist state, bringing Cuba closer to the Soviet Union. After 1974 a political structure very similar to that of the Soviet Union's Eastern European satellites was set up in Cuba. It was the first non-neighboring state of the former Soviet Union (Vietnam joined in 1978) to become a member of COMECON, the Soviet-led organization to promote economic cooperation between communist states. Cuba's continuing financial dependence on the Soviet Union was very great. One estimate put the value of Soviet economic aid to Cuba during the early 1980s overall to be $3.5 billion a year.

In the early years of the regime Cuba was committed to extending the revolution throughout Latin America. Che Guevara (1928–67), the theoretician of the revolution, was killed by Bolivian troops while organizing a guerrilla base there. Cuba joined the nonaligned movement opposed to neocolonialism and imperialism and assistance was given to left-wing and anticolonial groups elsewhere in the world, particularly in Africa. In 1989 the Cuban troops withdrew from Angola as part of the agreement for South Africa's withdrawal from Namibia, negotiated through the United Nations. Nearer home, the provision of aid to the Marxist regime in Grenada (1979–83) and to the Sandinista government in Nicaragua increased tensions with the United States in the mid-1980s. Despite the cessation of Soviet aid and the end of the Cold War in the early 1990s, Cuba had not moved from its position of entrenched communism.

Conflict in Nicaragua

The establishment in 1979 of a socialist government in Nicaragua, after more than a decade of fighting, was seen by many people in the United States as a direct threat to its interests in the region. Nicaragua occupies a strategically pivotal position in the Central American isthmus, from which military control could be established over El Salvador, Guatemala and Honduras to the north, and over Costa Rica and Panama to the south.

The Sandinista government (which took its name from a guerrilla group headed by Augusto César Sandino that had opposed the establishment of US naval bases in 1912) under Daniel Ortega immediately set out to reverse the pro-United States policy of the authoritarian Somoza regime it overthrew. The new government soon introduced extensive economic and land reforms. The part of the economy (about 60 percent) that had been under the direct control of the Somoza family was nationalized. The rest, including many large estates on the Pacific coast, remained in private hands. Nicaragua continued to operate a state-led mixed economy.

Nicaragua's relations with the United States deteriorated rapidly after the election of Ronald Reagan as president in 1981. Alleging that the Sandinista government was supporting anti-government FMLN guerrillas in El Salvador, the Reagan administration immediately cancelled the supply of economic aid, linked to the restoration of human rights, that his predecessor Jimmy Carter (president 1977–81) had inaugurated.

The price of United States' involvement
Active support was given to the Contra guerrillas opposed to the Sandinista government, and in 1984 Nicaragua's harbors were mined by the CIA. Nicaragua was denounced as a Marxist regime, and attempts were made to isolate it diplomatically and financially, and by trade embargo. When the United States Congress refused to allow the continuation of aid to the Contras, other channels were found to direct funds to them. The Iran-gate scandal of 1986 revealed that cash resulting from secret arms deals with Iran to bring about the release of United States' hostages in Lebanon was intended to fund back-door military aid to Nicaragua.

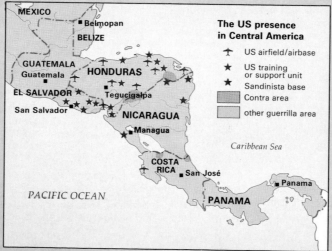

The United States exercises close guardianship over its Central American and Caribbean backyard. Its military support bases in the 1960s reflected its strategic goals of limiting the influence of the Cuban left-wing socialist regime, neutralizing Soviet influence, and protecting the free passage of the Panama Canal.

Nicaragua's civil war A government soldier shoulders his Soviet-supplied rifle. The successful seizure of power by the Sandinistas appeared to give Cuba and the Soviet Union the opportunity to extend their influence on to the Central American mainland. Fearing that one state after another would fall under communist control (the "domino theory"), the US government stepped up its support for the right-wing Contra guerrillas after 1981, and backed the establishment of anti-Sandinista bases in Honduras. An economic blockade and years of civil war caused havoc to the Nicaraguan economy, leading to the rejection of the Sandinista government by voters in 1990.

Nicaraguan "mothers of the disappeared" hold up pictures of their sons who vanished during the bitter years of fighting.

The United States' action against Nicaragua was the latest in a series of interventions in the region, since the onset of superpower hostility after World War II. These included support of the Bay of Pigs invasion of Cuba in 1961, and direct military intervention in the Dominican Republic in 1965 and Grenada in 1983 to prevent the threatened installation of Marxist governments.

Such action has traditionally been defended by invoking the Monroe Doctrine (1823), which stated that interference by any outside power in the affairs of the newly emerging independent states of the American continent would be regarded as an unfriendly act toward the United States itself. In 1904 this was extended by President Theodore Roosevelt (1858–1919) to allow the United States to exer-

cise police power in the region as necessary to defend its interests, and was subsequently used to justify direct action on many occasions. The United States' policy against Nicaragua had repercussions throughout Central America. To assist the Contra rebels United States' bases were maintained in Honduras, from which surveillance planes operated throughout the region.

Nicaragua responded to the hostile and destabilizing actions of the United States by developing its links with Cuba and the Soviet Union, which before 1987 supplied all its oil requirements. Conscript forces were employed to defend the Atlantic coast and the northern border from guerrilla activity, and by the mid-1980s the war against the Contras was absorbing more than half the national budget.

Governing the Bahamas

The Bahamas are typical of the smaller member states of the Commonwealth. Governed by the British from 1783 to 1973, it consists of some 700 islands between Florida and Haiti, 30 of which are inhabited. The total population is just under a quarter of a million.

When the Bahamas achieved independence the British monarch remained as head of state, represented by a governor general. The constitution is broadly modeled on that of Britain, with a two-chamber legislature consisting of an elected house of assembly and an appointed senate. The chief executive is the prime minister, who is appointed by the governor general from the party that controls the house of assembly.

Although the legacy of imperial rule lingers on in political institutions and traditions, the economic influence of the neighboring United States is becoming increasingly important. The economy is linked to that of the United States through its dependence on tourism and international banking. The changing situation is symbolized by the fact that the pound sterling is no longer accepted as general currency on the islands, and has been replaced by the US dollar.

A soldier on guard in Nassau, the islands' capital, provides tourists with a reminder of the Bahamas' former links with the British empire.

DICTATORS AND DEMOCRACIES

THE STRUGGLE FOR POWER · CONSTITUTIONALISM IN A TIME OF CHANGE · GETTING ON NEIGHBORLY TERMS

Within two years of the discovery of the New World by Christopher Columbus, Pope Alexander VI had divided South America between Spain and Portugal (in the Treaty of Tordesillas, 1494). This arrangement gave Spain the right to all land west of a line approximating to the 50 degree line of longitude, and Portugal the land to the east of it. By 1800 a vast Spanish empire had been created in the west and center of the subcontinent, stretching from Venezuela to Argentina, and a Portuguese empire in Brazil, by far the largest state in the region. Britain, France and the Netherlands had established colonies on the coast south of the river Orinoco (today Guyana, French Guiana and Surinam). Large tracts of the Amazon basin and areas farther to the south were still uncolonized, though Christian missions had been established.

COUNTRIES IN THE REGION

Argentina, Bolivia, Brazil, Chile, Colombia, Ecuador, Guyana, Paraguay, Peru, Surinam, Uruguay, Venezuela

Island territories Easter Island, Juan Fernandez (Chile); Galapagos (Ecuador); Tierra del Fuego (Argentina/Chile)

Disputed borders Guyana/Venezuela; Peru/Ecuador

Dependencies of other states Falkland Islands, South Georgia, South Sandwich Islands (UK); French Guiana (France)

STYLES OF GOVERNMENT

Republics All countries of the region

Federal states Argentina, Brazil, Venezuela

Multi-party states All countries of the region except Paraguay

One-party states Paraguay

Military influence Chile, Paraguay

CONFLICTS (since 1945)

Coups Argentina 1955, 1966, 1976; Bolivia 1964, 1980; Brazil 1960, 1964; Chile 1973; Paraguay 1954, 1989; Peru 1958, 1969, 1975, 1992; Surinam 1980–83 (six coups); Uruguay 1973

Civil wars Argentina 1974–83; Bolivia 1946, 1952, 1967; Colombia 1948–58, 1984–; Peru 1965, 1980–

Interstate conflict Argentina/UK 1982

MEMBERSHIP OF INTERNATIONAL ORGANIZATIONS

Caribbean Community (CARICOM) Guyana

Latin American Integration Association (ALADI) All countries of the region except Guyana and Surinam

Organization of American States (OAS) All countries of the region except Guyana

Organization of Petroleum Exporting Countries (OPEC) Ecuador, Venezuela

Note: Argentina and Chile have territorial claims in Antarctica.

THE STRUGGLE FOR POWER

The movement for independence in South America took place between 1808 and 1826, encouraged by the example given by the War of Independence in North America against Britain (1775–83). In Brazil, demands for political autonomy were met by the Portuguese royal family, who had fled there from Napoleon's armies, and an independent empire was established in 1822.

By contrast, Spain responded to its South American colonies' desire for freedom with violent repression. Politically weak at home, however, it was unable to resist the strength of the movement that swept across the subcontinent. By 1830 two colonial empires in South America had become 10 independent states.

Three colonies still remained. British Guiana (which became a British colony only in 1814, having been seized from the Dutch) gained its independence as the Cooperative Republic of Guyana in 1966. It is a member of the British Commonwealth. Neighboring Surinam, formerly Dutch, became fully independent in 1975. French Guiana is an overseas department of France. The Falkland/Malvinas Islands, with its dependencies of the South Georgia and Sandwich islands, is a British colony. Argentina has long contested claims to sovereignty there.

Power and patronage

Independence brought no significant change in political structure to the former Spanish and Portuguese colonies. Indeed, divisions that had existed earlier between a privileged minority and an impoverished peasant and worker population, often racially and linguistically separated from the ruling class, became even sharper after independence. The South American rulers handed out patronage in the form of land or position, and whole countries were run like private estates.

Political instability was always endemic – Bolivia had some sixty revolutions in the first century of independence, and Colombia twenty civil wars – and remains so to this day. Toward the end of the 19th century the scramble by European and North American investors to share in the exploitation of South America's agricultural and mineral resources created ever more dramatic extremes of wealth and poverty, increasing the tendency toward instability, especially during times of economic recession. When revenues fell compliance could not be rewarded, and rulers either lost their political support, or resorted to force to stay in power.

Those states whose economies were more stable did see some attempt to introduce democratic forms of government. By the beginning of the 20th century, for example, universal male suffrage and a secret ballot were operating in Argentina, but real political power remained in the hands of the great landowners. In Brazil, beneath the trappings of democracy, the army provided the ruling force. In Venezuela there was no pretense – the wealth generated there by oil flowed straight into the hands of a dictator, Juan Gomez, whose regime (1908–35) was noted for the savagery of its

Congress Building, Brasília, one of architect Oscar Niemeyer's innovative designs for Brazil's new capital, begun in 1960. Some 960 km (600 mi) inland, the city was planned as a monumental symbol of national unity.

Abuse of human rights is high in the region. The "mothers of the disappeared" appeal to the Argentinian government for news of their children who vanished during the "dirty war" of 1976–83, when thousands were killed.

States of South America

1812 date of independence

colonial power
- British
- Dutch
- French
- Portuguese
- Spanish

Independent states Most of the region had freed itself from European rule by the early 19th century. Many territorial boundaries are still disputed.

secret police. Only in Uruguay – where the population was mostly of European origin – and Chile did democracy become more than a form of words.

The rise of populism

Whatever their style of government, all the states of South America were increasingly dependent on the United States for economic support. The devastating effects of the Great Depression of 1929 on their export-based economies, followed by World War II, led to the rise of populist leaders, such as Getulio Vargas (1883–1954) in Brazil and Juan Perón (1894–1974) in Argentina. Having broken the power of the landowning aristocracy, they won popular support by offering paternalistic programs that rewarded the workers with higher wages.

Perón was elected president of Argentina in 1946. The enormous popularity of his wife Eva (1919–52), a former actress, was to contribute to his own charismatic leadership. Industrialization was encouraged under his regime, and social welfare programs were introduced. Banks, transport and public utilities were nationalized. Perón's populist policies worked only

in favorable economic circumstances. As expansion declined, prices fell, and his regime become more repressive: political opponents were imprisoned, and civil liberties suppressed. In 1955 he was overthrown by a military coup, backed by the church and the landowners.

After 1959, following the success of the Cuban revolution, Marxist-led guerrilla groups pressed for social change throughout South America. The armed forces, in the interests of national security, seized control of the state apparatus set up by the populists. In 1964 there were army coups in Brazil and Bolivia, followed by Argentina (1966) and Peru (1968). Finally, in 1973 the military seized control in the two South American states in which democratic institutions had seemed most firmly planted, Chile and Uruguay. In Chile, President Salvador Allende Gossen (1908–73) – the world's first democratically elected Marxist leader – was murdered in a military overthrow engineered by General Augusto Pinochet Ugarte. In Uruguay an increasingly rightwing and repressive government was replaced by the army, which instead of restoring a democratic form of government imposed its own brutal rule.

CONSTITUTIONALISM IN A TIME OF CHANGE

The 1970s were a decade of militarist rule in South America: only Colombia and Venezuela were without military regimes during this period. Perón was allowed to return briefly to power in Argentina in 1973, but the attempt to restore populist policies, carried on by his second wife Isabel after his death in 1974, was once again ended in 1976 by a military coup and Congress was dissolved.

South America's governments were in many respects merely a continuation of the traditional ruling interests in the region, combining political conservatism and repressive methods of government with economic policies that encouraged foreign capital investment and increased state expenditure. The military rulers themselves thus laid the foundations of the crippling debts that were to contribute to their downfall.

The return of constitutionalism
Most of South America had returned to civilian rule by the mid 1980s. Ecuador (1979), was followed by Peru (1980),

Bolivia (1982), Argentina (1983), Uruguay and Brazil (1985), and Paraguay (1993).

In 1988 55 percent of Chile's electorate voted in a referendum to reject a further term of office by General Pinochet leading to the election of a Christian Democrat, Patricio Aylwin as president in 1990. Eduardo Frei succeeded Aylwin as president in 1993 in the first democratic handover of power since 1970.

When Carlos Menem was sworn in as president of Argentina in July 1989, it marked that country's first democratic change of government in over sixty years. Argentina's economic crisis (including a 30-day state of emergency, after rioting following price measures and high inflation) forced him to take office five months sooner than the constitution allowed for the transfer of power. In 1990, his decision to restore diplomatic relations with Britain, ruptured since the Falklands/Malvinas war in 1982, was the cause of an unsuccessful army revolt.

The last remaining military dictatorship in South America, Paraguay, was superseded in 1993 when the country's first fully democratic elections resulted in victory for the ruling Colorado Party headed by Juan Carlos Wasmosy.

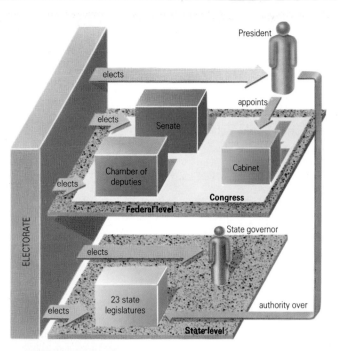

The trappings of democracy The lists of voters are closely checked, but only one name – that of Alfredo Stroessner – appears on the ballot paper to elect Paraguay's president. After winning power in 1954 Stroessner was "re-elected" seven times before his removal by the army in 1989. At that time only one party – the Colorado Party – was officially recognized. It remained the ruling party, though a multiparty system was restored in 1989.

A federal constitution Brazil is a federal republic, with 23 constituent states, 3 territories and a federal district (Brasília). The two-chamber national assembly consists of a 69-member senate elected for eight years and a 487-member chamber of deputies elected for four years. Executive power is exercised by the president, elected directly by universal suffrage for a 5-year term, who appoints and leads the cabinet.

The transition to democracy Crowds in Brazil demonstrate in support of Tancredo Neves, leader of the Liberal Party Front (PFL) in 1985. Elected by Congress as Brazil's first civilian president for 21 years, Neves died before taking office. He was succeeded by the vice president, José Sarney. In 1987 the PFL ended its coalition with Sarney.

THE DRUG BARONS

The Indians of the Andes have long chewed coca leaves, from which cocaine is produced, to reduce hunger pangs and as a source of vitamins and minerals. Only recently have they grown these plants to supply the international market for illicit drugs – a development that must be looked at in the context of South America's economic recession and the decline in trade.

The opening up of land in the 1970s on the eastern slopes of the Andes provided isolated, fertile areas where the coca bush could be secretly planted and cultivated. By the end of the 1980s Colombia had about 25,000 hectares (62,500 acres) given over to its production, Bolivia 50,000 hectares (125,000 acres) and Peru 100,000 hectares (250,000 acres). Profits from illicit drug exports make a major contribution to the economies of these countries; the laundering of drug money involves many of their financial institutions.

The growth of the trade in narcotics has created a new oligarchy that undermines the authority of national governments. In Peru the left-wing terrorist group known as the Shining Path strengthened its grassroots support by negotiating with the drug traffickers to obtain better prices for the peasants whose livelihood is dependent on coca-growing. The investment of drug profits in real estate by the powerful drug barons in Colombia means that one-twelfth of the country's productive land is in their hands, and they have won popular support by building housing for the poor.

Recognition that this was an international problem came in 1989 when the United States and other world governments responded to Colombia's desperate appeal for financial and policing aid to curb the activities of the barons. Their intimidatory campaigns, with bombings and assassinations, which undercut the ability of government to protect its people, continued.

Street sellers Bolivian farmers sell their crop of coca leaves at a local market.

Strains and stresses

South America's new democracies are fragile. They are often threatened by the economic strains imposed by international debt repayments. These have led to the imposition of austerity programs that have increased the unpopularity of governments. The pressures are particularly severe in the rapidly growing major cities of the region, where organized labor is strong, and rising prices and unemployment have led to strikes and food riots. In Argentina, a 30-day state of emergency was declared after price measures to curb soaring inflation led to violent rioting.

In Brazil, the right-wing policies of José Sarney failed to cope with the country's social and economic problems. It led in 1988 to the election of a woman member of the Socialist Workers Party as mayor of São Paulo, the largest city in the southern hemisphere. In 1989 Fernando Collor won power in the first direct presidential elections since the military coup of 1964 with a promise to renegotiate the foreign debt and root out corruption. However, his anti-inflationary measures incurred even greater unpopularity, and in 1992 he was charged with corruption and removed from office.

The left-wing Shining Path guerrillas posed an increasing threat to order in Peru. In 1990 the Social Democrat Alan García Pérez (president since 1985) was defeated in elections by Alberto Fujimori, who suspended the constitution in 1992. The leader of the Shining Path was arrested, yet terrorist attacks did not abate, and the army was implicated in its war against the guerrillas in deals with the barons of the cocaine trade to supply weapons and airstrips. As in Bolivia and Colombia, the drug barons have created an alternative elite, and attempts to control them are met with intimidation and bribes.

In the newest independent state of the region, Surinam, democracy is also very fragile. A military coup in 1980 was soon followed by successive coups and counter-coups (six in all between 1980 and 1983). After the deaths in custody of 15 opposition leaders in 1982, the Netherlands' government suspended grant aid to its former colony. Civilian government was restored with the holding of elections in November 1987 but guerrilla warfare continued and the army still exercised a powerful control.

GETTING ON NEIGHBORLY TERMS

During the 19th century, and well into the present one, the international relations of the South American states were governed by its economic role as a world supplier of raw materials – food and grain from Argentina, silver from Bolivia, rubber from Brazil, nitrates from Chile. Foreign investment, particularly from Britain, northwestern Europe and the United States, into South American mines, agriculture, railroads and ocean transport was extensive. But the rapid expansion of its export-based economies, which were often reliant on a single product, made the region particularly vulnerable to fluctuations in world demand. The effects of the Great Depression of 1929 were devastating here as elsewhere, severing supplies of foreign capital and lowering prices obtained for many products.

The pattern of external dependency has continued. Rapid industrial growth during and after World War II, when South America was cut off from foreign sources of consumer goods, increased reliance upon imported capital, technology and finance, with the effect of creating enormous foreign debts. The penetration of foreign subsidiaries into national industries, and the rise of the multinational corporations, caused many governments

to redefine their international relations in the 1970s and 1980s. The presence of foreign interests also increased domestic social tension. In Brazil, for example, the destruction of the rainforest by multinational companies began to arouse protest, particularly from threatened Indian communities.

The best hope of breaking the cycle of dependency and of speeding up social and economic development was by creating closer links between the states of the region. Cooperation between them had always proved difficult in the past, largely because settlement was generally limited to coastal areas.

Indians in the Amazon protest in 1989 against plans to build a massive dam for hydroelectricity that will destroy their lands and livelihood. In 1992, Rio de Janeiro hosted the UN-sponsored Earth Summit that sought international cooperation on the environment.

Obstacles to integration

Overland transport routes between the countries of the region are still few and far between, and communication remains easier with traditional trading partners overseas than with neighbors. Additional friction arises from long-standing rivalries between states for leadership of the region, and jealousy of the largest, non-Spanish speaking state, Brazil. Border disputes were, and still are, frequent.

THE FALKLANDS/MALVINAS WAR

British claims to the remote and sparsely populated Falkland Islands (named after Lord Falkland, treasurer of the British navy at the end of the 17th century), 480 km (300 mi) off the coast of Argentina, date back to the earliest British settlement in 1765. They are vigorously contested by the Argentinians, who know them as the Islas Malvinas. During the 1960s discussions were held over the sovereignty issue, but Britain insisted on the islanders' right to self-determination: they have always opposed cession.

On 22 March 1982 a group of Argentinian scrap metal merchants raised the Argentinian flag on South Georgia, a dependency of the Falkland Islands. On 2 April Argentinian troops occupied the Falkland Islands themselves, expelling the British governor.

Britain at once brought economic, diplomatic and military pressure to bear against the Argentinians. Resolu-

Alfredo Astiz, the Argentinian commander in South Georgia (later tried for human rights abuses) signs the document of surrender.

tion 502 of the United Nations Security Council called for the withdrawal of the Argentinian troops and peaceful settlement of the dispute. Almost every state of South America supported the Argentinian action, but Argentina failed to win the backing of the United States, despite invoking the Rio Treaty, which

provides for members of the OAS to give military assistance to one another. By 22 April, when the British naval task force reached the area, United States' support had tilted decisively in favor of its "special relationship" with Britain.

Military operations began at the beginning of May with the sinking of the Argentinian cruiser *General Belgrano*. On 21 May British troops landed at San Carlos, and by 15 June the Argentinian garrison at Port Stanley, the islands' capital, had surrendered.

Argentina's action in initiating the conflict was taken by General Leopoldo Galtieri, who at that time headed the military regime, to divert attention from economic crises and to bid for popular support. Defeat discredited the government: Galtieri was removed from power and later sentenced to 12 years' imprisonment for his mismanagement of the war, which led to the loss of over a thousand lives.

A nation unites National flags wave proudly as people in Buenos Aires take to their cars to celebrate Argentina's invasion of the Falkland/Malvinas Islands in 1982. The war diverted attention from the military junta's problems, but defeat hastened its downfall.

The Spanish empire in South America had been divided into administrative regions (viceroyalties and captaincies). These formed the territorial bases of the new states at the beginning of the 19th century. Boundaries were drawn through unpopulated areas and were often based on inaccurate maps. Subsequent border disputes, particularly between Argentina and Chile, and Ecuador and Peru, frequently erupted into armed conflict. After losing a war with Chile (1879–84) Bolivia had to give up its coastal territories and valuable mineral-rich lands, and further territory was lost to Paraguay after the Chaco war of 1932–35.

Many borders are still unsettled. In 1981 Peru and Ecuador almost went to war over disputed territory. Five-eighths of Guyana's territory (the province of Essequibo) is claimed by Venezuela.

Closer ties

It is against this background that the states of South America have tried to develop closer regional ties. The Latin American Free Trade Area (LAFTA), set up by the Treaty of Montevideo in 1960, aimed to encourage industrial development on a coordinated regional basis. Talks began between Argentina, Brazil, Chile and Uruguay in 1958, and the other countries of the region (plus Mexico) soon joined. Venezuela was the last, in 1972.

Trade between LAFTA states still amounts to only about 10 percent of total regional trade. The more developed economies of Argentina and Mexico, followed by Brazil, have benefited most from membership of LAFTA, and this was resented by the poorer countries. The Andean Pact, created in 1969 between Bolivia, Colombia, Ecuador, Peru and Venezuela within the framework of LAFTA, foresaw the creation of a real economic union with not merely the liberalization of trade, but also joint industrial planning and the harmonization of economic policies.

In 1980 the Latin American Integration Association (ALADI) replaced LAFTA; this change was an attempt to revitalize efforts to bring about the economic integration of the region.

All the states of the region are members of the Organization of American States (OAS). This was founded in 1948, though its pan-American origins go back to the 19th century. Its aims include the improvement of both social and economic cooperation thoughout Latin America, but with its quarters in Washington DC it is seen to be heavily influenced by United States' interests, and its neutrality has been called into question. Its weakness and disunity were demonstrated during the Falklands/Malvinas conflict in 1982.

Chile: the suppression of human rights

In 1973 General Augusto Pinochet Ugarte replaced the democratically elected government of Salvadore Allende Gossen in Chile with a military regime that was authoritarian and brutal in its repression of discontent. Allende's victory three years earlier had brought into power one of the world's first elected Marxist governments, and implementation of its radical program of economic and social change, including nationalization of the country's largest industries, agrarian reform and a program of exchange and commodity control, had led to an increase in unrest and growing violence.

Pinochet was determined to destroy the very fabric of democracy that had allowed the Allende government to come to power. This involved destroying or re-organizing anything considered to be an expression of support for the previous government. Estimates of the numbers of political killings in the first few months of the Pinochet regime range from 5,000 to 30,000; political detentions numbered about 65,000. A state of emergency was declared, which allowed the establishment of military courts and detention without trial.

As a result, Chile came to symbolize human rights atrocities in South America. Yet its record was no worse than those of the suppressive regimes in power in Brazil (1964–85) or Uruguay (1972–85), nor of the perpetrators of the so-called "dirty war" in Argentina (1976–83), in which thousands of men, women and children – adjudged left-wing opponents of the military junta – "disappeared". What was unusual in the case of Chile was the amount of information about state repression that was circulated at the time, not only within the country but also throughout the world by church groups opposed to human rights abuses – particularly the "Vicarate of Solidarity" and the "Relatives of the Detained".

Despite the government's use of terror tactics, opposition and protest grew throughout the 1980s, to be greeted with mass arrests. In November 1984 Pinochet decreed a state of siege and used government forces to smash demonstrations and arrest opposition leaders. The campaign of protest and bombings continued, and in September 1986 an assassination attempt on Pinochet brought brutal government response: strict censorship was reintroduced and right-wing death squads again became active.

The publicity surrounding Chile's use of repression reached new heights, and even began to damage the government's international reputation so that it was forced to comply with international safeguards on human rights. In September 1987 Pinochet signed a United Nations convention outlawing torture. Despite this, it was only a year later that the UN general assembly condemned the Chilean government for its violation of human rights for the thirteenth consecutive year.

The "transition to democracy"
The constitution approved by Pinochet in 1980 set the framework for Chile's "transition to democracy" by allowing for presidential elections to be held every eight years. In addition, no president could serve more than one term. In 1988 a referendum was held to extend Pinochet's term as president by a further eight years: the answer was a decisive rejection by the Chilean people.

The first election under the new constitution took place in 1989 and was won

General Pinochet had been appointed army commander in chief only 18 days before the coup that overthrew Allende in 1973. Damage to Chile's international reputation caused him later to moderate the worst excesses of military rule, but he demonstrated a great reluctance to relinquish power.

Beneath the heel of authority For nearly two decades the army in Chile wielded unlimited power and political protest was severely restricted. In the early days of its repressive regime thousands of people were killed and tortured, and many more detained without trial.

A resounding No Given the chance to decide whether or not to extend Pinochet's term of office in 1988, 55 percent of the 7 million voters taking part in the referendum turned him down. A year later they rejected his choice of candidate to succeed him and voted in a center-left president.

by Patricio Aylwin, the leader of the Christian Democrats who headed a center-left coalition, and who pledged to restore democracy. Before the election, Pinochet had moved to secure a role for himself in any future elected government by confirming his intention of remaining as commander in chief of the army for an indefinite period, and announcing conditions with which the civilian government would be expected to comply. These included the right of the military to oversee political development through a strong National Security Council and to control general defense policy.

Many feared that the presence of Pinochet and his supporters in high military posts would give only the semblance of democracy to Chile's elected government and that the military fist would remain within the civilian glove. However, in 1993 Eduardo Frei, the presidential candidate of the ruling coalition achieved a landslide electoral victory, winning 58 percent of the vote, the largest popular backing of any Chilean leader since 1931. Frei promised to introduce constitutional reform, including recovering for the president the power to appoint and remove senior military officers.

NORTHERN DEMOCRACIES

A SHARED PAST · GOVERNING BY CONSENSUS · KEEPING A NORDIC BALANCE

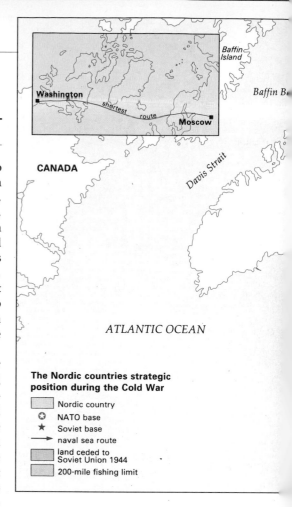

The five Nordic states have evolved over a thousand years from tribes who gradually recognized the overlordship of rulers based in central Norway, eastern Sweden and eastern Denmark. During the migrations of the Viking period people from these areas settled in the Faeroe Islands and Iceland. Finland became a duchy of Sweden; it was later annexed by Russia. For a time all the states of the region recognized the sovereignty of Denmark. Sweden was the first to become independent of the Danish crown, in the 16th century. Today all are parliamentary democracies. Denmark, Norway and Sweden are monarchies, Iceland and Finland republics. The largest island group – the Faeroes – is part of the Danish state; Jan Mayen and the Svalbard islands belong to Norway, the Åland Islands to Finland.

COUNTRIES IN THE REGION

Denmark, Finland, Iceland, Norway, Sweden

Island territories Åland Islands (Finland); Faeroes (Denmark); Jan Mayen, Svalbard Islands (Norway)

Territories outside region Greenland (Denmark)

STYLES OF GOVERNMENT

Republics Finland, Iceland

Monarchies Denmark, Norway, Sweden

Multi-party states Denmark, Finland, Iceland, Norway, Sweden

One-chamber assembly Denmark, Finland, Norway, Sweden

Two-chamber assembly Iceland

CONFLICTS (since 1945)

Interstate conflicts Iceland/UK 1956–58, 1974–78 (Cod Wars)

MEMBERSHIP OF INTERNATIONAL ORGANIZATIONS

Council of Europe Denmark, Iceland, Norway, Sweden

European Community (EC) Denmark

European Free Trade Association (EFTA) Finland, Iceland, Norway, Sweden

North Atlantic Treaty Organization (NATO) Denmark, Iceland, Norway

Nordic Council Denmark, Finland, Iceland, Norway, Sweden

Organization for Economic Cooperation and Development (OECD) Denmark, Finland, Iceland, Norway, Sweden

Notes: Iceland has no military forces and is not a member of NATO Military Command.

Norway has a territorial claim in Antarctica.

A SHARED PAST

The modern Nordic states began to emerge at the beginning of the 19th century. The Treaty of Vienna (1815), which concluded the Napoleonic wars, confirmed the separation of Finland from Sweden and its new status as a grand duchy of the Russian empire; it was allowed a measure of self-government. The Norwegians set up an independent parliament in 1814 but were forced into an unequal union with Sweden under a common monarch. External affairs were controlled by Sweden.

The Faeroes, Iceland and the recently acquired colony of Greenland remained with Denmark, but the disputed German-speaking territories of Schleswig and Holstein in the south, acquired in the 15th century, were ceded to Germany in 1864: in 1920, after a plebiscite, the Danish–German border was moved about 50 km (30 mi) to the south in order to bring areas with a Danish-speaking majority within Denmark.

Norway separated wholly from Sweden in 1905, and Iceland from Denmark in 1944. Following a referendum, Greenland was granted full self-government in 1981. Finland declared its independence from Russia during the revolution of 1917–18. In 1939 the two countries were involved in the 15-week "winter war", which Finland lost. It later joined Germany in attacking the Soviet Union, but agreed a separate armistice in 1944, with the loss of territory in Vyborg and southern Karelia, northwest of Leningrad.

About 33,000 Lapp (Sami) people are scattered over northern Norway and Swedish and Finnish Lapland. State boundaries in Lapland were marked out between 1751 and 1826, but those Lapps who herded reindeer were allowed to cross frontiers during their migrations. The establishment of a common labor market among the Nordic states in 1954 extended these rights to all workers.

Parliamentary constitutions

As a result of their shared history, all five states have many political features in common. All have long traditions of the rule of law and of the making and enforcement of the law at representative assemblies (the Icelandic parliament, the Althing, was first established in 930). Although the crowns of Denmark and

---

Cold War vulnerability The Nordic Countries – on a direct route between Washington DC and Moscow – were vulnerably positioned between the two superpowers of the Cold War, and awareness of the threat of nuclear war was high. Soviet nuclear submarines, leaving their bases in the Baltic and White Seas, passed close to the shores of all five countries.

Sweden had accumulated power over state and church, which were very closely linked, amendments to their 19th-century constitutions gradually established the supremacy of parliaments. By the early 20th century the decisions of government required the support of a majority in elected parliaments.

Norway chose a king at independence in 1905, but Finland (1918) and Iceland (1944) chose presidents. The Finnish president remains a powerful constitutional figure. Elected every six years, the president can dissolve parliament, conduct foreign relations (subject to review by parliament), and delay legislation. The Icelandic president may exercise some influence on the formation of governments. The same is true of the Danish and Norwegian monarchs, but in Sweden this function devolves on the speaker of parliament, and the role of the crown is virtually ceremonial.

The island territories

Denmark, Finland and Norway's island territories are now largely self-governing

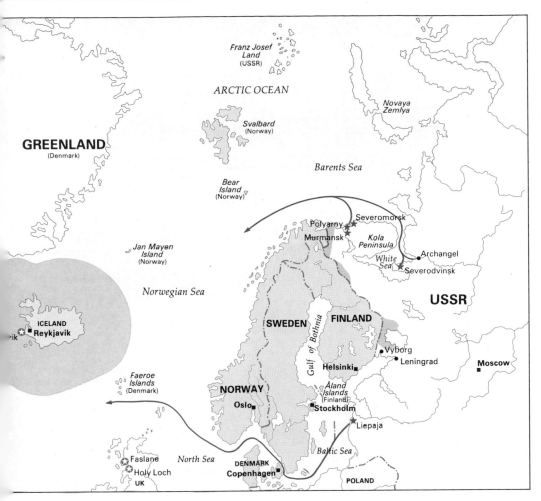

GOVERNING BY CONSENSUS

Nordic society is structured to encourage shrewd appraisal, caution, and conservative attitudes toward change, and the close cultural and linguistic links between the states of the region have allowed the development of similar political institutions and parties in each one. Political interest is high, with over 80 percent turnouts normal in general elections, which are held at fixed four-yearly periods (three years in Sweden): the Norwegian parliament cannot be dissolved between elections. Parliaments are elected by proportional representation, and the prime minister and cabinet drawn from the party, or group of parties, commanding a majority of votes.

Parliaments can modify the constitution, as happened in Sweden in 1971 and 1975 during the premiership of Olof Palme (1927–86). However, they are limited by the constitutionally defined powers of many administrative agencies that are not subject to direct intervention by ministers, though they may enjoy informal links with them. These agencies are responsible to the cabinet, and some of them, such as Swedish Railways, may operate as commercial enterprises, deriving part of their income from the state. Political parties and relevant interest groups, such as trade unions and employers' organizations, are represented on their governing boards.

The machinery of consultation

Parliaments have one chamber, though in Norway the elected members divide into two groups to discuss legislation. They are more concerned with the discussion than the formulation of policy, and a great deal of parliamentary work, especially the drafting of new laws, is done by standing committees with all-party membership proportionate to the number of votes cast. Widespread consultation is normal, and sometimes mandatory, before new policy can be made, and the administration agencies are strongly represented on the commissions of inquiry. Interest groups and other organizations may also be involved; consultation thus imposes a lengthy time-scale on policy change.

National referendums may also be held to decide policy on such issues as whether Norway and Denmark should join the European Community (EC) (in

island territories. The Faeroe Islands have a local assembly but are also represented by two members in the Danish parliament, which legislates for foreign affairs, law, social affairs and education. The Åland Islands, confirmed as a part of Finland by a League of Nations agreement of 1921, constitute a self-governing, Swedish-speaking province of Finland with its own assembly, which agrees local legislation with the Finnish president.

The Svalbard islands, of which Spitsbergen is the largest, were unpopulated until 1906. In 1920 Norway was granted sovereignty over Spitsbergen, but seven other states share the right to exploit minerals there, and coal is still mined by Norwegian and Russian corporations. Today the islands have a population of approximately 1,400 Norwegians and 2,500 Russian citizens. No military installations are allowed on the islands.

Celebrating independence On Independence day (17 May) Norwegians, many wearing national costume, parade down Oslo's Parliament Hill. A young nation by European standards, Norway did not become fully independent until 1905. It remains fiercely proud of its national identity and traditions.

1972) or Sweden abandon nuclear power (in 1980). The press has constitutional guarantees that enable it to act as a watchdog over government. The institution of an independent investigator of complaints against maladministration (the ombudsman), first introduced in Sweden in 1713, has spread to the other countries of the region and farther afield as well.

The party line-up is not identical in all five states, but parliaments may include – from left to right across the political spectrum – communists, socialists, social democrats, a center party (formerly representing agrarian interests), a Christian party, liberals and conservatives. In the late 1950s "progressive" parties were formed with programs to reduce taxation, which has been at a high level in order to maintain the comprehensive systems of state welfare that are characteristic of

Policy making in Sweden The Swedish parliament (Riksdag) discusses policy, which is decided by the prime minister and cabinet in consultation with the political parties and other interest groups inside and outside parliament. Execution of policy is in the hands of about a dozen administrative agencies.

WOMEN IN POLITICS

Women in political systems throughout the world have a disproportionately small role in government. Scandinavian countries were among the first to give women the vote at the beginning of the 20th century, and Scandinavian women now achieve a higher share of parliamentary seats than in any other of the world's democratic assemblies, ranging from 20 percent in Iceland to 37 percent in Sweden. The only other states to approach this were some communist regimes, for example East Germany and the Soviet Union. Women continue to be poorly represented in major democracies such as Britain, France, Japan and the United States, where representation may be below 6 or 5 percent.

Between 1945 and 1979 individual women had achieved a major position of power in only four states – Britain (Margaret Thatcher), India (Indira Gandhi), Israel (Golda Meir) and Sri Lanka (Sirimavo Bandaranaike). Women generally remained underrepresented in their national assemblies. Iceland elected a woman president, Vigdis Finnbogadottir, in 1980 and Norway a woman prime minister, Gro Harlim Brundtland, in 1981 and 1986. Half the Norwegian cabinet ministers are women. A European Commission survey in 1987 found that only 7 percent

A woman prime minister Gro Harlim Brundtland, Norway's former prime minister, with some of her male colleagues in the cabinet.

of Danish men (compared to 30 percent in Germany) expressed greater confidence in men as political leaders.

Ten major political parties in Scandinavia and Finland have a quota (usually 40 percent) of party offices for women. A Women's Alliance party was set up in Iceland in the belief that traditional political parties incorporate women within male political structures. In 1983 it became the first specifically women's party to sit in a national parliament, and in 1987 won six seats. Women are also well represented on county and district (municipal) councils. In 1971 Oslo city council had a majority of women members; parity of male and female members is not unusual on local councils.

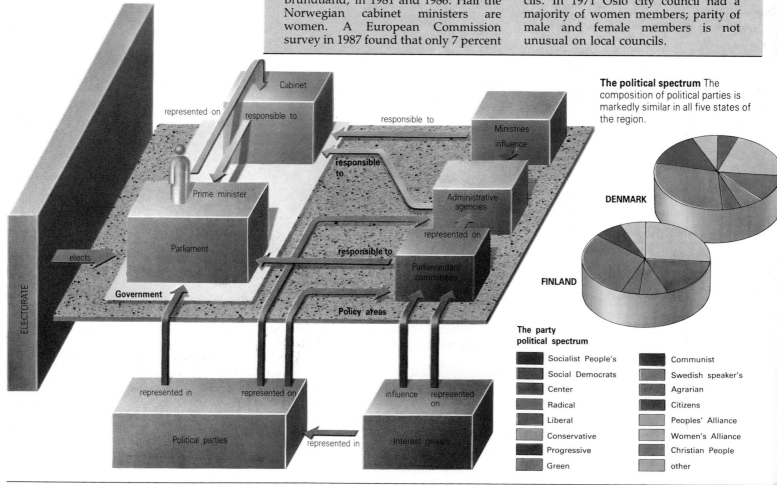

The political spectrum The composition of political parties is markedly similar in all five states of the region.

DENMARK

FINLAND

The party political spectrum

- Socialist People's
- Social Democrats
- Center
- Radical
- Liberal
- Conservative
- Progressive
- Green
- Communist
- Swedish speaker's
- Agrarian
- Citizens
- Peoples' Alliance
- Women's Alliance
- Christian People
- other

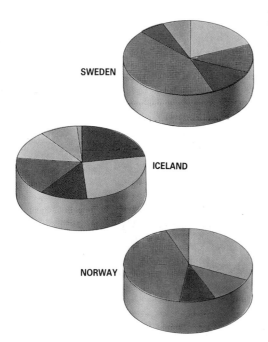

Swedes mourn Flowers are piled at the place where Olof Palme (1927–86), twice Labor prime minister and a champion of Third World policies, was shot dead in an apparently motiveless crime.

SWEDEN

ICELAND

NORWAY

Scandinavia. Some of the progressives have subsequently opposed liberal immigration policies.

Party loyalty is very strong among voters, and the proportion of seats held by each party changes only slightly from one election to the next. The number of seats held by the parties of the left and right are evenly balanced, but the parties on the right find it more difficult to form a common program than do those on the left. These circumstances enabled the Social Democratic Labor Party of Sweden to form the government (usually in alliance with some of the smaller parties) from 1933 to 1976 and again from 1982 to 1991, though it has rarely enjoyed an absolute majority of seats.

Sectional interests

A minimum-vote threshold excludes the smallest parties from parliament but, outside Sweden, small parties – often representing limited sectional or regional interests – have survived to influence policies by supporting larger parties that

lack an overall majority. In Iceland, for example, there is a women's party, and in Finland a Swedish-speaker's party. In 1988 the Green party was represented for the first time in the Swedish parliament. Representation on administrative agencies and local government bodies is also usually proportionate to party voting strength, and so party leaders have strong incentives to reach compromise agreements with each other.

The larger political parties are pragmatic rather than ideological, always seeking to widen their appeal to voters in the middle ground. Social Democratic governments have been primarily concerned with the welfare of individuals and regions rather than with state ownership of industrial firms. Coalition governments of the right have continued to support state welfare provision. Unlike many other Western European states, where each government may counter the programs of its predecessor, the changes implemented by Scandinavian governments are less dramatic and wideranging.

KEEPING A NORDIC BALANCE

Cooperation between the neighboring states of Scandinavia and Finland has always depended upon achieving a balance between the common interests that pull them together and the need to maintain independent relations with other powers, which may drive them apart. For most of the period since World War II each has sought close cooperation with the others through a series of cultural, political and economic alliances. The 1970s and 1980s, however, saw some attempt to assert individual state interests more strongly.

The chief instrument of cooperation is the Nordic Council, which was established in 1951. Finland joined in 1955. The representatives from all five parliaments and cabinets who comprise the council consider many issues of common interest,

Saying no to Europe In summer 1992 Danish voters confounded politicians moving toward greater integration within the European Community by voting in a national referendum to reject the terms of the Maastricht treaty, which would have taken that process a stage further. Here campaigners against the treaty are canvassing support in Copenhagen for a No vote. Ratification of the treaty by all the member states was necessary to implement its clauses, and so the decision by Denmark – one of the smallest countries in the EC – caused a major political upset.

World leaders at Reykjavik The Icelandic capital was chosen in 1986 as the meeting place for President Ronald Reagan of the United States and Mikhail Gorbachev of the Soviet Union. Following the success of their preliminary discussions here, the first nuclear arms reduction treaty between East and West was signed a year later in Washington DC.

excluding national defense and relations with non-Nordic countries. It has set up planning regions across state boundaries and initiated research to explore common social and cultural problems. Considerable progress has been made in harmonizing social legislation.

There has been less success in developing common economic policies, though steps toward this have included the establishment of the Scandinavian Airlines System (SAS), comprising the national airlines of Denmark, Norway and Sweden, in 1950 and a combined railway tariff system in 1953. Proposals for a Northern Economic Union (Nordek) with a common external tariff and trade policy were abandoned in 1970 when Finland felt unable to join. Iceland chose not to participate in proposals for economic union because of concern over markets for its principal export, fish.

Intense political debate was generated by Denmark's and Norway's applications to join the European Community (EC). Denmark joined in 1973, but farming, fishing and some rural manufacturing interests united to defeat Norway's application. Swedish business interests have always favored joining but successive governments felt that the provisions for entry laid down in the Treaty of Rome (1957), which established the EC, would breach Sweden's longstanding and highly regarded policy of political neutrality.

Despite these fears, in July 1991 Sweden applied to join the EC, and Finland and Norway later indicated that they would also apply. However, the ambivalent feelings that the Nordic Countries have for the EC were shown yet again in June 1992 when the Danish people voted not to ratify the Maastrict Treaty, setting out the ground for still greater integration within the EC, which needed the approval of all member states.

Questions of defense

Apart from Sweden, which has not been involved in war since 1806, the region was drawn into the conflicts of World War II: Finland fought the Soviet Union twice; Denmark and Norway were invaded by Germany; the Faeroe Islands and Iceland were occupied by British and US forces. Immediately after the war Sweden failed to persuade its neighbors to set up a common defense policy. Finland signed a non-aggression pact with the Soviet Union in 1948; Denmark, Iceland and

ICELAND'S COD WARS

Traditionally, the "high seas" were held to be under no national jurisdiction, and the claims of maritime states over coastal waters extended only to 5 km (3 nautical mi). The development of huge factory ships to operate in distant waters and other modern methods of fishing, as well as the exploitation of fuel and other mineral resources, made states more conscious of the potential value of their coastal waters. Fishing territories were ever more aggressively protected and extended, leading to a series of disputes between states, the most famous of which were the so-called "Cod Wars" (1956–58 and 1974–78) between Iceland and Britain and other Western European fishing nations. Its total dependence on fish exports led Iceland to extend its territorial fishing waters, and to protect them with gunboats. In the mid-1970s several clashes took place in the North Atlantic when British naval frigates were sent to support trawler fleets fishing in the disputed waters.

It was largely as a result of pressure from Iceland that the United Nations Conference on the Law of the Sea (UNCLOS) in 1982 recognized a territorial waters limit of 22 km (12 nautical mi) for coastal states, with further provision for an exclusive economic zone of 370 km (200 nautical mi) in which fishing and mineral rights are protected.

Conflict at sea An Icelandic gunboat alongside a British trawler fishing in the North Atlantic at the height of the Cod War in the 1970s.

Norway joined the North Atlantic Treaty Organization (NATO) a year later. Except for US-maintained NATO bases at Thule in Greenland and Keflavik in Iceland they do not allow the permanent stationing of foreign troops or nuclear weapons on their territory. Supporters of the concept of "Nordic Balance" argue that the limited commitment to NATO of Norway and Denmark helps to maintain the political independence of Finland.

Finland's postwar treaty (the Mutual Assistance Pact or YYA Treaty) with the Soviet Union bound it to repel attacks on the Soviet Union across Finnish territory. By its terms Finland's neutrality was respected. While it retained its capitalist economy, parliamentary democracy and trading links with Western Europe, the defense clauses were designed to prevent the smaller power from becoming a springboard for an attack on Leningrad by Western powers. The Soviet government dissuaded Finland from joining Nordek in 1970, but refrained from overt interference in internal Finnish politics.

In the early 1980s there were suspicions that Soviet submarines were lurking around Swedish naval bases. These appeared to receive confirmation when one ran aground in 1982. This led to a substantial increase in the Swedish defense budget for 1987–92, and heightened enthusiasm among the Swedish people for the establishment of a nuclear-free zone in northern Europe that would encompass not only the Nordic states but also the heavily militarized Kola Peninsula.

The Nordic Countries played a prominent role in the process that led to the separation of the Baltic states from the Soviet Union in 1991. Iceland was the first country to recognize the new states on 26 August 1991, and Sweden was the first to open an embassy in Vilnius, capital of Lithuania.

The role of the peacekeeper

The Nordic states have a long history of working to promote international peacekeeping operations and of fostering peace-seeking conferences and institutions. In 1917 the kings of Denmark, Norway and Sweden made the first proposal for a League of Nations, an idea that had originated with Fridtjof Nansen (1861–1930), the Norwegian polar explorer. Scandinavians have played an active part in the work of the League and its successor, the United Nations, providing many of its diplomatic personnel and high-ranking officials, including the first two general secretaries of the UN, the Norwegian Trygve Lie (1896–1968) and the Swede Dag Hammarskjöld (1905–61).

All the Nordic states support capitalist economic systems yet have enjoyed long periods of socialist-dominated politics. This made their governments particularly acceptable as intermediaries between the superpowers of East and West: they condemned those policies of both superpowers that appear to increase international tension. They perform a worldwide role in attempting to bring a speedy end to armed conflict wherever it may happen to take place, and offer negotiating facilities to warring sides once other aspects of international diplomatic machinery has been found to fail.

Troops from the region have served in UN peacekeeping forces from the Congo in the 1960s to Cyprus, Lebanon and Namibia in the 1980s and Bosnia in the 1990s. In addition, Nordic governments have tried to promote economic stability in the Third World to reduce the danger of conflict there, and have consistently supported programs for dialog between North and South with the ultimate objective of a new international economic order. Gro Harlim Brundtland chaired a UN Committee on Environment and Development (1983–87), which published a report, *Our Common Future*, listing threats to the environment and calling for development policies to feed the world's increasing population. Nordic governments have strongly opposed the apartheid policies of South Africa and many other violations of human rights.

Peace conferences

Several peace conferences have been held in the region. Finland hosted the first Strategic Arms Limitation Treaty talks (SALT I) between the United States and the Soviet Union (1969–71), as well as

A prize for peace The Nobel peace prize is awarded at an annual ceremony every December in Oslo. It is the world's most prestigious international award, and is given to an individual who is considered to have best promoted and pursued the interests of peace. It has often been the cause of political controversy. Here

Oscar Arias Sanchez, the president of Costa Rica, holds his citation and medal after receiving the award in 1987. His plan to bring about a ceasefire between guerrilla groups led to an accord being signed by five Central American presidents, but it failed to win international backing, particularly from the United States.

the Conference on Security and Cooperation in Europe (1975). Attended by the leaders of 33 European states, as well as those of Canada and the United States, this affirmed the post World War II boundaries of Europe, and led to the Helsinki declaration on human rights. An associated foreign ministers conference on confidence, security-building and disarmament in Europe took place in Stockholm in 1985. In 1986 the Icelandic capital of Reykjavik was chosen by US

Two Swedes killed in peace missions Count Folke Bernadotte (1895–1948), nephew of the Swedish king, was murdered in Jerusalem by members of the Jewish Stern Gang while acting as United Nations mediator in Palestine (*top*). Dag Hammarskjöld (1905–61), secretary general of the UN, was killed in an air crash in Northern Rhodesia (now Zambia) while trying to bring peace to the newly independent state of the Congo (Zaire), divided by civil war (*above*).

Keeping peace around the world Troops from the region have served with United Nations security forces in many major trouble spots throughout the world. These Finnish soldiers are engaged in keeping the peace between warring factions in south Lebanon.

President Ronald Reagan and Soviet Secretary General Mikhail Gorbachev as the venue for the arms reduction talks that preceded the Intermediate Nuclear Forces (INF) treaty signed in Washington the following year.

The international prizes awarded each year by the Nobel Foundation of Sweden to commemorate achievement in various fields of human activity have included since their inception in 1901 a prize for achievement in promoting peace. Given on the advice of the Norwegian parliament (chosen because it did not then have responsibility for foreign policy), it has been given seven times to Scandinavians, including the Norwegian Fridtjof Nansen for his work for refugees.

Recent recipients have been Soviet human rights protestors Alexander Solzhenitsyn and Andrei Sakharov (1970 and 1975), President Anwar Sadat of Egypt and Prime Minister Menachem Begin of Israel (1978), and Costa Rican President Oscar Arias Sanchez (1987). In 1966, to commemorate 150 years of peace, the Swedish parliament established the Stockholm International Peace Research Institute, whose international staff publish studies on disarmament and arms regulation. The willingness of Cold War world leaders to welcome detente and look for practical solutions to reduce international tension owed more than a little to the peace-seeking efforts of Nordic leaders and people.

THE SOVEREIGNTY OF PARLIAMENT

GOVERNING BRITAIN · ELECTING THE HOUSE OF COMMONS · DESCENT FROM POWER

The United Kingdom of Great Britain and Northern Ireland (commonly called Britain) consists of the three countries of England, Scotland and Wales, and the separate province of Northern Ireland. It evolved through the expansion of its largest part, England. Wales was absorbed by conquest in 1277; following the succession of the Scottish king to the English throne in 1603, Scotland and England were formally united by the Act of Union in 1707. From 1801 the neighboring island of Ireland was joined in a political union with Britain. Its independence, except for the six counties of Northern Ireland, was secured in 1922 and it is today the Republic of Ireland. The Isle of Man, off the northwest coast of England, and the Channel Isles, lying close to the French mainland, are self-governing under the British crown.

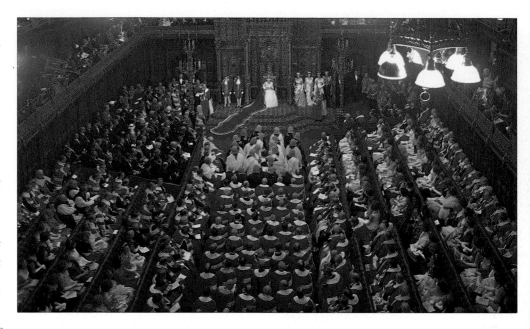

The mother of parliaments The Houses of Parliament have served as the model for governments in many parts of the world that formerly belonged to the British empire. At the ceremonial State Opening of Parliament, which takes place every fall, the monarch addresses members of both houses, and outlines the government's business for the forthcoming session in a speech written by the prime minister of the day.

COUNTRIES IN THE REGION

Ireland, United Kingdom

Island territories   Channel Islands, Isle of Man (UK)

Territories outside region   Anguilla, Ascension, Bermuda, British Indian Ocean Territory, British Virgin Islands, Cayman Islands, Falkland Islands, Gibraltar, Hong Kong, Montserrat, St Helena, South Georgia, South Sandwich Islands, Tristan da Cunha, Turks and Caicos Islands (UK)

STYLES OF GOVERNMENT

Republic   Ireland

Monarchy   UK

Multi-party states   Ireland, UK

Two-chamber assembly   Ireland, UK

CONFLICTS   (since 1945)

Nationalist movements   UK (Scotland, Wales)

Civil war   UK (Northern Ireland) 1969–

Colonial wars   UK/Malaya 1947–57; UK/Cyprus 1955–60; UK/Egypt 1956; UK/Aden (Southern Yemen) 1963–67

Interstate conflicts   UK/Iceland 1956–58, 1974–78 (Cod Wars); UK/Argentina 1982; UK (with US and others)/Iraq 1991

MEMBERSHIP OF INTERNATIONAL ORGANIZATIONS

Council of Europe   Ireland, UK

Colombo Plan   UK

European Community (EC)   Ireland, UK

North Atlantic Treaty Organization (NATO)   UK

Organization of Economic Cooperation and Development (OECD)   Ireland, UK

Notes: Hong Kong ceases to be a UK dependency in 1997. The UK has a territorial claim in Antarctica.

GOVERNING BRITAIN

Britain's system of parliamentary government evolved over many centuries. Parliament itself had its origins in the council of rich landowners (the barons), which advised the medieval kings of England and raised money and armies for the crown. During the 13th century representatives from the counties and the boroughs (local administrative divisions) were also summoned to parliament, and began to meet separately from the barons, thus forming the House of Commons. The assembly of barons was known as the House of Lords.

Over the next centuries a long struggle for power and privilege took place between the monarchy and parliament, particularly the Commons. Civil war between royalists and parliamentarians (1642–52) led in 1649 to the execution of King Charles I and to the establishment of a parliamentary Commonwealth. The monarchy was restored in 1660, but the Glorious Revolution of 1688 resulted in further checks to the monarch's authority, and finally abolished all royal claim to legislate or raise taxes without parliamentary consent.

Britain today is a constitutional monarchy. Government operates through a number of widely accepted conventions and rules whose framework was set in 1688. Formal sovereignty is held to reside with the "Crown in Parliament".

Though often viewed as a balanced constitution, in which the two chambers of parliament and the monarch hold each other in check, greatest power has come to rest with the second, elected chamber, the House of Commons. The leader of the party with most parliamentary seats is invited by the monarch to become prime minister, and to form a government whose policies are presented to parliament as bills to be debated by both chambers. On being passed, they become acts of parliament: these become law on being signed by the monarch, who in practice has no powers to change them and cannot refuse to sign.

The upper chamber, the House of Lords, consists of senior bishops of the established Church of England, law lords (judges), the hereditary peers (nobles) and government-appointed life peers. It scrutinizes legislation but has the power only to delay its passing into law: it cannot exercise even this power where financial bills are concerned.

The British prime minister selects the Cabinet through which executive power is controlled, sets the legislative program, and appoints party officers (whips) to safeguard its passage through the House of Commons. The British prime minister has greater untrammeled power than the leader of any other Western democracy.

Scottish and Welsh nationalism

The decline of the British economy after World War II and the increasing centralization of government in the same period, helped to foster nationalist feeling in Scotland and Wales. During the 1960s there was a growing demand for greater autonomy. The most important trigger for this in Scotland was the discovery of oil in the North Sea off the northeast coast; Scots argued that this potential increase in wealth belonged to them. Welsh nationalism has been more cultural, focusing on the survival of the Welsh language. Support for the Scottish and the Welsh nationalist parties reached its height in the general election of 1974. Referendums held in 1979 on whether directly elected assemblies should be granted, failed to gain the necessary support, but the nationalist parties continue to win support at local and national elections.

A separate state in Ireland

In Ireland, the nationalist movement demanding autonomy from Britain became increasingly militant during the 19th century as successive British governments failed to legislate for home rule. This was conceded in principle in 1914, but the outbreak of World War I delayed its implementation.

In 1922 the self-governing Irish Free State came into being in the south of the island, with dominion status within the British Commonwealth. In 1937, as Eire, it was established as a sovereign state; in 1949 it declared itself the Republic of Ireland on leaving the Commonwealth. It is a parliamentary democracy with an elected president and a two-chamber parliament headed by a prime minister.

The predominantly Protestant population of the six counties that make up the province of Northern Ireland (commonly known as Ulster) chose in 1922 to remain under British rule. It is entitled to send 17 Members of Parliament (MPs) to the British parliament.

Northern Ireland normally administers its own affairs, but the increase of political violence both by Roman Catholic nationalists and by those opposed to them led to the imposition of direct rule from London in 1972. The terms of the 1985 accord between the British and Irish governments (the Anglo–Irish Agreement) confirmed that the status of Northern Ireland cannot be changed without the consent of the majority of the people.

Traditional rivals Tartan-clad Scottish soccer supporters flaunt the blue and white flag of St Andrew and other national emblems at an encounter with England. Although the two countries have been united for nearly 400 years, rivalry is still fierce. Within Scotland there is a strong nationalist movement to secure political independence.

Ireland's two governments Irish premier Garret FitzGerald and Britain's Margaret Thatcher at the signing of the Anglo-Irish Agreement in 1985. This provided for the regular exchange of information on political, legal, security and cross-border matters, and recognized the principle of self-determination in Northern Ireland. It was fiercely criticized.

ELECTING THE HOUSE OF COMMONS

Elections to the House of Commons must be held at least once every five years. If the government loses a vote of confidence in the House of Commons, and no other party commands a majority, the monarch dissolves parliament and calls an election. More usually, however, the prime minister advises the monarch (who cannot disregard the request) to call an election, choosing a time within the five-year term when the government is popular and can expect to be reelected.

There are 650 parliamentary constituencies in Britain (525 in England, 72 in Scotland, 36 in Wales and 17 in Northern Ireland); their boundaries are revised every 10 to 15 years. There is one Member of Parliament (MP) for each constituency. Every citizen over the age of 18 is entitled to vote; the candidate winning the most votes in each constituency is elected as the MP (the plurality or "first-past-the-post" system). This system of voting discriminates against the smaller parties: the number of MPs elected is proportionately smaller than the number of votes obtained. In the 1987 general election, for example, the Liberal–Social Democratic Alliance achieved 22.6 percent of the votes but won only 22 seats.

A two-party system
Parliament's present domination by two parties – partly the result of its electoral system – has its roots in the past. The House of Commons is even designed to accommodate two parties only, the ruling party sitting on one side of the chamber, and the "opposition", as it is known, facing it on the other. Until World War I the contest for government was carried on between the Conservative (Tory) Party and the Liberal (Whig) Party. After 1918 the Labor Party, created in 1900 as the political wing of the trade union movement, came to command the second largest number of seats in the Commons, and took over the role of the official opposition party from the Liberals. Since then power has been contested between the Conservative and Labor parties. Since 1945 there have been seven Conservative and four Labor governments.

The Labor party bases its policies on principles that emphasize the role of the state as a redistributor of wealth. Most of

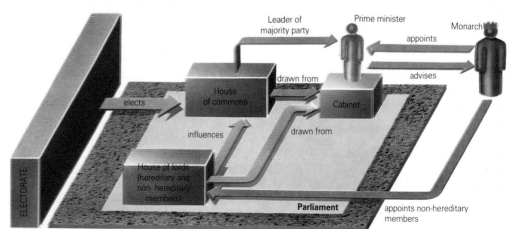

A constitutional monarchy Britain has no written constitution: the relationship between the monarch as head of state and the prime minister as head of government is based on rigid convention. The prime minister heads the majority party in the elected House of Commons, and selects a cabinet of ministers to form the government. Government is subject to the laws made in parliament, as interpreted by the courts.

The party political contest Conservative supporters rally behind Margaret Thatcher during the 1987 general election campaign, which resulted in her third successive victory. Sufficient Conservative MPs failed to support her during a leadership contest in 1990, and she resigned as party leader and prime minister.

its support has traditionally come from the working class, and has been geographically based in industrial, mining and shipyard areas. The Conservative party, on the other hand, represents property and business interests, and draws the bulk of its support from people occupying professional and managerial positions, though it has always attracted some working-class voters. It has been the party of rural areas, small towns and the affluent suburbs.

LOCAL GOVERNMENT

In Britain some major functions of local government are carried out by elected county and district councils (in Scotland, regional councils). These include education and some social services, the renting of public housing, and the maintenance of land-use planning controls. Other areas of local administration are also undertaken by the regional authorities of some central ministries, as well as by specialist bodies.

In the 1980s in most of Britain's major cities Labor-controlled councils clashed with the Conservative central government's policy of public spending cuts. This led to conflict over what services were to be provided and how they were to be financed. Central government was able to use its majority in the House of Commons to overcome resistance in a number of ways. Limits were set on local budgets; the councils of England's six metropolitan counties (including Greater London) were abolished; some of their local government functions were placed in the hands of smaller district councils. London is the only major capital city in the world without its own city government.

Elected councils derive their income from two main sources: a central government grant, and local taxation. In 1990 a direct personal tax (the poll tax), introduced by the Conservative government as the means of raising local taxes, proved highly controversial and was a major factor in eroding Margaret Thatcher's popularity.

The north–south divide

Generally speaking, patterns of voting showed little overall change between the 1920s and the 1970s. In recent elections there have been unmistakable signs of sharp polarization in electoral behavior between the north and south of Britain. In part this reflects a weakening of the link between occupational class and voting habits. In the 1987 general election the Conservatives won a swathe of constituencies across the whole of prosperous southern England, with the exception of inner London, winning significant support from large sections of an increasingly affluent working class. Labor consolidated its hold on those regions in the north of England, south Wales and Scotland whose traditional heavy industries had been hardest hit by the economic recession of the 1970s. It also did well in the depressed inner cities. In Scotland it achieved its best ever election results, winning 50 constituencies; the Conservatives were reduced to only 10 seats.

This new polarization in British elections provided an opportunity for other parties to challenge the domination of the two main parties. There was a revival of the Liberal Party in the 1970s, and in 1981 a new central party, the Social Democratic Party (SDP), broke away from the Labor Party. The Liberals and the SDP fought the elections of 1983 and 1987 as the Alliance. They came third in terms of votes, but did not win many seats. They achieved their greatest successes in the south of England, where the Labor vote had collapsed. The growth of the Scottish Nationalist Party (SNP) affected the strength of the Conservative vote there. In the 1992 general election the north–south pattern continued, though support for Labor rose slightly in the south and the Conservative vote increased in Scotland. The Liberal Democrats (former Alliance) lost votes across the country.

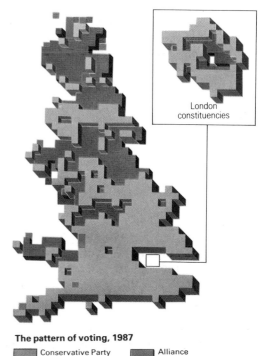

London constituencies

The pattern of voting, 1987

- Conservative Party
- Labor Party
- Alliance
- Nationalist parties

Two nations? The pattern of voting in the 1987 general election shows the blue Conservative seats concentrated in the affluent south and central areas, Labor's in the industrial areas of the north, Scotland and Wales, and in the inner cities – a pattern that was repeated with little variation in 1992.

DESCENT FROM POWER

In the 19th century Britain was the greatest power on Earth. As the pioneer of the Industrial Revolution it was the "workshop of the world". As an exporter of industrial goods and an importer of food and raw materials it was the largest trading nation in the world, and owned the world's largest merchant fleet. The City of London was the financial center of the world.

In 1900 Britain ruled an empire that covered about a quarter of the world's total land area and population. It included Australia, Canada and New Zealand, the Indian subcontinent and Burma, much of the Middle East, a substantial part of Africa, islands and territories in the Caribbean and Central America, and in the Pacific Ocean. By the end of the 20th century all this had gone. New roles have had to be found as Britain's influence has waned.

Churchill's circles of influence

In the crucial period just after World War II Sir Winston Churchill (1874–1965) located Britain at the center of three interlocking circles or power structures – empire, Atlantic and Europe – a position that would give Britain a continuing role as a major world power. In practice, each of Churchill's circles has represented an unconnected, alternative strategy for for-mulating British foreign policy.

In the first place, Churchill and many others did not foresee the rapid disintegration of Britain's empire after World War II. The centerpiece of the empire, India, became independent in 1947, followed by Burma and Ceylon (Sri Lanka). In 1957 Ghana became the first African colony to gain independence. By 1966 12 African colonies had become new states, and the pattern was repeated in the Caribbean, the Middle East and Southeast Asia in the same decade. Churchill's empire option was never a realistic role for Britain after 1947.

Changing relationships

As a victorious ally Britain was accorded "Big Three" status alongside the United States and the Soviet Union in the years after 1945. But the war had taken a heavy economic toll. Britain was virtually bankrupt, dependent on the United States for loan facilities to keep its economy going. In effect the United States had taken over Britain's position of a century earlier as the greatest political and economic power in the world. Britain adjusted to this by searching for a role as the United States' deputy in the new world order then being created. The result was the development of the "special relationship" between Britain and the United States.

In 1947 and 1948 Britain took the lead in Western Europe in organizing its response to the United States' aid package to Europe (the Marshall Plan). Britain was also instrumental in drawing the United States back into defense commitments in Western Europe through the formation of the North Atlantic Treaty Organization (NATO) in 1949, and fought alongside the United States in the Korean war (1950–

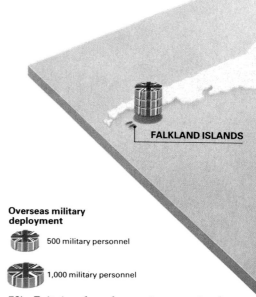

FALKLAND ISLANDS

Overseas military deployment

500 military personnel

1,000 military personnel

53). Britain played a major part in determining the division of Europe into Western and Eastern blocs, and in the events that led to the "Cold War" period of hostility between the superpowers.

After 1960 the special relationship soured as Britain's power declined and the United States looked to other states in Europe, notably West Germany. An economic crisis in 1967 led to the withdrawal of most of Britain's armed forces from commitments east of Suez. Britain had changed from a world power to a European power. After two failures to join, Britain finally entered the European Community (EC) in 1973.

After joining the EC, Britain's relations with the rest of the community were uneasy. Britain's contribution to the EC budget, considered excessive by British governments – the original bone of contention – was successfully renegotiated. However, vital differences remained over the future direction of the EC. A strong body of opinion in Britain, especially the Conservative party, opposed to greater integration has influenced Britain's erratic stance on Europe. Although Britain supported the removal of internal trade barriers, Margaret Thatcher (prime minister 1979–90) firmly rejected moves toward

---

THE COMMONWEALTH

The Commonwealth is a voluntary association of 49 independent sovereign states from every continent and ocean of the world. Its members range from India, with over 750 million people, to Tuvalu in the southwest Pacific, with only 8,000. They vary greatly in terms of culture, race, religion and language. The only common denominator is that they are all former colonies of the British empire.

The Commonwealth originated in the Statute of Westminster (1931), which recognized the British "dominions" (Australia, Canada, Ireland, New Zealand and South Africa) as independent and equal to Britain under the crown. In 1950 India, as a republic, was able to join the Commonwealth by recognizing the British monarch as symbolic "head of the Commonwealth", a formula that allowed most colonies to join on independence. The exceptions were Burma (1947), Sudan (1956), the South Arabian Federation (1967), and the Gulf States (1971). In addition, Ireland (1949), South Africa (1961) and Fiji (1987) have left the Commonwealth, and Pakistan did not belong from 1972 to 1989.

There is a permanent secretariat in London whose major activity is to organize the biennial heads of government meeting. The first of these to take place outside London was in Singapore (1971). Since then it has met in Australia, Bermuda, Canada (twice), Jamaica and Zambia. The 1971 Singapore Declaration identified common interests in seeking peace and cooperation, reducing inequalities between states and combating racial prejudice. Implementation of the last of these aims has dominated recent meetings. In 1977 members assented to the Gleneagles Agreement, which discouraged sporting links with South Africa.

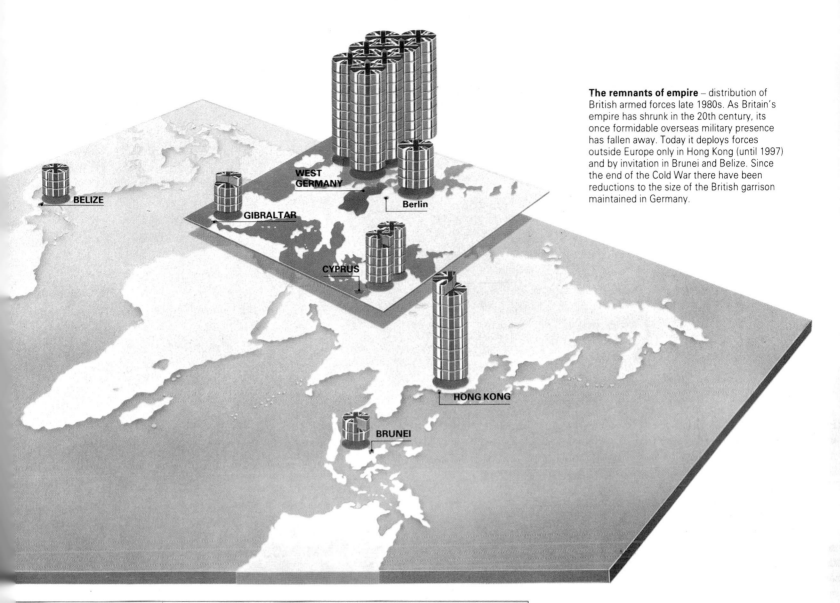

The remnants of empire – distribution of British armed forces late 1980s. As Britain's empire has shrunk in the 20th century, its once formidable overseas military presence has fallen away. Today it deploys forces outside Europe only in Hong Kong (until 1997) and by invitation in Brunei and Belize. Since the end of the Cold War there have been reductions to the size of the British garrison maintained in Germany.

BELIZE

WEST GERMANY

Berlin

GIBRALTAR

CYPRUS

HONG KONG

BRUNEI

Returning from victory Celebrating crowds welcome HMS *Invincible* home from the Falklands/Malvinas war. Argentina disputes sovereignty of the sparsely populated islands with Britain, and invaded them in April 1982. British forces recaptured them within weeks, at a cost of nearly $4 billion.

political union. Her "Euroskeptic" views were a factor in her downfall in 1990. Her successor John Major accepted the plans formulated at Maastricht in 1991 for closer economic union and a common defense policy but pressure from his party led him to delay ratification of the agreement.

Close similarities between the political and economic policies of Britain and the United States, helped by a personal affinity between Margaret Thatcher and president Ronald Reagan, brought the two countries closer than at any time since the 1940s. Britain received diplomatic and logistic support from the United States in its war with Argentina over the Falklands/Malvinas islands in 1982, and in 1986, against the wishes of its European allies, Britain allowed the United States' air bases on its territory to be used for a bombing raid by the United States on Libya. In 1991 Britain contributed a substantial land, sea and air force to the multinational United Nations force led by the United States in the Gulf War against Iraq.

The politics of a divided Ireland

Ireland's long history of conflict reaches back to its early colonization by English settlers, first under the Norman kings and then (with systematic confiscation of land) during the 16th and 17th centuries. Following the crushing of Irish revolt by King William III in 1689–91, there were two important Protestant minorities in the largely Roman Catholic country: a landed aristocracy, and the Ulster "plantations" of small-scale agricultural settlers, many from Scotland, in the north of the island. This basic religious divide has underpinned much of the violence that has afflicted Ireland ever since.

The exclusion of the six counties of Ulster from the terms of the 1922 treaty that created the Irish Free State was bitterly opposed in the south. Its acceptance by the Irish Republican leader Michael Collins (1890–1922) led to further civil war. Collins was shot by a Republican opponent, and Fianna Fail ("warriors of destiny"), the party led by Eamon de Valera (1882–1975) who resigned from the Irish parliament (Dail Eireann) over the treaty, took over the mantle of republicanism.

It was a Fianna Fail government that drew up the 1937 constitution. In it the refusal of the Irish state to regard the partition as permanent was made explicit: "The national territory consists of the whole island of Ireland, its islands and territorial seas." A Fianna Fail government also proclaimed Ireland a republic outside the Commonwealth in 1949.

In electoral terms, Fianna Fail has remained strongest in western Ireland, the traditional base of republican support. The Fine Gael party ("the tribe of Gaels") is less anti-English, and has built its power base on a middle-class core of support in Dublin. The Labor Party has never matched the other two parties, but has entered several coalition governments with Fine Gael.

Two religions, two communities

The Northern Ireland province retained its representation in the British parliament and was given its own assembly at Stormont Castle in Belfast. The devolution of much domestic policy to Stormont meant that the Protestant majority was able to exercise discrimination against the Roman Catholic community for over half

A thankless task A British soldier in Belfast's Falls Road, which runs through the Roman Catholic area of the city. At first British troops seemed to contain the sectarian violence, but later they aggravated it. Between 1969 and 1972 deaths in the province rose from less than 10 to over 450 a year.

a century, particularly in the labor and housing markets. In the late 1960s a civil rights movement grew up.

Confrontation with the police became increasingly explosive and in 1969, following the violent dispersal of a People's Democracy march from Belfast to Londonderry, sectarian fighting broke out and the British army was sent in to keep the two sides apart. The Irish Republican Army (IRA) intensified its campaign to bring about a unified Ireland; the British army – originally welcomed as protectors by the Catholic population – came to be seen as a foreign occupation force. Violence continued unabated. By 1992, the troubles had claimed over 3,000 victims.

Both communities in Northern Ireland are politically divided. The Catholic, or nationalist, community is mainly represented by the Social Democratic and Labor Party (SDLP), which emerged from the civil rights movement. This has been more willing to negotiate both with local Protestants and with the British government than has Sinn Fein, which supports the IRA's military campaign. In the Protestant community the division is less marked. It lies largely between the traditional party of the Protestants, the Official Unionists, and a new party, the Demo-

cratic Unionists, which emerged in the current troubles under the leadership of Dr Ian Paisley, Protestant Ulster's most popular politician.

The most far-reaching attempt to achieve political compromise in Northern Ireland came in 1985, when the Anglo-Irish Agreement was signed. This was supported by all three main parties in the south, but only by the SDLP in the north. The Unionist parties were united in their opposition to the Irish Republic being allowed a say in Northern Ireland's affairs; their MPs boycotted the British parliament from 1985 to 1987. Sinn Fein continued to back the IRA; the level of violence did not diminish in the way supporters of the agreement had hoped.

Orange Day parade
Protestants in distinctive bowler hats and sashes march every year to commemorate the Battle of the Boyne on 12 July 1690, in which William III of Orange defeated the Catholic Irish forces of James II. Marches like these keep alive bitter sectarian differences.

The six counties of Ulster
The minority Roman Catholic population (about one-third of the total) is most heavily distributed in rural areas, where it forms a majority except in the north and east. Effective policing of the border with the south, which runs through remote country, is a perennial problem.

Roman Catholics in Northern Ireland

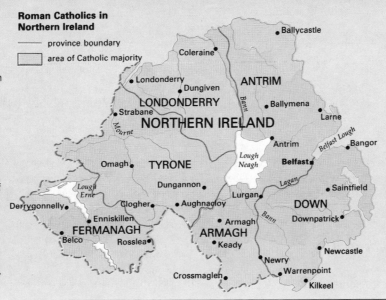

- province boundary
- area of Catholic majority

Ballycastle
Coleraine
Londonderry
Dungiven
ANTRIM
LONDONDERRY
Strabane
Ballymena
NORTHERN IRELAND
Larne
Bann
Mourne
Omagh
TYRONE
Lough Neagh
Antrim
Belfast Lough
Bangor
Belfast
Dungannon
Lagan
Saintfield
Lough Erne
Lurgan
Derrygonnelly
Clogher
Aughnacloy
DOWN
Enniskillen
Bann
Downpatrick
FERMANAGH
Armagh
ARMAGH
Belco
Keady
Rosslea
Newcastle
Newry
Warrenpoint
Crossmaglen
Kilkeel

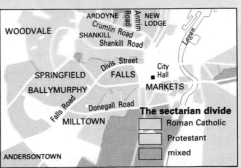

WOODVALE
ARDOYNE
NEW LODGE
Crumlin Road
Antrim Road
SHANKILL
Shankill Road
Divis Street
SPRINGFIELD
FALLS
City Hall
BALLYMURPHY
MARKETS
Falls Road
Donegall Road
MILLTOWN
Lagan
ANDERSONTOWN

The sectarian divide
- Roman Catholic
- Protestant
- mixed

Segregated communities Most of Northern Ireland's towns have Protestant majorities: in Belfast less than 30 percent of the population is Roman Catholic. Industrial East Belfast is split into two highly segregated communities. The Falls Road bisects the Roman Catholic district and the Shankill Road the Protestant district. In the violence of the 1970s disturbances were greatest in these areas. Segregation is less apparent in the middle-class suburbs.

The power of the unions

Trade unions have played an important role in the development of British political life in the 20th century. In 1900 they were instrumental in forming the Labor Representative Council, which became the Labor Party in 1906. Most trade unions remain formerly affiliated to the Labor Party, and have been a powerful influence on their policies both in and out of government.

Trade unions emerged in the 19th century through the struggle of working people to win the right to band together against employers in negotiations over wages and conditions at work. The rise of trade unionism coincides with other political gains, notably the extension of the franchise: the first annual Trades Unions Congress (TUC) and the move to give working men the right to vote both took place in 1868. The TUC has been active in supporting the worldwide trade union movement.

Strikes and demonstrations have been the major direct actions of trade unions. The latter have been formalized as annual parades and galas to display strength and solidarity. These have become important political events – the Labor Party leader is expected to address the annual Durham Miners Gala, for instance – and help to proclaim Labor's power in the community. The Conservative government of Margaret Thatcher in the 1980s directly confronted the strength of the unions by putting through a legislative program to limit strike action.

The colorful banners of local trade union branches add a festival atmosphere to a miners' demonstration in London during the year-long strike against coalpit closures in 1985.

THE ONE AND INDIVISIBLE REPUBLIC

THE MAKING OF THE FIFTH REPUBLIC · EXECUTIVE POWER-SHARING · IN SEARCH OF A WORLD ROLE

The modern French state is the direct heir of the French Revolution of 1789. This brought into being a republic based on the universal principles of liberty, equality and fraternity, incorporated in a written constitution that has served as a model for other nation-states throughout the world. The territorial expansion of the French state was begun by the Capetian kings at the end of the 10th century. Over the centuries they gradually extended their rule into the lands south of the river Loire (Languedoc), the west of France (Aquitaine), over the duchies of Brittany and Burgundy, to the Alps, and northeast to the upper Rhine (Alsace and Lorraine). When the republic replaced the monarchy the territorial boundaries of France had more or less reached their present extent, and included many ethnic and linguistic groups.

COUNTRIES IN THE REGION

Andorra, France, Monaco

Island territories Corsica (France)

Territories outside region French Guiana, French Polynesia, Guadeloupe, Martinique, Mayotte, New Caledonia, Réunion, St Pierre and Miquelon, Wallis and Futuna (France)

STYLES OF GOVERNMENT

Republic France

Principalities Andorra, Monaco

Multi-party states France

States without parties Andorra, Monaco

One-chamber assembly Andorra

Two-chamber assembly France, Monaco

CONFLICTS (since 1945)

Nationalist movements France (Alsace, the Basque country, Brittany, Corsica, Languedoc)

Internal unrest France 1958, 1968

Colonial wars France/Indochina 1946–54; France/Algeria 1954–62

Interstate conflicts France (with US and others)/Iraq 1991

MEMBERSHIP OF INTERNATIONAL ORGANIZATIONS

Council of Europe France

European Community (EC) France

North Atlantic Treaty Organization (NATO) France

Organization for Economic Cooperation and Development (OECD) France

Notes: France withdrew from NATO Military Command in 1966 but remains a member of the alliance.

France has a territorial claim in Antarctica.

THE MAKING OF THE FIFTH REPUBLIC

"The one and indivisible Republic" has undergone many vicissitudes in the two centuries since it was established by the Revolution. The early years of revolutionary government gave way to the dictatorship of Napoleon Bonaparte (emperor of France 1804–15), and there have been subsequent periods of monarchy and empire, as well as two revolutions (1830 and 1848). The constitution has been rewritten on several occasions.

During the German occupation of France in World War II, a puppet government was set up under Marshal Pétain (1856–1951). Alsace and Lorraine, which had been acquired by Germany after the Franco-Prussian War (1870–71) and then handed back to France at the end of World War I, were reannexed by Germany. After the liberation of France by the Allies in 1944 a provisional government was set up under General Charles de Gaulle (1890–1970), who had headed the wartime Free French government exiled in Britain, based in London, until the Fourth Republic was established in 1946.

The years of the postwar period were troubled ones for France: the process of decolonization in Indochina and North Africa brought political instability and conflict – between 1947 and 1954 France had 15 governments. It was the crisis over Algerian independence that precipitated the downfall of the Fourth Republic in 1958. De Gaulle was called back from retirement to head a government of national unity and supervise the framing of a new constitution, which was voted in by national referendum. In 1959 de Gaulle became president of the Fifth Republic.

The constitution of the Fifth Republic (modified in 1962) separated the executive branch of government from the two-chamber legislature – the national assembly and the senate – whose importance was downgraded. The role of the president was greatly enhanced: as guarantor of the continuity of the French state the president was given wide powers in order to preserve national independence and territorial integrity, including the right to dissolve the national assembly and propose a referendum.

The de Gaulle era was a time of economic growth, accompanied by tight censorship and strong centralization. In

1968 students' demonstrations in Paris and elsewhere sparked off a series of strikes and protests that briefly threatened the government. In the elections called immediately afterward, however, de Gaulle won a landslide victory, but resigned a year later after he had been defeated in a referendum to approve his proposed reforms for regional devolution.

Separatist pressures on the state

Regionalist and nationalist movements in the 1960s offered a strong challenge to the highly centralized French state. Separatist groups were most active in Brittany, Occitanie (the area of the south that includes Languedoc and Provence), among the Basque people of the French Pyrenees, and on the island of Corsica (an administrative department of the French Republic). Demands centered on the unfair distribution of resources for regional economic development, on state interven-

A symbol of France's past The Republican Guard, in ceremonial uniform, parading along the Champs-Elysées in Paris. The Guard, an elite force of the French army, is responsible for the security of national buildings, and for carrying out ceremonial duties.

Rapport with the people General de Gaulle gained support for his policies by appealing directly to Frenchmen and Frenchwomen. In a characteristic gesture, his finger is poised to drive a point home in one of his many broadcast speeches.

tion in regional affairs, and on the loss of cultural identity, particularly the decline of minority languages. The German dialect spoken in Alsace and Lorraine, for example, has no official status, giving rise to some demand for autonomy. The fiercest campaigns were waged in respect of Breton and Corsican, and since 1975 it has been possible to study both these languages in schools.

A program of decentralization put into effect by the "left coalition" government led by President François Mitterrand (elected 1981) partially defused separatist demands, though the Corsican National Liberation Front (FLNC) still engaged in violent attacks against government officials. It was banned in 1983, a year after Corsica had been awarded special status with its own regional assembly, albeit with somewhat ill-defined powers. Social and political conflict continued.

Two mini-states

The region includes two of Europe's tiny sovereign states that have escaped absorption by their stronger neighbors. The joint heads of state of Andorra in the Pyrenees are the bishop of Urgel in Spain and the president of France. Its unwritten constitution is based on long-established custom and privilege. The hereditary principality of Monaco, just within the French frontier with Italy, has been ruled by the house of Grimaldi since 1308: its current ruler (since 1949) is Prince Rainier III. According to agreements made between France and Monaco, in the event of a reigning prince having no male heir the principality will be fully incorporated into the French state.

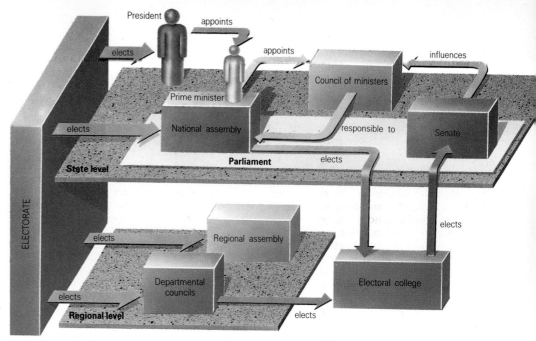

EXECUTIVE POWER-SHARING

The constitution of the Fifth Republic shares the functions of the executive between the president and the prime minister. In addition to his formal role as head of state and commander-in-chief of the armed forces, the president of France appoints the prime minister, presides over cabinet meetings and is responsible for negotiating treaties.

Ultimate policy control rests with the prime minister, who is responsible for appointing members of the council of ministers. If elected members of the national assembly or the senate (the second chamber of the legislature) are appointed to government office they must give up their parliamentary seat, but ministers are often drawn from outside the political sphere, from commercial or technocratic backgrounds.

During the early years of the Fifth Republic, when there was a secure majority of Gaullist representatives in the national assembly, the power of the president was virtually unchallenged. In the 1970s majorities became less stable and supportive; under the socialist administration of 1981–86, and even more so between 1986 and 1988 when a socialist president was paired with a right-wing prime minister, the balance of power within the shared executive came to depend on whether the president or the prime minister commanded greater political support within the assembly.

The prime minister Jacques Chirac used his majority in the national assembly to claim legitimacy for a program of legislation that reversed earlier policies by returning state-owned enterprise to private ownership and removing statutory controls. He moderated his program only when he ran into serious opposition from students and workers.

The party political system
There are four main political parties in France. Until the 1978 election the Communist Party was the largest left-wing party, but it gave way to the Socialist Party, led by President Mitterrand, which in 1981 became the largest parliamentary party. The right-wing Republican Rally Party (RPR) was established by Jacques Chirac as the inheritor of the Gaullist tradition. The Union for French Democracy (UDF) is a federation of liberal

France's system of government Under the constitution of the Fifth Republic the president appoints the prime minister, who then advises him on appointments to serve on the council of ministers. This is responsible to the national assembly, which can censure its handling of the administration.

THE RISE OF THE FRENCH NATIONAL FRONT

France now has the largest National Front (right-wing nationalist) party in Western Europe. After 1981, when it had insufficient support to put forward a presidential candidate, its rise under the leadership of Jean-Marie le Pen was extremely rapid. Under the system of proportional representation used in the 1986 national assembly elections it won 35 seats and 9.6 percent of the votes cast. However, with the return to the two-round majority ballot system in the national assembly elections in June 1988, it won only one seat, in the department of the Var (Provence), though its share of the vote had risen to 14.39 percent two months earlier in the first round of the presidential election.

The range of its support is diverse, from those demanding repatriation of immigrants and tighter law and order policies, including a return to capital punishment, to those protesting against the political system in general, and the more prominent role it is predicted France will play in the wider European Community (EC) after 1992. Its original strongholds were in the Paris region and the Mediterranean coast (where it achieved 30 percent of the vote in 1988), but it made significant inroads in the east of the country in particular, and generally performed well in rural areas throughout France.

Mounting guard in Monaco Though an independent state, the principality's affairs are closely tied to France. Its council of government is headed by a French civil servant, chosen by the prince.

Rallying the right Jean-Marie le Pen, leader of the National Front, campaigning in 1988. His party's rise was the latest sign of a strong right-wing tradition.

right and center parties set up in 1978 by President Valéry Giscard d'Estaing (president 1974–81).

The 577 (since 1986) deputies of the national assembly are elected every five years from single-member constituencies. A two-ballot "run-off" majority system is used: if no candidate wins a clear majority in the first ballot, a second ballot is held in which candidates with more than 12.5 percent of the first ballot may stand. This has led since the 1970s to an increasing tendency for electoral pacts to be concluded, allowing the strongest candidate from left or right to go forward. A system based on proportional representation was introduced for the 1986 national assembly elections, but was not repeated in 1988. The president has the right to call new elections to create a national assembly likely to support his policies. This President Mitterrand did in 1981 and 1988, when he would otherwise have been confronted with right-wing majorities in the assembly.

Presidents are elected for seven-year terms by direct universal suffrage on the basis of a single constituency for the whole country, also using the two-ballot "run-off" system. Former voting patterns based on geographical areas of traditional right- and left-wing support have altered in recent years, and electoral success is secured by capturing a greater proportion of the central vote, as President Mitterrand did in 1988.

Local government

Below the elected bodies at national level are three further levels of government, each of which carries specific, though not totally exclusive functions. The first level consists of the communes (36,383 of them in 1978), dating from 1789 but given electoral representation only in 1884. They are now responsible for land-use planning, urban policies and local public works. Elections are held every six years and have become, since the 1970s, highly politicized and increasingly interwoven with national interests.

Above the communes are departments (départements), the administrative areas into which France was first divided in 1790, of which there are now 96 in metropolitan France. The departments were the main financial beneficiaries of the socialists' decentralization legislation in the early 1980s, and are responsible for local social services and health, school transport and buildings.

In 1982, 22 regions were established above the departments, with the aim of directing regional economic planning away from central government. With the exception of Corsica, which had already held elections in 1982, the first elections for regional assemblies took place in March 1986. All but two (Limousin and the Nord) were gained by the right.

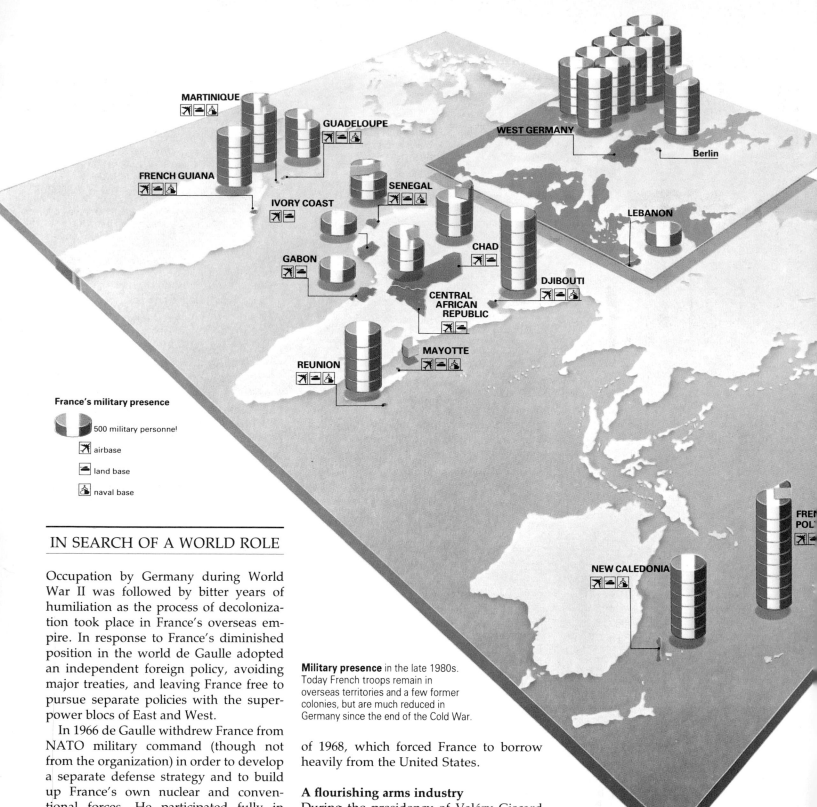

France's military presence

⬤ 500 military personnel

✈ airbase

🚢 land base

⚓ naval base

MARTINIQUE
GUADELOUPE
FRENCH GUIANA
IVORY COAST
SENEGAL
GABON
CENTRAL AFRICAN REPUBLIC
CHAD
DJIBOUTI
REUNION
MAYOTTE
WEST GERMANY
Berlin
LEBANON
NEW CALEDONIA
FRENCH POLYNESIA

IN SEARCH OF A WORLD ROLE

Occupation by Germany during World War II was followed by bitter years of humiliation as the process of decolonization took place in France's overseas empire. In response to France's diminished position in the world de Gaulle adopted an independent foreign policy, avoiding major treaties, and leaving France free to pursue separate policies with the superpower blocs of East and West.

In 1966 de Gaulle withdrew France from NATO military command (though not from the organization) in order to develop a separate defense strategy and to build up France's own nuclear and conventional forces. He participated fully in developing agricultural and industrial policies with France's partners in the European Community (EC), of which France was a founder member in 1957. Suspicious of "Anglo-Saxon" influence in Europe, he repeatedly blocked Britain's requests for entry. Links were revived with Eastern European states, but the initiatives were destroyed after the Soviet Union's brutal invasion of Czechoslovakia in 1968. French independence was further dented by the domestic unrest

Military presence in the late 1980s. Today French troops remain in overseas territories and a few former colonies, but are much reduced in Germany since the end of the Cold War.

of 1968, which forced France to borrow heavily from the United States.

A flourishing arms industry

During the presidency of Valéry Giscard d'Estaing (1974–81), France's nuclear strike power was quadrupled, and defense spending had increased to 3.8 percent of the GNP by 1980. It continued at around this level through the 1980s.

It is through its arms industry that France has most effectively extended its influence in the postwar world. By the end of the 1970s French arms sales were the third largest in the world, after the United States and the Soviet Union, having grown by 22 percent a year between

1972 and 1982. Most of this trade was with the oil-rich states of the Middle East.

Under President Mitterrand the struggle to maintain an independent world role continued, for example through the pursuit of a progressive Third World policy that attempted to transcend the rigid North–South division. However, France's ability to remain nonaligned was

A small piece of France in the Indian Ocean President Mitterrand's photograph is held aloft as he is welcomed to Réunion during the 1988 presidential election campaign. As an overseas department the island of Réunion, with a population of 530,000, is administered on the same basis as departments in mainland France, and takes part in national elections. It has its own elected general and regional assemblies. It has been a colony since the 17th century (though it belonged to the British from 1810 to 1815) and became an overseas department in 1946. Government of the island is a heavy expense, but independence movements have nevertheless so far made little progress there.

eroded after 1982 in the face of its weak economy and the realities of international politics, and it came closer to NATO. It supported the siting of Pershing 2 and Cruise missiles by the United States in Western Europe, as long as they were not sited on French soil. It sent peacekeeping troops to the Lebanon, and a naval force to the Gulf to protect international shipping threatened by the Iran–Iraq conflict. In 1991, France contributed military contingents to the multinational force in the Gulf War against Iraq.

The need to comply with a European consensus in world affairs and build a unified Western Europe, which preoccupied France from the mid-1980s, played a key role in advancing the single European market of 1992.

Remnants of empire

Overall, however, France's zone of influence remains limited to its colonial territories, past and present. It has supplied economic aid to many of its former colonies in Africa, and has maintained military connections with many of them, including Benin, Cameroon, Chad, Congo, Gabon, the Ivory Coast, Senegal and Togo. There is a military force of 5,000 permanently stationed in Djibouti, and 1,200 French military advisors are deployed in 17 African states. A number of authoritarian regimes in Africa have been maintained in power as a result of French military intervention; for example, Chad in 1984 and Togo in 1986.

Scattered remnants of the former French empire still remain. The islands of Guadeloupe and Martinique in the Caribbean, French Guiana on the South American mainland, the island of Réunion in the Indian Ocean, and Saint-Pierre and Miquelon off the coast of Newfoundland have the status of overseas departments (*départements d'outre-mer*). This means that they are regarded as an integral part of France, with the same laws and currency. They send deputies to the national assembly and have their own regional assemblies with wide-ranging powers.

In addition, there are the overseas territories (*territoires d'outre-mer*): French Polynesia, New Caledonia and the Wallis and Futuna islands in the South Pacific Ocean, and Mayotte in the Indian Ocean. These enjoy a variety of relationships with France. All French nuclear tests since 1975 have taken place in French Polynesia on the Mururoa atoll. France's determination to prevent surveillance of these tests led to the scuttling of the *Rainbow Warrior*, the boat belonging to the environment pressure group Greenpeace, in New Zealand on 10 July 1985.

The problems and conflicts of colonialism are posed most acutely in New Caledonia, where a large proportion of the population (over 33 percent) is of French origin and there have been violent clashes with the indigenous Kanak population and other ethnic groups. An agreement made in 1988 with the French government allows for a referendum on independence to be held there in 1998. Elsewhere, independence movements have not gained significant strength. All the overseas territories and departments are heavily subsidized.

THE ENDING OF FRENCH IMPERIALISM

In 1945 France possessed the world's second largest colonial empire, after Britain, with important colonies in Indochina (Cambodia, Laos and Vietnam) as well as Africa. The French colonies in Indochina had been occupied by the Japanese in World War II. When the French returned to Vietnam after the war they found a nationalist regime well entrenched in the north. A seven-year struggle against the communist Viet Minh rebels led by Ho Chi Minh (1892–1969) finally ended in defeat for the French in 1954 at Dien Bien Phu. Though the French sustained only 7,000 casualties (compared to 20,000 on the other side), the defeat dealt a humiliating blow to national morale. As a result the French withdrew from Vietnam, and the country was divided between the communist government in the north, with its capital in Hanoi, and pro-Western South Vietnam, with its capital in Saigon.

In Africa, Morocco and Tunisia gained their independence in 1955 and 1956. Then, in 1960, French West Africa and French Equatorial Africa were divided into 11 separate sovereign states. The most bitter struggle took place in Algeria. From 1954 to 1962 the National Liberation Front (FLN) waged a drawn-out war against a million French settlers (*pieds-noirs*) and 400,000 French troops. The cost of fighting the guerrillas was enormous, amounting to 28 percent of the annual national budget for the years 1954–60. The consequences for France were momentous. The uprising of the settlers in May 1958 brought General de Gaulle back to power and eventually led to an independent Algeria in 1962.

France and Europe

The European Community grew out of moves to rebuild the economies of Europe shattered at the end of World War II, and to prevent such a war happening again. It originated in the 1950 Coal and Steel Plan to unify the industries of France and Germany, and was the brainchild of two Frenchmen, Jean Monnet (1888–1979) and Robert Schuman (1886–1963), who had been born in Lorraine, which was then part of Germany.

By June 1950 the Benelux countries (Belgium, Luxembourg and the Netherlands) and Italy had expressed an interest in joining, and a year later the European Coal and Steel Community (ECSC) was formally instituted. Very soon proposals were made for more generalized economic cooperation, and in March 1957 the Treaties of Rome agreed the establishment of the European Economic Community (EEC) and the European Atomic Energy Commission (Euratom), which came into being the following year.

The EEC became instrumental in France's efforts to maintain its independence, especially from the United States. As first among equals France would be able to play a leading role in a strong Europe, bolstered by its special relationship with West Germany. The treaty of cooperation concluded by de Gaulle and West German Chancellor Konrad Adenauer (1876–1967) in January 1963 covered foreign affairs, defense and education, and has withstood all the vicissitudes of the ensuing years.

Though appreciative of the economic benefits of European union, de Gaulle (initially hostile to the EEC) opposed closer integration that would override national interests. In 1967 France gained acceptance for the Luxembourg Accord, by which any decision of the council of ministers may be vetoed by a member whose national interests are at stake.

De Gaulle consistently vetoed Britain's applications for entry, first in 1963 and again in May 1967, and it was not until after his death that Britain, together with Denmark and the Republic of Ireland, finally joined the European Community (as it now became known) in 1973. A national referendum held in Norway rejected entry. Later entrants were Greece (1981) and Portugal and Spain (1985). A decision on Turkey's request for entry was delayed, depending on its government's willingness to improve democratic and human rights.

The EC is governed through a complicated structure of interacting institutions. The extent to which each represents separate national or common European interests varies considerably. The council of ministers, consisting of representatives of the governments of the constituent states, is the decision-making body of the EC: it is here that competing national interests most often come into conflict.

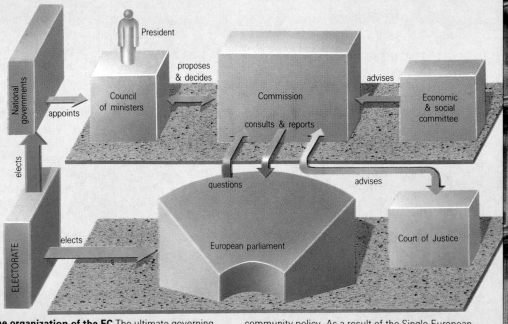

The organization of the EC The ultimate governing authority is the council of ministers, the presidency of which rotates at six-monthly intervals among member states. In the past, unanimous agreement was necessary for council decisions to gain acceptance as community policy. As a result of the Single European Act, majority (two-thirds) decisions are now accepted.

First in line The flag of the EC heads the line of national flags outside community headquarters.

The commission provides administrative backup, but is also responsible for putting forward proposals for integrated action. It consists of 17 commissioners. The economic and social committee, made up of 189 members appointed for four-year terms, represents employers, workers and other interest groups. It advises the council of ministers, and has to be consulted on certain issues. The European Court of Justice, which sits in Luxembourg, is responsible for interpreting community laws. Its decisions have moral force only.

The European parliament is essentially a consultative body, with few legislative powers. Those it does have – to dismiss the commission and reject the budget – have never been called into effect. Since 1979 members of the European parliament (MEPs) have been directly elected from each country. Turnout levels at these elections are usually lower than in national elections.

The future of the EC

During the 1980s President Mitterrand was a leading supporter of initiatives to increase and reinforce European unification through the EC. These were not always nationally popular – the inclusion of the Mediterranean states, for example, provoked massive demonstrations from French producers of wine, fruit and vegetables. In 1986 it was agreed to create a single European market, which came into effect in 1992, eliminating existing restrictions on the circulation of capital and labor between member states.

In 1991 the heads of state of all 12 member countries, meeting at Maastricht in the Netherlands, produced an agreement that would lead to a more integrated Community. The treaty included plans for a single currency, a common defense policy, common citizenship and the granting of more power to the European Parliament. However, in referendums in 1992 the Danes voted to reject it and the French voted to ratify it by only a narrow margin, throwing some doubts on its future implementation.

At the same time, there were indications that the Community would be widened, with EFTA countries (Sweden, Norway, Austria and Finland) due to start talks to join the EC in the mid 1990s and the possibility of Poland, Hungary, the Czech Republic and Slovakia entering in the late 1990s.

Anniversary of a Revolution

For 200 years the French Revolution, with its call for *liberté, égalité et fraternité* (liberty, equality and brotherhood), has had repercussions far beyond the boundaries of France. Since 1789 all political movements against traditional powers have based their ideas on this greatest of all revolutions, after which the world could never be the same again.

In 1989 President Mitterrand chose to celebrate the 200th anniversary of the fall of the Bastille on 14 July 1789, which began the chain of events leading to the overthrow of the French monarchy and the establishment of the revolutionary government, in lavish style. More than $100 million was spent on the two-day-long activities, which included a huge parade, the opening of the new Opera House in the Place de la Bastille and the inauguration of the American architect I. M. Pei's controversial pyramid in front of the Louvre.

The celebrations were not all play. The 15th economic conference of the seven major industrial countries (Britain, Canada, France, Italy, Japan, the United States and West Germany) was held in Paris to coincide with the bicentenary, and President Mitterrand also convened a special meeting of the leaders of 11 Third World states. In all, 34 heads of state came to Paris, including President George Bush of the United States and Soviet President Mikhail Gorbachev.

Many of Mitterrand's domestic political opponents boycotted the events, which they felt had been organized as a propaganda exercise on the president's behalf. Parisians complained that the tight security arrangements needed to protect so many world leaders brought the city to a halt for a week. Nevertheless, the bicentenial celebrations fixed the attention of the world on the achievements of France's Revolution in a year that would later see its own revolutions in the communist states of Eastern Europe.

A fantastic show The climax of the bicentennial celebrations was a gigantic fireworks display on 14 July, Bastille Day, which lit up the familiar landmarks of Paris.

EUROPEAN PARTNERS

TWO CONSTITUTIONAL MONARCHIES · GOVERNING BY COALITION · PEACE AND PROFITS

A number of small feudal lordships in the region of the Low Countries passed in the 15th century to the possessions of the dukes of Burgundy and thence in 1504 to the Spanish royal house of Habsburg. The Protestant population of the Netherlands, in the north, rebelled in the 16th century against Catholic Spain; its independence was won, after a hard-fought struggle, in 1648. Belgium and Luxembourg remained under Habsburg rule until the whole region was drawn into the orbit of revolutionary France. In the peace negotiations of 1815 that concluded the Napoleonic wars the Low Countries were reunited under the crown of the Netherlands, but a rising in 1830–31 among the French-speaking Belgians led to the creation of a separate kingdom. Belgium was formally recognized as an independent sovereign state in 1839.

COUNTRIES IN THE REGION

Belgium, Luxembourg, Netherlands

Territories outside region Aruba, Netherlands Antilles (Netherlands)

STYLES OF GOVERNMENT

Monarchies Belgium, Luxembourg, Netherlands

Multi-party states Belgium, Luxembourg, Netherlands

One-chamber assembly Luxembourg

Two-chamber assembly Belgium, Netherlands

CONFLICTS (since 1945)

Internal unrest Belgium 1960s, 1980, 1983 (language riots); Netherlands 1975–78 (South Moluccans)

Colonial wars Netherlands/Indonesia 1945–49; Belgium/Congo (Zaire) 1960

MEMBERSHIP OF INTERNATIONAL ORGANIZATIONS

Council of Europe Belgium, Luxembourg, Netherlands

European Community (EC) Belgium, Luxembourg, Netherlands

North Atlantic Treaty Organization (NATO) Belgium, Luxembourg, Netherlands

Organization for Economic Cooperation and Development (OECD) Belgium, Luxembourg, Netherlands

---

TWO CONSTITUTIONAL MONARCHIES

Belgium and the Netherlands are two of Europe's surviving monarchies. Both are parliamentary democracies. The hereditary monarch, as head of state, performs a ceremonial role with agreed constitutional functions, but with limited powers.

In both countries the monarch has the power to appoint all ministers and to dissolve parliament, whereupon new elections have to be called. When necessary, the monarch also sets in motion the process of forming coalition cabinets. A wide range of political advice is taken on these occasions, but they allow the monarch a small degree of personal impact on the political process.

Contrasting styles

Despite these similarities, the political evolution of the two states means that there are differences in the way the role of the monarch is perceived. In the Netherlands the Prince of Orange (William the Silent, 1533–84) led the Dutch revolt against Spanish rule, and his descendants became the hereditary *stadholders* (chief magistrates) of the independent Dutch Republic, and the monarchs of the kingdom after 1815. The royal family remains a potent symbol of Dutch independence and is a curious link with the republican past of the country.

Monarchy came later to the Belgians, and their royal family was a foreign import. On breaking away from the Netherlands in 1830–31 they elected to become a parliamentary monarchy, and chose as their king a German prince, Leopold of Saxe-Coburg. His descendants remained more aloof from the people than their neighbors in the Netherlands did. They have experienced periods of great unpopularity.

Both states, too, are former colonial powers, but here again there are marked differences in the way they acquired and ran their empires. In the 16th and 17th centuries Dutch maritime enterprise laid the foundations of a trading and commercial empire that made the Netherlands for a time the preeminent state in northern Europe in the practice of arts and sciences. The Dutch East India Company established trading links that stretched from the Dutch settlement on the Cape of Good Hope (southern Africa) to Nagasaki

in Japan, and founded a rich colony in the Dutch East Indies (Indonesia). The Dutch also competed for commerce and prizes with the Spanish and the British in the Caribbean and along the American seaboard, eventually emerging with a colony on the South American mainland (Surinam) and possessions in the Caribbean (Curaçao and neighboring islands – today, the Netherlands Antilles).

Belgium acquired its colonial empire in Africa during the 19th century. For 20 years the Congo Free State (now Zaire) in central Africa – 80 times the size of Belgium and including the rich copper-mining area of Katanga – was under the personal rule of Leopold II (1835–1909) before it was annexed by the Belgian state in 1908.

Toward decolonization

The path to decolonization in the period since World War II did not go smoothly for either the Dutch or the Belgians. The Dutch East Indies were occupied by the Japanese from 1942 to 1945. On reassuming control, the Dutch rulers were involved in three years of bitter and costly conflict with Indonesian nationalist fighters, until in 1949 they were reluctantly compelled to withdraw from a colonial role that they could no longer sustain economically or politically.

In 1960 – the year the French abandoned their empire in Africa, creating 13 independent states – the upswell of nationalist feeling in the Congo proved too violent for the Belgian colonial rulers to withstand. Their precipitate granting of independence to the Congo left the new state unprepared for self-government, and a brutal civil war followed in which UN troops intervened unsuccessfully to try and achieve peace.

Some 300,000 immigrants of Dutch–Indonesian descent entered the Netherlands in the years after 1949. On the whole they were smoothly integrated into Dutch society, but one small group, the South Moluccans, for the most part ex-soldiers who had served with the Netherlands Indies army, chose to live with their families in separated camps and apartment blocks. In the mid 1970s, when independence was granted to Surinam, a further 150,000 people of varied origin (about 40 percent of the population), descendants of those who had originally been taken to the South American colony by the Dutch as slaves or contract laborers to work the sugar and banana plantations, emigrated to the Netherlands.

The period since World War II has seen a considerable influx of migrant laborers from Mediterranean countries into both Belgium and the Netherlands, many of whom have settled permanently with their families. The assimilation of all ethnic minority groups – accounting for a little over 5 percent of the population in the Netherlands, and 10 percent in Belgium – has proved by no means easy. In the Netherlands indications are that while definite progress has been made in the housing market, their position is much more precarious in the labor market and in gaining access to education.

In Belgium the existence of two distinct language blocs in the north and the south of the country is an ever-present threat to stability. Constitutional changes were put into effect in the 1980s in an attempt to halt demands for the complete political separation of these communities in two distinct states.

A cause for national celebration For Dutch people of all ages the monarchy has remained a popular source of national pride.

Train hijack in the Netherlands In the 1970s young exiles from the South Moluccas (formerly part of the Dutch East Indies) committed a number of acts of violence in demand of South Moluccan independence.

A patchwork of history The intricate pattern formed by Belgium and the Netherlands' provincial boundaries, and tortuous twists of the frontier, are a legacy of the region's feudal past.

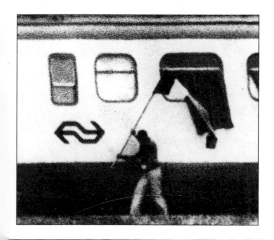

Provinces and languages of the Low Countries

—— province boundary

languages
- Dutch
- French
- French and Dutch
- German
- Lëtzebuergesch and German

117

GOVERNING BY COALITION

The political structures of Belgium and the Netherlands, two of the most densely populated countries in western Europe, have much in common. Both have two-chamber parliaments elected by universal suffrage. In Belgium the chamber of representatives, elected for four years by a system of proportional representation, has 212 members and the senate has 182, of which the majority are directly elected, 50 are drawn from the nine provincial councils and 24 are coopted. The role of the senate is at present under discussion and will be subject to redefinition if Belgium moves further in the direction of federalization.

The states-general of the Netherlands

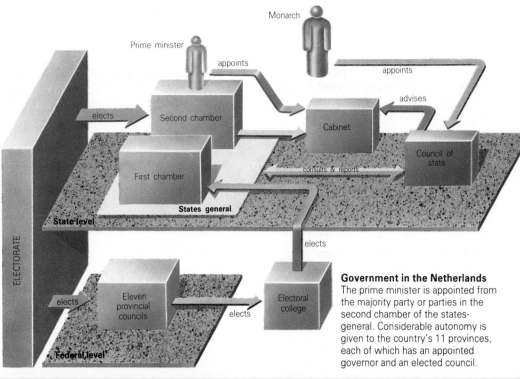

Government in the Netherlands
The prime minister is appointed from the majority party or parties in the second chamber of the states-general. Considerable autonomy is given to the country's 11 provinces, each of which has an appointed governor and an elected council.

consists of the first chamber, whose 75 members are indirectly elected by the 11 provincial councils for a six-year term. The 150 members of the second chamber are elected by proportional representation for four years. The first chamber has the right to approve or reject the legislation introduced in the second chamber.

The executive of both these countries is headed by a prime minister, normally appointed from the party commanding the most votes. In the Netherlands cabinet members are not permitted to be members of the legislature, but they may attend its meetings and take part in debates, and they are collectively respon-

Political choices Voters in Belgium's elections are wooed in two languages. The three major parties all have Dutch- and French-speaking wings to accommodate the country's language divide.

sible to it. There is also a council of state, the government's oldest advisory body, whose members – intended to represent a broad cross-section of national life – include former politicians, judges and retired civil servants, all of whom are appointed for life.

Party loyalties

In both countries a multiplicity of parties reflects religious, social and (in Belgium) linguistic divides. Since the 1960s there have been significant shifts in traditional party alignments in response to altered social and economic circumstances.

In the Netherlands political attachments in the past were formed primarily on the basis of religious affiliation, with regional and class interests also playing a significant role. More recently, the in-

LUXEMBOURG: ONE OF EUROPE'S MINI-STATES

The history of the grand duchy of Luxembourg is closely linked with that of its neighbors, Belgium and the Netherlands. Previously part of the Habsburg possessions, the grand duchy was created at the Treaty of Vienna (1815) under the sovereignty of the kingdom of the Netherlands. It was given to the Dutch crown in return for their hereditary territories in Germany, which were handed by the same settlement to the kingdom of Prussia. Half of its territory in the west was lost to Belgium in 1839, and it finally achieved independent status within the European state system in 1867.

Luxembourg is a parliamentary democracy; the hereditary grand duke acts as the formal head of state. It has a single-chamber parliament, the chamber of deputies, which is elected by universal suffrage. There are four main political parties, and proportional representation has resulted in habitual coalition government.

Despite its size, Luxembourg's prosperity, derived originally from iron and steel manufacture and now from banking, ensures its importance in European affairs. The customs union it formed with Belgium and the Netherlands (Benelux) in 1948 was the forerunner of wider European cooperation, and it is a founder member of many international organizations, including the European Community (EC). Though a founder member of NATO, it has only token defense forces of 800, of whom one is always on guard outside the ducal palace.

creasing secularization of society, and the movement of large sections of the population to the suburbs, has blurred these distinctions. The Roman Catholic and Protestant parties have been most affected. They have lost about 40 percent of their traditional vote in recent years, and have merged to form the Christian Democratic Appeal (CDA). The two other main parties are the Labor Party (PvdA) and the liberal People's Party for Freedom and Democracy (VVD).

None of these parties has traditionally been able to command a majority of votes. There is no clearly discernible left-wing or right-wing movement, and every government since 1945 has been a coalition, with the parties differing mainly over the implementation of economic and environmental policies. Christian parties in both countries have been the pivots of these coalitions. The effect of successive coalition governments has been to provide political stability, which has not been undermined by the increasing volatility of the electorate.

The language factor

In Belgium the situation is complicated by the language divide. In the north (Flanders) Dutch is the predominant language; in the south (Wallonia) French is spoken. The three major political parties – the Social Christian Party, the Socialist Party and the Liberal Party – are split on linguistic lines, with the result that there are separate French- and Dutch-speaking parties for each half of the country. The Social Christians are strongest in the Dutch-speaking north and the socialists in the south. The liberals have their electoral power base in Brussels. There are also a number of smaller parties; the ecology (Green) parties, in particular, have begun to attract electoral support.

As in the Netherlands, the formation of governments depends on coalitions being made, and post-election negotiations between the parties are often complex and protracted. In 1987 discussions to establish a ruling coalition took nearly six months. One effect of the language division has been to devolve greater power to the regions. In 1980 five regional assemblies were set up for Flanders, Wallonia, the French and German communities, and for Brussels, with powers to spend up to 10 percent of the national budget on cultural facilities, health, roads and urban development.

PEACE AND PROFITS

Fought over for centuries by contending European powers and nicknamed "the cockpit of Europe", each state in the Low Countries pursued a policy of strict neutrality throughout the 19th century. The creation of the unified German empire in 1871 on their eastern boundaries heightened their vulnerability. World opinion was outraged by Germany's breach of Belgian neutrality when it launched its attack on France across Belgian territory at the start of World War I. Barely a quarter of a century later Belgium, Luxembourg and the Netherlands again fell victim to German expansionism: all three were occupied from 1940. Belgium and Luxembourg were liberated in 1944, the Netherlands in 1945.

Building a new Europe
Even before the end of World War II the governments of the Low Countries exiled in Britain were having discussions in London aimed at preventing German expansion again in Europe. The first step toward the postwar reconstruction of Europe was the Benelux union (1948). This was soon superseded by more wide-ranging forms of cooperation in which the governments of Belgium, Luxembourg and the Netherlands were prime movers. All three were founder members of what was then the European Economic Community (EEC) in 1957, and their contribution to its organization has been of pivotal importance since its inception: they are all committed to the plans for a more integrated Community. The headquarters of the European Community (EC), as it is now known, is in Brussels, and Luxembourg is home to the European Court of Justice.

The Benelux countries were all founder members of the North Atlantic Treaty Organization (NATO), set up in 1949 to provide for the collective defense of Western Europe, Canada and the United States in the face of the perceived threat from Soviet domination in Eastern Europe. Since 1967 the headquarters of NATO have been in Brussels, and both Belgian and Dutch personnel have held key positions within the organization. Before the implementation of the Intermediate Nuclear Forces (INF) treaty (1987) there was disagreement over the deployment of Cruise missiles: the

CONTRIBUTING TO WORLD AID

Together with Scandinavian governments, the Dutch were foremost among Western governments in acting on United Nations initiatives in the 1960s to increase the transfer of funds from developed countries to Third World states to stimulate economic and industrial growth. There has for a long time been widespread popular support for the idea of development cooperation.

Jan Tinbergen, the Dutch economist and winner of the first Nobel prize awarded for economics in 1969, has played a key role in leading this campaign in the UN and in the Netherlands. One of his pupils, Jan Pronk, was the cabinet minister responsible for development assistance in the mid-1970s, when the Netherlands' contribution to development aid attained 0.7 percent of GNP, the target set for all developed governments by the UN, but scarcely ever reached. (This figure has been surpassed by Kuwait and Saudi Arabia; the only other European countries ever to attain it are Denmark, Norway and Sweden.)

The Dutch aid program receives broad support from all four main political parties, and humanitarian concern to redress imbalances between rich and poor states (the North–South divide) is high among the motives guiding aid policy. However, the program is not undertaken wholly without self-interest. There is no doubt that Dutch companies worldwide, particularly consultancy agencies, have greatly benefited from the implementation of the aid program.

Wider interests

The acrimonious winding up of Dutch colonial power in Southeast Asia left a bitter legacy of hostility, from which Dutch economic influence in Indonesia never quite recovered. The Netherlands government approached the granting of independence to its South American and Caribbean possessions with this lesson in mind, but with not much more success. In 1954 Dutch Guiana and the six islands of the Netherlands Antilles were made equal members of the kingdom of the Netherlands, with internal self-government. Full independence for Dutch Guiana as the state of Surinam followed in 1975. The new state was endowed with a substantial development fund by the Netherlands government, but further aid was withdrawn for a time in the early 1980s following a succession of military and revolutionary coups and the breakdown of effective government.

The six islands of the Netherlands Antilles are among the most prosperous in the Caribbean. Nevertheless their economies are dependent on the injection of grant aid from the Netherlands government. One island, Aruba, has seceded from the other five and is moving toward full independence and separate status.

Despite the years of administrative chaos and civil war left in the wake of Belgian withdrawal from the Congo (Zaire since 1975), the Belgian government has been able to retain an influence there by the provision of development aid, particularly in relation to mining interests in the Shaba province, formerly known as Katanga.

Until the 1960s the conduct of foreign policy was very much the unchallenged domain of a restricted circle of professional diplomats who were left more or less free to interpret questions of national interest as they wished. Since then, the policy-making arena has widened considerably, particularly in the Netherlands. Pressure is brought to bear on government by public protest and by lobbies from special interest groups such as those concerned with environmental issues.

The approach of the Belgian and Dutch governments to foreign policy is broadly similar in outline, but the aims of the Netherlands government are generally more idealistic. The Belgians remain committed to carrying out all their treaty obligations, though their day-to-day approach is more pragmatic.

The weight of public opinion Anti-Cruise weapon protestors in the 1980s graphically display the horrors of nuclear war. Policy on defense, as well as on such issues as the South African apartheid regime, development and environmental questions are all matters of public concern. Government decisions in these areas have more and more to take account of popular opinion.

Farmers protest The presence of the headquarters of the EC in Brussels brings many to agitate against Community decisions. These farmers are from France, protesting against EC agricultural policies.

Netherlands government, responding to public pressure, delayed for as long as possible the decision to site them on Dutch soil. Tactical nuclear weapons were accepted in the late 1950s. In the debate within NATO in the late 1980s over the decision to modernize these weapons, the Belgians sided with West Germany in pushing for postponement.

Belgium's language question

For historical reasons, French was for long the single official language of Belgium, though it was spoken by a minority of the population (at present 44 percent), living mainly in the southern part of the country, Wallonia. Dutch (or Flemish, as the local variant is commonly called) is spoken in the north. There is also a small German-speaking minority in the eastern part of the country.

Complicating the issue is the fact that Brussels, the center of diplomacy and government, is in the north. French was therefore the language of the ruling and social elite imposed on a local Dutch-speaking population, and it is not surprising that feelings on language traditionally ran high here. The Flemish campaign to confer equal status on their language – and thus to provide equal opportunities in society – began before World War I. Proposals that Belgium should become two separate language communities were first voiced by Dutch speakers seeking to enhance their economic chances.

In the period after World War II, however, the language dispute took a new direction as the coal and steel industries of the south went into decline at the same time as the development of service industries and the growth of Brussels as a major international center were bringing new employment opportunities in the north. French speakers in Wallonia began to demand greater regional powers to tackle their economic problems.

The language question came to a head in the 1960s, particularly among protest movements in Wallonia. There were frequent demonstrations, leading to clashes with the police. There were many proposals for political and constitutional change, and governments began to implement reforms. By the early 1970s linguistic parity in government appointments had been achieved.

Toward federalization?

In 1977 a coalition government led by Leo Tindemans of the Dutch-speaking Social Christian Party proposed the creation of a federal Belgium based on Flanders, Wallonia and Brussels. The plan was not adopted and Tindemans resigned. Four successive coalition governments under Wilfried Martens struggled to find a solution to the problem, and in 1980 it was eventually agreed to set up executive councils to administer the Dutch- and French-speaking communities.

The Flemish community council consists of 221 members who are Dutch-language representatives of both chambers of parliament. The French community council has 171 members, and there is additionally a Walloon regional council whose members all represent the Walloon provinces in parliament. The German community council has 25 members, and there is a three-member executive for Brussels. The councils have powers to spend up to 10 percent of the national budget on cultural facilities, health, roads and urban projects.

Plans for further constitutional change, devolving greater powers to the regions, were accepted by the coalition government formed by Martens in 1987, but no date was fixed for their implementation.

Pointing the way in two languages
Brussels (Bruxelles in French, Brussel in Dutch) is officially a bilingual city – every street sign, by law, must be in both languages.

Rioting communities During the 1960s the streets of Brussels were the scene of frequent violent clashes sparked off by the language question. In this incident in 1962 demonstrators had poured into the capital from all over the Dutch-speaking regions of Belgium to reaffirm their determination to gain equal status for their language. French-speaking Walloons, including students from Brussels' universities, had gathered to disrupt their march. Petrol bombs were thrown, and police were called in to quell the riots. It was incidents such as this that eventually forced Belgian governments to implement reforms.

OLD STATES, NEW DEMOCRACIES

THE ROAD TO DEMOCRATIC CHANGE · CONSTITUTIONS AND ELECTIONS · OUT OF AFRICA, INTO EUROPE

Two separate states began to emerge in the Iberian peninsula only when all the lands in the south had been reclaimed from the Islamic Moors, established there from the 8th and 9th centuries. Portugal became an independent kingdom in the 12th century. In Spain the last of the Moorish strongholds, Granada, fell in 1492, and a period of expansion began under the united crowns of Aragon and Castile. By the mid-16th century Spanish rule extended over much of Europe, including Portugal from 1580 to 1668, and in addition a new empire had been won in Central and South America (except for Brazil, which was colonized by Portugal). The Napoleonic wars (1803–15) highlighted Spain's decline in Europe, and in America independence movements had ended both Spanish and Portuguese rule by 1830.

THE ROAD TO DEMOCRATIC CHANGE

The Spanish state formed by the marriage of Ferdinand and Isabella in 1479 brought together the two separate kingdoms of Aragon, in the northeast corner of the peninsula and including the Pyrenean region of Catalonia, and Castile, which occupied the central plateau. With the royal capital at Madrid, in the center of the country, it was Castile that came to dominate the union.

At the height of its power the centralizing ambitions of the Spanish crown provoked an unsuccessful revolt by the Catalans (1640). The boundary with France, fixed in 1659, bisected both the country of the Basque people (Pais Vasco), at the Atlantic end of the Pyrenees, and the Catalan lands at the eastern end. The Basques, Catalans and Galicians (in the extreme northwest) have always presented a challenge to the Spanish concept of a single nation-state.

The territorial sovereignty of Spain today extends beyond its mainland frontiers to the Balearic Islands and a number of smaller islands in the Mediterranean, the Canary Islands in the Atlantic, and the ports of Ceuta and Melilla on the northern coast of Morocco. Its major possession in Africa, the Spanish Sahara, was transferred to Morocco and Mauritania in 1976. One part of the mainland itself, Gibraltar, ceded to Britain after its capture in 1704 during the Spanish wars of succession (1702–4), remains a British crown colony.

Following the loss of its African colonies in the mid-1970s, the only overseas territories remaining to Portugal are the Atlantic islands of the Azores and Madeira, which form an integral part of the Portuguese state. Unlike Spain, it has no ethnic minority peoples to threaten national cohesion.

Europe's last dictators

During the 19th century Spain and Portugal were characterized by weak governments. Politics became polarized between monarchists and liberal constitutionalists. A republic was temporarily established in Spain in 1873–74; the Portuguese monarchy was abolished in 1910 following the assassination of Carlos I two years earlier. In the 20th century both countries have experienced long periods of authoritarian military rule.

In Spain the military government of Miguel Primo de Rivera (1870–1930) lasted from 1923 to 1930. A year later King Alfonso XIII abdicated and a republic was declared. The center-left Popular Front alliance, which included both anarchists and communists, introduced a series of reforms that precipitated a right-wing revolt in 1936. General Francisco Franco (1892–1975) emerged victorious from the civil war that followed (1936–39). As Spain's undisputed ruler he established a fascist dictatorship. He was named head of state for life in 1947; in fact he relinquished the post of premier in 1973.

In Portugal António de Oliveira Salazar (1889–1970) exercised a virtual dictatorship as prime minister from 1932 to 1968. Social conditions improved at the expense of personal liberty, but Portugal

remained one of Europe's poorest states. Salazar's last years were troubled by the beginnings of armed revolt in Portugal's African colonies.

Changing the system

Salazar's successor, Dr Marcello Caetano (1906–80), failed to deal with the increasingly costly anticolonial wars. He was overthrown on 25 April 1974 by a group of radical army officers calling themselves the Armed Forces Movement.

The next two years saw the formation of a number of provisional governments under the authority of a military junta. Coups were attempted by right-wing and left-wing elements in the army before a constituent assembly was elected in 1975, paving the way for parliamentary elections the following year.

In Spain, Juan Carlos, Alfonso XIII's grandson, was named as Franco's successor in 1969. He became king on Franco's death in 1975, guiding the country toward democratically elected government. He appointed Adolfo Suárez, a civil servant, to introduce a number of reforms, including the legalization of political parties. A referendum of December 1976 gave massive approval to democratic change. Two years later a new constitution, which all the major parties had helped to formulate, received public assent.

Defenders of the Spanish state
The civil guard (*Guardia civil*) belongs to Spain's oldest national police force. It is part of the military establishment, and in peacetime is responsible to the Ministry of the Interior for guarding ports, airports, frontiers and prisons, and for maintaining civil order.

Victory parade, Madrid 1939
General Franco gives the fascist salute to Italian troops. Military success in the civil war was backed by Europe's fascist leaders – Adolf Hitler in Germany and Benito Mussolini in Italy – to whom Franco gave initial support at the start of World War II.

Spain's 17 autonomous communities have their own elected assemblies and regional governments. Its three largest national minorities – the Basques, Catalans and Galicians – were the first to be granted self-governing status under the 1978 constitution. Portugal is divided into 18 administrative districts.

Administration in Iberia
—— province boundary

national minorities in Spain
- Basque
- Catalan
- Galician

CONSTITUTIONS AND ELECTIONS

Portugal's new constitution, approved in 1976 and modified in 1982, provides for a president who is elected for five years and appoints the prime minister. He in turn nominates other members of the council of ministers. The single-chamber legislative assembly consists of 250 deputies, who are elected for four years under a system of proportional representation.

The new constitution perpetuated the limited autonomy of the Azores and Madeira, but divided mainland Portugal into 18 administrative districts governed by assemblies. Below these are 305 municipal councils (*concelhas*). Each level has its own executive body.

Portugal's first elected president was the army chief General António Ramalho Eanes. Tensions during his term of office (1976–86) meant that no single political party was able to command a stable majority; there were five prime ministers in under three years. Following the elections of 1976 the Socialist Party (PS), led by Mário Soares, formed a minority government. Increasingly split by internal dissent, it remained precariously in power until its defeat by the right-wing Social Democratic Party (PSD) in 1983. The PSD achieved a notable electoral success in 1987, and its leader, Anibal Cavaco Silva, was able to form Portugal's first majority government. In the presidential elections the previous year Mário Soares had become Portugal's first civilian president for 60 years.

The Spanish constitution

Unlike Portugal, Spain has a two-chamber national assembly (the Cortes Generales), consisting of a congress of deputies and a senate. The congress has 350 members, elected by proportional representation every four years. The senate is elected by a majority system that returns four members from each of the 50 provinces, with the exception of the islands and North African towns. The regional parliaments also send members to the senate, making 250 in all.

Numerous small political parties were formed to contest Spain's first parliamentary elections, held in June 1977. Adolfo Suárez, leader of the Democratic Center Party (UCD), a grouping of diverse right-wing and center parties, became the

country's first elected prime minister. Divisions within the alliance brought about its downfall in February 1981, and Suárez resigned.

As his deputy, Calvo Sotelo, was being sworn in as the new prime minister, an attempt was made by right-wing army officers to engineer a military coup in the Cortes. It failed; the king, as head of the army, was firm in repudiating the coup. The episode illustrates the sort of difficulty that societies accustomed to authoritarian control may encounter in adapting to the the habits of democracy.

Spain's Socialist Party (PSOE), led by Felipe González, a rousing and popular orator, won a clear electoral victory the following year, securing 48 percent of the vote. In his first years in office he decentralized the government, extended welfare

A separate homeland for the Basques An ETA mural portrays Spain and France's thumbs-down approach to demands for independence. Isolated in the Pyrenees, the Basques have preserved their cultural uniqueness, centered on the *caserio* or traditional farmstead although more than four-fifths of Basques now live in the industrialized coastal region around Bilbao.

benefits and led Spain into the European Community (EC). Support for his government later declined. In the 1989 elections the PSOE had an absolute majority of only two seats.

Regional autonomy

The 1978 constitution allowed for the possibility of self-governing regional governments with their own elected assemblies. These were quickly granted in the Basque country, Catalonia and Galicia; in Andalusia discussions were

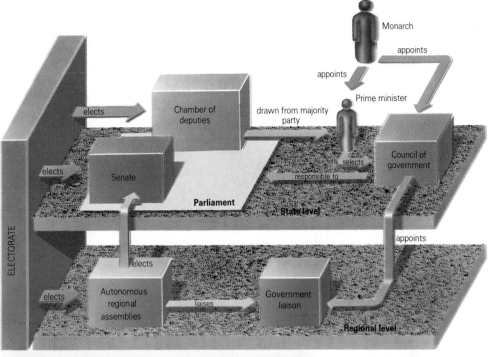

Spain's constitutional monarchy
The 1978 constitution made the hereditary monarch the formal head of state. He appoints the prime minister (called the president of government), who takes office only upon a vote of confidence from the chamber of deputies and appoints ministers (the council of government). The congress of deputies votes on new legislation. The senate, the upper house, has limited power to amend or veto bills.

A respected politician Mário Soares, leader of the Socialist Party, who headed Portugal's first civilian government in 1976, was elected the country's president in 1986 and again in 1991.

REGIONAL CHALLENGES TO THE STATE

Separatist demands from Spain's three historic national minorities increased during the late 19th and 20th centuries, a period of rapid economic growth and cultural renaissance in both the Basque Country and Catalonia. Regional autonomy was granted to the Catalans under the Second Republic in 1932, and to the Basques and Galicians in 1936.

All nationalist movements were banned during Franco's dictatorship, but they began to reappear in the 1950s. Particularly prominent was the ETA (*Euskadi ta Askatasuna*: Basque Homeland and Liberty) movement, whose activities against the Spanish state became increasingly violent.

One of the earliest actions of Spain's elected government was to carry out the promises in the 1978 constitution with regard to regional assemblies: the first elections (in the Basque Country and Catalonia) were held in 1980.

The regional parties gained significant electoral strength. In Catalonia the nationalist party Convergence and Union (CIU) won power in 1980, and after 1986 the Basque Nationalists ruled in coalition with the Socialists. Their influence in national politics is growing. Some Basques continue to press for full independence by nonviolent protest, such as abstaining from elections. A few still engage in direct action, including bombings, but most Basques now repudiate such acts of terrorism. An ETA truce ended in 1989 when talks with the government collapsed.

more protracted. The remaining 13 regions were created together in 1983. This different process accounts for variations in the status of the different regions. All have to conform, to differing degrees, to policy frameworks that are laid down by the government in Madrid, but they nevertheless have wide-ranging responsibilities – for agriculture, urban and rural planning and development, tourism, housing, health and social services. Elections to the regional assemblies are held every four years.

Below them are the 50 provinces, each with its own council (Diputación Provincial). Like the other levels of government, elections to them are held every four years. After years of underfunding from Madrid, the larger cities faced increasing financial burdens in the 1980s.

OUT OF AFRICA, INTO EUROPE

As long as Portugal and Spain remained under fascist dictatorships, which ruthlessly suppressed all political dissent at home, neither country was able to share fully in the processes that reshaped Europe after World War II. However, intensification of the Cold War tempered hostility from other Western powers, and a treaty signed in 1953 allowed four US bases to be set up in Spain.

Both regimes retained colonies in Africa. When Spain rid itself of the Spanish Sahara in 1976 it did not consult the wishes of the people, but simply ceded the colony to Morocco and Mauritania, leaving them to decide how to divide it. The independence movement, known as the Polisario Front, then demanded self-

government. In 1979 Morocco annexed Mauritania's Saharan territory. It refused to negotiate directly with the Polisarios, who continued their guerrilla campaign.

An expensive war of attrition
Portuguese colonies (designated "overseas provinces" in 1951) still remained in Asia (Goa, Damao and Diu, annexed by India in 1961, Macao and Timor) and in Africa (Angola, the Cape Verde islands, Guinea, Mozambique, São Tomé and Principe). These last were the scene of increasingly expensive anticolonial wars. Salazar, and then Caetano, refused to countenance any form of self-government for Portugal's African dependencies, but the effort of clinging on was an enormous drain on its resources – and Portugal was already one of the poorest countries in Europe. Some 45 percent of its GNP was

GIBRALTAR: A DISPUTED COLONY

Gibraltar is the smallest colony in the world. The rock on which it stands is only 5 km (3 mi) long and 1.2 km (0.75 mi) wide. Its population of just less than 30,000 is governed by an assembly with 15 elected and 2 nominated members. The governor (representing the British crown) has executive authority. The colony was assigned to Britain by treaty in 1713, following its capture in 1704, and has remained a British garrison ever since. Its position at the entrance to the Mediterranean gives it strategic importance today as an air and naval base. It also serves as an international transit port for oil.

Spain's claims to sovereignty over Gibraltar are strongly pressed, but a referendum in 1967 confirmed the overwhelming wish of the local population to remain in association with Britain.

As in the Falkland Islands, Britain gave assurances that the colony's right to self-determination would be upheld: a demand from the United Nations that Britain should withdraw by 1969 was consequently rejected.

On Spain's entry into the EC fresh proposals were made for Gibraltar's return. Guarantees were given that it would be treated as one of Spain's autonomous communities: it would continue to be administered by British subjects, and its current laws, rights and economic organization would be recognized. These promises did not satisfy the local population, who demonstrated in 1987 to reject them. The sovereignty issue remains as far from resolution as ever, but cooperation has increased in such matters as shared use of Gibraltar's airport.

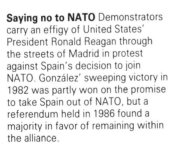

Saying no to NATO Demonstrators carry an effigy of United States' President Ronald Reagan through the streets of Madrid in protest against Spain's decision to join NATO. González' sweeping victory in 1982 was partly won on the promise to take Spain out of NATO, but a referendum held in 1986 found a majority in favor of remaining within the alliance.

Cultural ties In the post-Franco years Spain has worked actively to promote better relations with the Spanish-speaking states of Central and South America, to which it is tied by a common language and culture. Its socialist government has also extended its diplomatic and commercial links with communist countries throughout the world. Here premier Felipe González is seen with Cuba's longstanding revolutionary leader, Fidel Castro.

spent on maintaining military forces in Africa. Emigration was encouraged to bolster white rule.

The officers' coup of 1974 set Portugal on the path of negotiation in Africa, and opened the way to fuller cooperation with its neighbors in Europe. Although Portugal had been a member state of NATO since its inception in 1949, and was a founder member of the European Free Trade Association (EFTA), set up in 1960, it had not shared in the increased prosperity of its neighbors. So long as their undemocratic systems remained, neither Spain nor Portugal was able to join fully in the closer ties of the European Community (EC). Adolfo Suárez and others guiding Spain's transition to democratic government argued that membership of the EC (which Spain had first tried to join in 1962) and of NATO was an inherent part of the process of democratization, and was the only way to enhance Spain's economic prospects.

Joining the European Community

Both Spain and Portugal were admitted to the EC in 1986. As part of its preparation for entry Spain sought to regulate and clarify some areas of conflict that still remained in its relations with other European states, including the reopening of the frontier with Gibraltar, closed since 1969. Bilateral agreements were also reached with France over the extradition of suspected ETA terrorists, paving the way for closer political cooperation.

Membership produced a number of immediate benefits. Spain's economic growth rate was the most rapid in Western Europe (though it still lagged behind in industrial investment and modernization). Communications and transport were improved, and both countries received substantial subsidies from the European Regional Development Fund. The presence of Spain and Portugal in the deliberations of the EC helped to focus attention on the Mediterranean region as a whole, creating a bloc with France, Greece and Italy to counterbalance the dominant influence of its northern members.

In the late 1980s Spain was involved in negotiations with the United States over the presence of four US bases in the country. The treaty of 1953 that had originally agreed to them expired in 1987. Spain wanted a substantial reduction in the numbers of United States personnel attached to these bases (from 12,000). In particular, the size of the Torrejon base – the scene of many pacifist demonstrations was questioned. It was considered to be too close to Madrid.

Spain has sought to strengthen its role in Central America, to which it is linked by ties of language, history and culture. In conjunction with other EC countries, it backed the peace initiatives to bring an end to the civil wars in El Salvador and Nicaragua, proposed in 1987 by the Contadora group of countries (Colombia, Mexico, Panama and Venezuela).

Withdrawal from empire

Portugal was the earliest European colonial power in Africa, and the last. It was here that the heart of its empire (which at its height stretched from Brazil in South America to Macao in Southeast Asia) was to be found. The foundations of its African possessions were laid in the 15th century when, under the patronage of Prince Henry the Navigator (1394–1460), Portuguese expeditions were made down the west coast of Africa, reaching the Cape of Good Hope (1488) and traveling eastward on to India (1497).

Portugal's early colonization of west and east Africa was essentially commercial. A series of trading posts, protected by fortified bases, was set up to exchange goods for ivory and other resources: Portugal inaugurated the trade in African slaves as enforced labor in the new colonies of America, though other European powers were not slow to follow.

In the 19th century rampant European imperialism, fueled by rivalries at home, exploded into Africa. Britain, France and Germany proceeded to divide up the continent between them, and Portugal's colonial ambitions were revived. From its long-established coastal trading settlements in Angola and Mozambique, military expeditions were made into the interior during the 1870s.

Portugal's hold over these new colonies was never very firmly secured. From the beginning a considerable military establishment was needed to maintain power. Its policy of ruling by fragmenting indigenous opposition and exploiting rivalries between different ethnic groups (a common enough colonial strategy) served to prolong resistance and spark off sustained civil wars. A substantial anti-colonial uprising in Angola was put down in 1913, and there was later resistance in both Guinea and Mozambique.

By the 1960s independent states were being created throughout the African continent as European colonial power disintegrated. Most of these new states supported independence movements in Portugal's African territories. The refusal to allow even limited self-government led to a long and increasingly bitter war. A further element of conflict arose from the fragmentation of the anti-Portuguese opposition into rival guerrilla groups.

Revolution in Lisbon

The revolution launched by the radical young army officers who overthrew Caetano's government in Lisbon on 25 April 1974 had a dual aim. Their purpose was to bring a halt to Portugal's increasingly unpopular colonial wars. One in four men of military age were in the armed services, and many were leaving the country in order to escape military call-up). They also set out to "save the nation from government", and to establish some form of democracy in Portugal.

The leaders of the coup installed a Junta of National Salvation under General António de Spínola, a former governor-general of Guinea who advocated a diplo-

Portugal's worldwide empire was built on trade and commerce. A chain of coastal stations was established the length of the route round the Cape to trading centers in India and Southeast Asia. Not until the late 19th century did it consolidate its hold in Africa, in emulation of other European powers then engaged in dividing the continent between them. But it clung on longer to its colonies than did countries elsewhere.

Portraits of Marx and Lenin and a mural showing the revolutionary struggle, dating from 1974, decorate a Lisbon wall. The group of army officers who drove Spínola out of office in September 1974 attempted to impose a program of Marxist reforms, which won little public support. Farms were seized, and key industries placed under workers' control. Assembly elections in April 1975 marked the first step toward parliamentary democracy.

matic solution in Africa. Ceasefires were immediately brought into effect, but Spínola, a conservative, was in favor of a neocolonial arrangement of an "imperial federation of democratic Portuguese-speaking states", and stifled radical reforms within Portugal itself.

Map caption (lower left):

Azores
PORTUGAL SPAIN
Madeira
CAPE VERDE 1975
GUINEA-BISSAU 1974
SAO TOME & PRINCIPE 1975
CABINDA (Angola)
ATLANTIC OCEAN
ANGOLA 1975
INDIAN OCEAN
MOZAMBIQUE 1975

Last strongholds in Africa

Portugal
former Portuguese colony
1974 date of independence

An alliance of disaffected army officers and socialist politicians forced him into exile in September 1974. Guinea was given its independence as Guinea-Bissau, followed by Angola, Mozambique and the offshore African colonies a year later. In Southeast Asia the island of Timor was annexed to Indonesia, and Macao became a "special territory" with its own governor (in 1999 sovereignty returns to China under an agreement signed in 1987). As a result of these arrangements Portugal ceased to be a colonial power.

Some 650,000 people (known as *retornados*) returned to Portugal from its former colonies after 1976, some 90 percent of them from Angola. This very large number, in a country of only 9 million inhabitants, put further strain on inadequate urban resources and an economy that was already under pressure.

NEW STATES IN THE CLASSICAL WORLD

THE QUEST FOR NATIONAL IDENTITY · TWO PARLIAMENTARY DEMOCRACIES · CARVING AN INDEPENDENT ROLE

Although they occupy the part of the Mediterranean that is regarded as the cradle of Western European civilization, Italy and Greece are relatively modern states. The classical world of Greece and Rome fragmented in the 5th century AD under pressure from nomadic invaders. In Italy, Rome became the center of the medieval Christian church under the spiritual leadership of the pope. The Italian peninsula, a mosaic of small states, was gradually subordinated by its more powerful neighbors – France, Spain and Austria. From the 15th century Greece was a province of the Turkish Ottoman empire. Only when the Greeks had won their independence in 1830, and when a nationalist movement had brought about Italian unification between 1860 and 1870 (the Risorgimento), did today's states emerge.

COUNTRIES IN THE REGION

Cyprus, Greece, Italy, Malta, San Marino, Vatican City

Island territories Aegean Islands, Crete, Ionian Islands (Greece); Elba, Capri, Ischia, Lipari Islands, Sardinia, Sicily (Italy)

STYLES OF GOVERNMENT

Republics Cyprus, Greece, Italy, Malta, San Marino

City state Vatican City

Multi-party states Cyprus, Greece, Italy, Malta, San Marino

State without parties Vatican City

One-chamber assembly Cyprus, Greece, Malta, San Marino

Two-chamber assembly Italy

CONFLICTS (since 1945)

Internal unrest Greece 1965, 1967, 1974; Italy 1978

Coups Cyprus 1974; Greece 1967

Civil wars Cyprus 1963–64, 1974 (with Greek and Turkish involvement)

Independence war Cyprus/UK 1955–60

Interstate conflict Italy (with US and others)/Iraq 1991

MEMBERSHIP OF INTERNATIONAL ORGANIZATIONS

Council of Europe Cyprus, Greece, Italy, Malta

European Community (EC) Greece, Italy

North Atlantic Treaty Organization (NATO) Greece, Italy

Note: The Turkish Republic of Northern Cyprus was proclaimed in November 1983; it is not internationally recognized.

THE QUEST FOR NATIONAL IDENTITY

The boundaries of Greece and Italy underwent a handful of changes in the early years of this century. In 1912–13, as a result of the Balkan wars, Greece acquired Macedonia and the Aegean islands from Turkey; Thrace, in the northeast, was added at the end of World War I. In 1947 the Dodecanese islands in the southeastern Aegean, including Rhodes, were ceded from Italy to Greece. The Istrian zone around Trieste, on the north Adriatic coast, transferred to Italy from Austria in 1918, was claimed by Yugoslavia; the zone was divided between them in an agreement made in 1954.

Unstable pasts

Italy and Greece have both undergone periods of political instability during the 20th century. Greece has experienced dictatorship or military rule on a number of occasions, most recently from 1967 to 1974. In Italy the right-wing fascist dictatorship of Benito Mussolini (1883–1945) lasted from 1924 to 1943.

Today both countries are republics. They have had, and have abolished, monarchies. The Greek monarchy, of German origin, was never firmly or popularly established, and was overthrown after a military coup in 1967; a referendum seven years later rejected its restoration. In Italy, Victor Emmanuel II, king of Sardinia–Piedmont, and one of the leaders of the nationalist struggle against Austrian rule, was proclaimed king in 1861. The monarchy lasted until 1946, when it was abolished by a plebiscite.

Both are highly centralized states, though an important regional tier of government was introduced throughout Italy in 1970. (Several peripheral regions already had their own governments.) While both capitals, Athens and Rome, act as the foci for national life where the symbols of nationhood – such as the Victor Emmanuel monument in Rome – are housed, they are nevertheless widely associated with inefficient and corrupt central bureaucracies. Rome is also the center of a global church.

Strong cultural and political differences between Italy's regions have aided the growth of an entrenched provincialism. This has produced the need to create a sense of national identity. This was not so

necessary in Greece, where a history of opposition to the ruling Turks led to the easy development of a powerful and widespread Greek nationalism. The dilemma that faced Italy's unifiers – "Having created Italy, we must now create the Italians" – remains a major problem for the Italian state to this day.

The greatest challenge to national integration in Italy arises from the economic disparity that lies between the industrialized north and the underdeveloped south. This difference has persisted and deepened since unification. The stagnation of the south, in contrast to the startling economic growth of the north in the years before World War I, owed a good deal to the determination of the southern landowning class to protect its political privileges.

There are separatist movements in Sardinia and Sicily, as well as Aosta, Friuli and Süd Tirol in the Alpine areas of northern Italy. None, however, poses a strong threat to the territorial integrity of the state, with the possible exception of the German-speaking Süd Tiroleans in the Alto-Adige region. In northeast Greece the Turkish minority in western Thrace has not been well treated, but there is no organized demand for the return of the area to Turkey.

Other states of the region

The island of Cyprus had been colonized by successive empires, from that of the Assyrians in the 8th century BC to the Ottomans in 1571, when it came under British administration in 1878. By then a Greek-speaking majority was already seeking union with Greece, strongly resisted by a Turkish minority. Shortly after independence (1960) the Turkish community withdrew from power-sharing with the Greeks. In 1974 the ruling Junta in Greece was implicated in attempts to unite Cyprus with Greece; Turkey responded by invading the island and setting up the Turkish Republic of North Cyprus. No other state recognizes its sovereignty.

Malta has been a democratic republic within the British Commonwealth since 1964. San Marino, with a population of 22,000, claims to be the world's oldest republic. The sovereignty of the Vatican City, an enclave within the city of Rome, was first recognized by Mussolini's fascist government in 1929 and confirmed by the Italian government in 1947.

The statue of Victor Emmanuel II, first king of a united Italy, in front of his grandiose monument in Rome. It symbolizes the artificial nature of Italian unification rather than its ideological power.

Italian unification, led by the northern Piedmontese, was completed by 1870. It was the work of an urban minority in a largely rural society: to most peasants the state was a foreign intrusion, responsible for imposing taxes and conscription.

TRENTINO-ALTO ADIGE 1919

LOMBARDY 1859

VENETIA 1866

PIEDMONT 1860

Po

Adige

DUCHY OF PARMA 1860

DUCHY OF MODENA 1860

Arno

GRAND DUCHY OF TUSCANY 1860

PAPAL STATES 1860

Corsica (to France 1770)

Tiber

LATIUM 1870

SARDINIA 1860

▷ BENEVENTO (PAPAL STATES) 1860

The unification of Italy

— state boundary

☐ kingdom of Piedmont-Sardinia

1860 date of unification

KINGDOM OF THE TWO SICILIES 1860

VALLE D'AOSTA
• Aosta

TRENTINO-ALTO ADIGE
• Trent

FRIULI-VENEZIA GIULIA

LOMBARDY
• Milan

VENETO

• Trieste

Turin •

Po

Adige
• Venice

PIEDMONT

EMILIA-ROMAGNA

Genoa •

• Bologna

LIGURIA

Florence •

SAN MARINO

Arno

TUSCANY

• Ancona

Perugia •

MARCHES

UMBRIA

Corsica (France)

Tiber

LATIUM

• Aquila

ABRUZZI

■ Rome

MOLISE

• Campobasso

APULIA

CAMPANIA

• Bari

• Naples

SARDINIA

Potenza •

BASILICATA

Administrative regions of Italy

— regional boundary

☐ special status region 1948

Cagliari •

CALABRIA

• Catanzaro

• Palermo

SICILY

A league apart Supporters of the Northern League — a right-wing regional movement that won 80 seats in the national elections in 1992 — meet to declare their support for a separate state in northern Italy. The party's appeal draws on dissatisfaction with national politics in Rome and on resentment that the richer, industrialized north is called upon to subsidize the poorer regions of the south

The north–south divide The boundaries of Italy's 20 administrative regions roughly correspond to those of the past. Vast cultural and economic differences remain between the industrial and farming areas of the north and center and the much poorer Mezzogiorno, the area south of and including Naples and Campania, and between the mainland and Sardinia and Sicily.

133

TWO PARLIAMENTARY DEMOCRACIES

Parliamentary institutions are of quite recent date in both countries. Greece's bitter experiences in World War II – having resisted invasion by Italy in 1940, it was occupied by Germany from 1941 to 1944 – were followed by three years of civil war and subsequent political turmoil. Nine general elections were held between 1946 and 1964, and parliament was dominated by the right wing, though charges of intimidation and bogus voting were frequent. The role of parliament was additionally weakened by the persistent undermining action of the monarchy. This led to the eventual discrediting of parliament, and prepared the way for the military dictatorship of George Papadopoulous (April 1967 to July 1974).

After the abortive attempt to unite with Cyprus, democratic elections were restored in 1974; the present constitution was adopted a year later. A single chamber elects the president for a five-year term; he appoints a prime minister and cabinet, and ratifies all bills passed by parliament. His veto, however, can be overridden by an absolute majority of parliament.

The center-left, headed by the Panhellenic Socialist Movement (PASOK), came to displace the right as the dominant electoral bloc. Its national support increased from 13.6 percent in 1974 to just over 48 percent in 1981, when it won an absolute majority in parliament under its leader, Andreas Papandreou.

Between 1981 and 1985, when it was re-elected to office, the PASOK administration carried through a number of important social changes such as the lowering of the voting age to eighteen, the legalization of civil marriage and divorce, and an overhaul of the universities and the army. In the late 1980s it was shaken by a series of personal and financial scandals surrounding the leadership of Papandreou. Following PASOK's defeat in a series of elections in 1989–90, Xenophon Zelotas emerged as leader of an all-party coalition. Papandreou was charged with corruption and acquitted in 1992.

Postwar politics in Italy

Italy's present constitution was adopted in 1946, when the country became a republic headed by a president elected by

The heat of political battle A candidate for Greece's right-wing New Democracy party campaigns in an excited atmosphere amid a shower of thrown papers and the smoke of firecrackers let off among the crowd. Three general elections were held between June 1989 and April 1990 as Greece was plunged into a turmoil of political activity and confusion.

The expansion of Greece Since mainland Greece won its independence in 1830 it has more than doubled its territorial extent. Today it is divided into 10 administrative regions. Its many fragmented peninsulas and islands present problems of communication, and its long maritime boundary exposes it to many points of contact with its traditionally hostile neighbor, Turkey.

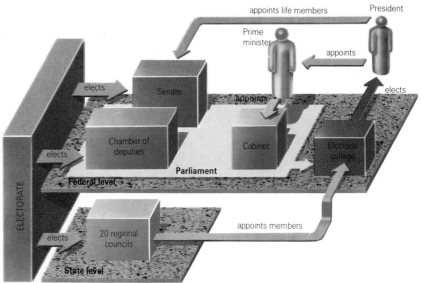

Italy's government The president, elected by parliament and the regional councils, has little actual power. He appoints the prime minister to head the government. The constitution is overseen by a constitutional court.

Red power A national rally of Italy's communist party, the largest in Western Europe. Its traditional strength is in the central areas of Italy and at local level: Bologna's municipal council has been ruled by a communist majority since 1946.

parliament. The two chambers of the parliament have equal powers; it is elected for five years by a system of proportional representation.

Italian elections from 1947 to 1963 produced shortlived center-right governments dominated by the Christian Democratic (DC) party in coalition with between three and five smaller parties. After 1963 the tendency of governments was toward the center-left, but the DC managed to maintain its dominant role, despite periods of stress such as that surrounding the proposed entry into government coalition of the neo-Fascist (MSI) party in 1960, the "hot autumn" of strikes in 1969, and the rampant terrorism of both the extreme left and the extreme right in the late 1970s.

After the Communist Party (PCI) won over a third of the votes for the chamber of deputies in 1976, an alliance with the DC – the "historic compromise" – was proposed, by which the PCI agreed to support the governing coalition. But it was rejected and, apart from a brief period between 1977 and 1978, the PCI was excluded from power sharing by the other parties.

In 1983 Bettino Craxi became Italy's first Socialist prime minister, and headed its longest-serving government since World War II: it lasted until 1987. He attempted to create a "third force" in Italian politics, lying between the DC and the PCI. However, Italian governments remained fragile coalitions between the DC, the Socialists (PSI) and small parties such as the Liberals, Republicans and Social Democrats. The entire political system became more fragmented as traditional

CONTRASTING STYLES OF COMMUNISM

Greece and Italy emerged from World War II with the two largest communist parties in Western Europe. The Greek Communist Party (KKE) had its own guerrilla army, and became embroiled in a civil war with royalist government forces, which were supported by the British. The Communists were eventually defeated when the United States' President Harry Truman (1884–1972), in an early stage of the Cold War, declared that a line must be drawn in Greece against communism (the Truman Doctrine, 1947), and lent massive aid to anticommunist forces.

For more than twenty years the KKE went underground, and pursued a clandestine strategy of uncompromising hostility to existing political institutions. It was legalized in 1974 when democracy was restored after Papadopoulous' downfall, and has since taken part in electoral politics, gaining about 10 percent of votes. It later split into two principal factions, the dominant one of which retains an old-fashioned Stalinist orientation.

By contrast, the Italian Communist Party (PCI) – which had played a

leading role in the prewar antifascist opposition – from 1945 committed itself to achieving social change through its various affiliated organizations such as cooperatives and trade unions, and by involvement in electoral politics. It flourished in Italian political life to become the second largest party, after the Christian Democrats, in numbers of votes and seats. However, after its high point in 1976, when it came close to sharing in national government its support declined, remaining strongest in *la zona rossa* (the red zone) of central Italy, where it built extensively on older traditions of anticlericalism and collectivism to produce a powerful communist subculture.

Its most significant losses after 1976 were in the industrial northeast and in some areas of the south, where it had made considerable inroads in the 1960s. In 1991, after the failure of communism in Eastern Europe, it reformed itself as the Democratic Party of the Left (PDS), and its policies bear a close resemblance to those of the social democratic parties of other European states, such as the Nordic Countries and Germany.

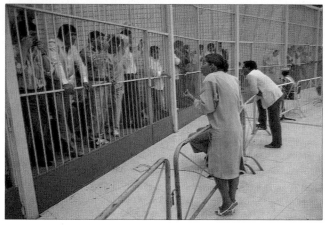

Urban guerrilla tactics Italy has suffered greatly at the hands of extremist groups of both left and right. Some 70 people were killed when neofascists exploded a bomb in Bologna station in May 1980. Among the outrages of the left-wing Red Brigades was the murder of former premier Aldo Moro in 1978.

Accused Mafia agents are visited by relatives during a mass trial in Naples. Organized crime in Italy directly challenges government, particularly in the south where the Mafia is able to influence local government by its control of political funds and contracts.

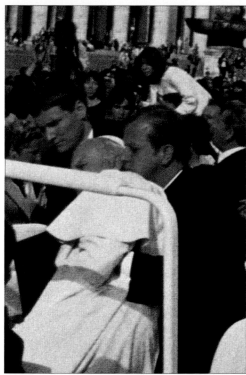

Pope John Paul II is held by aides after an attempt on his life in 1981. At the time it was believed that Bulgarian and Soviet agents were implicated in the plot to kill him, but the credibility of this theory was shaken by the increasingly erratic behavior of the would-be assassin, a Turkish right-wing fanatic, Mehmet Agca. Its credence rests on the pope's record as a leading anticommunist. In particular, he had given support to the Solidarity trade union in his native Poland.

party support declined and regional parties such as the right-wing Northern League increased in strength. In 1992, after an inconclusive general election, the president resigned and it took many weeks of crisis before a coalition government (under the premiership of the deputy leader of the reformed Communist Party, the PDS) was formed from among the warring parties.

Allegations of corruption

One reason for the rise of the regional parties is growing dissatisfaction with the highly centralized nature of the Italian state machine and the allegations of cor-ruption and criminality that surround the national government, which has been implicated in a series of scandals since the early 1980s.

The government has found itself unable to curb massive underground economies, estimated to equal perhaps a quarter of the official GNP of Italy. Despite substantial increases in state provision of social services and development planning, criminal organizations such as the Mafia have increasingly grown in strength. In 1992 the assassination of two anti-Mafia judges in Sicily resulted in the Italian army being deployed against the Mafia for the first time.

CARVING AN INDEPENDENT ROLE

After centuries of domination by their neighbors, Italy and Greece were late entrants to the European state system. As a result of their geographical position on the southeastern flank of Western Europe both countries were seen, and have continued to be seen, by the major powers of the day as potentially significant pawns in containing or restricting the influence and expansion of their rivals. Nonetheless both have tried to carve out a degree of independence for themselves in their relations with other states.

Expansionist ambitions

In the first half century and more of its existence Greece focused its efforts on expanding its territory to include all Greeks within its boundaries. In effect, this meant recovering the Greek-populated territories of the Ottoman empire, particularly Macedonia and Crete. At its most extreme, this policy was expressed in the *Megali idea* (Great Idea). This envisaged that all Greeks (and, inevitably, some non-Greeks as well) would be incorporated in an empire that would have as its capital Constantinople (Istanbul), the former center of the Greek Byzantine empire, lost to the Turks in the 15th century.

Pursuit of the *Megali idea* dominated Greek national politics for decades. It reached its height in 1919 when, following Turkey's defeat in World War I, Greece attempted to annex parts of Anatolia in western Turkey. It was defeated by the Turkish nationalists of Mustafa Kemal (Atatürk, 1881–1938), with the result that over a million Greeks were expelled from their longtime homes on the Turkish mainland to Greece – an influx that increased the population by 20 percent. Even today, the influence of the *Megali idea* is reflected in the central place that relations with Turkey, and the Cyprus issue, occupy in Greek thinking on foreign policy.

Italy's attempts to create an empire in Africa (Eritrea, Ethiopia, Libya, Somalia) and the Balkans (Albania, Greece) from the 1890s to the 1940s reflected a desire to acquire status in relation to Europe's established colonial powers, particularly Britain, France and Germany. Mussolini's ambitious plans to build a "new Roman empire" that would make Italy more respected, enabling it to compete more effectively with other states, brought about Italy's ruinous involvement in expensive colonial wars and in the Spanish Civil War. Entering World War II on the side of Nazi Germany, Italy suffered humiliating defeats, which ended its expansionist ambitions.

Postwar alignments

Italy was a founder member of the North Atlantic Treaty Organization (NATO), formed in 1949 to provide for the collective defense of Western Europe against the perceived threat of communist domination in Eastern Europe. Greece joined in 1952 (as did Turkey). Until the 1970s Italy and Greece tended to follow the policies of the United States very closely: these were linked to economic and financial aid, and the "communist issue" was still seen to be alive in domestic politics.

The growing importance of the European Community (EC) – of which Italy had been a member since its formation in 1957 (Greece did not join until 1981) – provided an alternative focus for external relations. The emergence of issues such as the Middle East conflict and oil supplies, détente with Eastern Europe and the Soviet Union and, for Greece, renewed conflict with Turkey, broadened both countries' scope for independent action.

As the 1980s progressed they came to identify less and less with the interests and policies of the United States.

The power politics of the Cold War seemed increasingly to be a thing of the past, and continuing membership of NATO was questioned in both countries, particularly in Greece. The radical socialist platform on which Papandreou was elected to power in 1981 included the withdrawal of US bases from Greece, but once in power he renegotiated Greece's defense agreement with the United States, and the bases – four major and twenty minor ones – remained. A survey made in 1985, however, showed that more than half the population felt that NATO membership and the alliance with the United States had been detrimental to Greece's interests.

Some politicians argued that national interests would best be served if Greece were to be included within the nonaligned network of Third World states, and many saw a renewed threat of external domination to lie in the closer ties of the EC that developed during the 1980s. By contrast, in Italy membership of the EC was increasingly prized, as the country enjoyed an unprecedented period of economic growth in the 1970s and 1980s. In Greece, as in Italy, European ties seemed likely to assume growing economic and practical importance after the creation of the single market in 1992.

THE *ACHILLE LAURO* AFFAIR

In the last quarter of the 20th century violence has increasingly been used to widen the arena of internal or local conflict by threatening the security of non-involved states. Italy and Greece, close to the Middle East area of conflict have been particularly vulnerable to attack by Palestinian and other groups. Both have suffered bomb attacks and hijackings. Such action subjects governments to enormous pressure: negotiation strengthens the hand of the extremists, but resistance to their demands risks creating targets for still further outrages.

The event that most clearly demonstrated this dilemma – and showed that Italy was prepared to act contrary to the interests and wishes of the United States – was the affair of the *Achille Lauro* cruise ship (1985). This was hijacked by terrorists of the Palestine Liberation Organization (PLO), and an elderly passenger, a citizen of the United States, was murdered. The atrocity enflamed public opinion in the United States. As the leaders of the hijack were being flown to Tunis United States' fighters forced their aircraft to land in Sicily. However, the Italian military asserted its right to arrest the hijackers on Italian territory; their ringleader, however, was allowed to leave the country.

The affair caused a major international storm, the repercussions of which threatened to unseat Craxi's coalition government. The year before Italy had signed a cooperative agreement against terrorism with six other states, including the United States. However, on this occasion Italy clearly felt that to hand the leader over to the United States' government, or to imprison him for any length of time, would increase the danger of further attack, and damage Italy's trade and economic links with Middle East states.

The Greece–Turkey conflict

Invasion troops are welcomed by the women of a Turkish Cypriot village. The Turkish invasion of 1974 – undertaken to preempt the threat of Cyprus's enforced union with Greece – marked a new stage in the long-running history of Greek–Turkish conflict.

A guard of honor at Gecitiikale (Lafkoniko) airport in the Turkish-controlled area of northern Cyprus. The declaration of the independent Turkish Republic in 1983 was condemned by the UN Security Council, which called unsuccessfully for the removal of Turkish troops.

Greeks and Turks have been feuding for centuries – the result of the scattering of Greeks in communities throughout the eastern Mediterranean, and of the attempt to create a state for this diffuse population. It was the desire for union (*enosis*) with the Greek mainland that fed growing nationalist fervor among the majority Greek population of Cyprus, then a British colony, in the 1950s. *Enosis* posed a threat to the separate identity of Cyprus's Turkish minority, which was scattered all over the island.

Deep-rooted animosities came to a head in the years after independence (1960). In 1963 the Turkish community withdrew from the Greek-dominated government headed by Archbishop Makarios (1913–77) who had earlier been exiled by the British for supporting armed action to achieve *enosis*. The following year a United Nations peacekeeping force was established on the island to keep the two sides apart.

After ten years of ethnic strife, the government of Turkey landed troops in Cyprus and took control of approximately a third of the island. In 1983 the Turkish Republic of North Cyprus was established, recognized only by Turkey. The Greek Cypriot government refuses to accept partition, and its claim to represent the whole island receives general international acceptance.

Cyprus has been a more or less permanent source of discord between Greece and Turkey over the decades. What added a new dimension to the quarrel was that both states were members of the NATO alliance, recruited in the early 1950s as bulwarks against Soviet expansionism in the eastern Mediterranean. Both, however, found it difficult to subordinate their historic enmity to the wider goals of NATO strategy.

The Aegean dispute
It is in the Aegean that the major causes of conflict have arisen. Greece's acquisition of the Aegean islands from Turkey in 1913 (the Dodecanese islands were added in 1947) created an exposed maritime boundary just a few kilometers off the coast of Turkey. The loss of the islands is still resented by Turks. One recent Turkish prime minister (Süleyman Demirel) declared that the Aegean islands were "in Ottoman hands for more than six hundred years. As a Turkish boy and the prime

Archbishop Makarios, who was president of Cyprus from 1960 to 1977, was briefly deposed by the military coup in June 1974, which was led by officers from mainland Greece and precipitated the Turkish invasion. The Greek ruling junta's involvement in the plot led to its downfall on 23 July, and its replacement by a democratic government.

minister of the Turkish Republic, nobody can want me to call the Aegean islands Greek islands".

At the core of recent disputes lies the demarcation of the continental shelf and the associated right to exploit mineral resources (especially oil) under the sea bed. This issue assumed great importance during the energy crisis of 1974, which affected both Greece and Turkey, shortly after significant amounts of oil had been discovered near the island of Thasos, off the coast of Thrace.

Exacerbating the conflict were two further issues: the question of air traffic control in the Aegean – organized from Greece until 1974 but challenged by Turkey since then – and the rapid militarization of Greece's islands off the coast of Turkey in the wake of the Turkish invasion of Cyprus.

Tensions eased in the 1980s, but in Greece the sense that NATO strategy – and United States' policies – favored Turkey led to widespread disillusionment with NATO as a guarantor of Greece's territorial integrity. The regional conflict with Turkey was considered more important than the increasingly irrelevant containment of a "communist threat".

A divided island Since 1974 Cyprus has been divided along the "green line" that marks the fullest extent of the Turkish advance at the ceasefire on 16 August. British sovereign bases remain in the Greek Cypriot area, and UN troops patrol the line. The conflict split communities and families. Many Greek Cypriots were living in the area of northern Cyprus occupied by the Turkish army; almost 200,000 fled as refugees. With the failure of talks in Paris in 1991 because of lack of agreement on basic issues, hopes of reunification seemed as distant as ever they have been.

The Cyprus conflict

→ Turkish advance 1974
— ceasefire line 16 August 1974
British sovereign base area
Greek Cypriot area before 1974

Mediterranean Sea

Kyrenia
Lapithos
Morphou Bay
Nicosia
Famagusta Bay
Famagusta
CYPRUS
Larnaca
DHEKELIA
Larnaca Bay
AKROTIRI
Limassol
Akrotiri Bay

The world's smallest sovereign state

In the heart of the city of Rome lies the Vatican City, an independent sovereign state in which absolute authority is exercised by the pope, the head of the worldwide Roman Catholic church. Rome has been the seat of the pope since the time of the Roman empire. The splendid Vatican palace, the pope's residence, dates from the 14th century.

The Vatican City became a sovereign state in 1929 when, under the Lateran Treaty, the then fascist government of Italy recognized the independence of the tiny enclave – only 0.44 ha (109 acres). Though there is no formal frontier with Italy, the Vatican's sovereignty is maintained through its own currency, postal service, newpapers and radio stations, and its army of more than a hundred Swiss Guards.

The Vatican City today yields an influence in international affairs out of all proportion to its size. Its population is about 1,000, but the pope's spiritual jurisdiction extends over the worldwide Roman Catholic congregation of nearly 800 million people. The Vatican enjoys full diplomatic relations with nearly a hundred states. Its civil service, the Curia, is responsible for maintaining links around the world and for handling the Vatican's complex and secret finances, the subject of much controversy in recent years.

The Swiss Guard, the pope's personal bodyguard, dates from the 16th century when Swiss mercenary troops came to the aid of Pope Julius II. Its officers and soldiers still come from Switzerland; their costume was designed by the Renaissance artist Michelangelo.

POLITICS AT THE HEART OF EUROPE

DIVISION AND REUNION · FEDERAL SYSTEMS · LOOKING EAST AND WEST

The four states of the region – Austria, Liechtenstein, Switzerland and Germany – share close historical ties. They formed the core of the medieval German empire (952–1250). This broke up into a number of small states under the political supremacy of the dukes of Habsburg, who by the 19th century controlled the Austro-Hungarian empire. Religious wars (the Thirty Years War, 1618–48) left the power of the Roman Catholic Habsburgs greatly reduced among the German states in the west of the region, where Prussia's rise as a dominant military power created a unified German empire in 1870. Both Austria and Germany were embroiled in the rivalries of World War I (1914–18). Defeat left Austria a small, landlocked republic, while Germany lost territory to France and Belgium, as well as its overseas empire.

COUNTRIES IN THE REGION

Austria, Liechtenstein, Switzerland, Germany

STYLES OF GOVERNMENT

Republics Austria, Switzerland, Germany

Monarchy Liechtenstein

Federal states Austria, Switzerland, Germany

Multi-party states Austria, Liechtenstein, Switzerland, Germany

One-chamber assembly Liechtenstein

Two-chamber assembly Austria, Switzerland, Germany

MEMBERSHIP OF INTERNATIONAL ORGANIZATIONS

Council of Europe Austria, Liechtenstein, Switzerland, Germany

European Community (EC) Germany

European Free Trade Association (EFTA) Austria, Switzerland

North Atlantic Treaty Organization (NATO) Germany

Organization for Economic Cooperation and Development (OECD) Austria, Switzerland, Germany

Note: Switzerland is not a member of the United Nations.

DIVISION AND REUNION

The period after World War I was a time of great political instability in Germany. Following the overthrow of the emperor, a democratic republic was established (the Weimar republic, 1918–33). This collapsed amid economic and social chaos caused by the world Depression of the late 1920s and early 1930s, enabling the Nazi party under Adolf Hitler (1889–1945) to seize power. In 1938 the Nazis seized control of Austria by military force (the *Anschluss*). Hitler's subsequent eastward expansion into Czechoslovakia and Poland led to World War II (1939–45).

Following Germany's defeat, it was divided into four zones controlled by the occupation forces of Britain, France, the Soviet Union and the United States. The former capital, Berlin, which fell within the Soviet Union zone, was also divided into four sectors. Heightened political tension between the Soviet Union and the other occupying powers, particularly over the future of Berlin, marked the beginning of the "Cold War" period of hostilities that divided Europe politically and militarily until the late 1980s.

In 1948 the Soviet Union cut off the road and rail links to the West from the western sectors of Berlin. A huge Allied airlift of essential supplies ensured that

West Berlin remained under the control of the Western powers. It was given special status within the Federal Republic of Germany, known as West Germany, which was created in 1949 in the British, French and United States occupation zones. The new state gained full sovereignty in 1954, and entered the United Nations a year later. In October 1949, a socialist German Democratic Republic (East Germany) was established in the Soviet zone. Political power was exercised through and by the ruling Communist Party, which exerted rigid control on all aspects of economic and cultural life.

Written into the Basic Law (*Grundgesetz*) that embodied the constitution of the new state of West Germany was the unwillingness of its politicians to give formal expression to the partition of Germany. The Basic Law was regarded as a temporary solution, adopted to give "a new order to political life for a transitional period", though for long it seemed unlikely that reunification would ever take place.

The ending of the two Germanies in 1989–90 was sudden and dramatic. The Berlin Wall came down in November 1989, and in July 1990 agreement was reached for monetary unification under the West German mark. Then on 3 October 1990 East Germany formally acceded to the Federal Republic to form a united

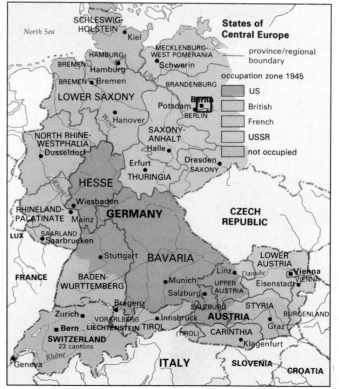

A divided city West and East Berlin were divided for nearly thirty years by the heavily fortified wall that was erected overnight through the center of the city by the communist regime of East Germany.

Two German states For 40 years, between 1949 and 1989, Germany was divided between the Federal Republic of West Germany and a separate communist state in the East. The neutral states of Austria and Switzerland lay between the Eastern European bloc and the West.

Germany. The first all-German elections were held in December 1990, producing a Christian Democrat victory, which was seen as a reward for chancellor Helmut Kohl, the architect of reunification. After intense debate, the German parliament voted to return the seat of government from Bonn to Berlin in a move that was planned to take 12 years to complete.

Stable democracies

Like Germany, Austria was divided in 1945 into four occupation zones. In 1942 the Allies had issued the Moscow Declaration, promising Austria's speedy restoration as a free and independent state. However, it remained under occupation until 1955, when the Soviet Union dropped its demands to maintain troops in eastern Austria. A treaty that reestablished its independence was agreed, and a constitution approved that guaranteed Austrian neutrality.

The rapid establishment of stable government in Austria and West Germany, devastated by war and with no history of democracy, was a remarkable achievement; Switzerland's democratic traditions stretch back several centuries. All three states have federal systems.

The political structure of Switzerland evolved from a league of three mountain cantons (local communities) formed in 1291 to defend their liberties against the Habsburgs. The mountainous terrain favored local rather than national development: it was not until 1848 that a federal constitution was introduced. It was amended in 1874 to give more authority to central government. Switzerland today has 20 cantons and 6 half-cantons.

Although the federal systems of Austria and West Germany were laid down in constitutions drafted by the occupying powers to discourage the sort of central government that was believed to have fostered the Nazi regime, they reflected existing local allegiances. Most of West Germany's 10 *Länder* or states (16 since reunification) were based on former principalities whose regional independence was jealously guarded. Austria's seven provinces conform to prewar regional divisions.

The principality of Liechtenstein, one of the five ministates of Western Europe, is a constitutional monarchy, independent since 1719. The Swiss represent its interests abroad and maintain border protection on its frontier with Austria.

A chink in the wall As the Berlin Wall, harsh symbol of partition, begins to come down in December 1989 an East German border guard peers through the gap into West Berlin. Within months of these events the two parts of Germany, separated since World War II, had reunited into a single political system

The Berlin airlift A United States' airplane flies in supplies to the besieged people of West Berlin in 1948. The Soviet blockade marked a hardening of hostility in the early stages of the Cold War and committed the United States to maintain its defense of Western Europe.

FEDERAL SYSTEMS

The organization of government in Austria and Germany is very similar. Both have a two-chamber federal parliament and multi-party representation. The executive head of government, the chancellor (prime minister) is chosen from the party holding the greatest number of seats in the lower house of the federal assembly, which is elected in both cases for four-year terms by voting systems based on proportional representation. The second, smaller chamber is elected from the state assemblies. It is able to review laws passed by the lower house, and has limited powers of veto.

In Austria the formal head of state, the president, is elected by popular vote every six years. The election of Kurt Waldheim, secretary general of the United Nations from 1972 to 1981, to the presidency in 1986 aroused great controversy because of his association with alleged war crimes in Yugoslavia during World War II. In Germany the president is elected every five years by an assembly consisting of the 662 deputies to the federal assembly and an equal number of members elected proportionally from the state assemblies. The position was intended to be mostly ceremonial, but the strong personalities of West Germany's postwar presidents gave the post greater influence than was indicated in the constitution.

At state level, the *Länder* all have their own elected parliament headed by a minister–president (in Austria, a provincial governor). The division of roles between the federal and state governments is defined in the constitution. States have care of education and cultural facilities, environmental protection, the police, public transport and the provision of public services. They are responsible for implementing federal legislation through their own civil services. They are assigned a share of federal revenues to supplement local taxes to meet all these responsibilities. Germany even provides for financial burden-sharing so that wealthier states subsidize poorer ones.

Coalition governments

The German constitution ensures that political parties have to win a minimum of 5 percent of the national vote in order to be represented in the federal assembly.

This hurdle was designed to hinder the electoral success of extremist parties of both left and right. Power is contested between two main parties, the conservative Christian Democratic Union (CDU) in permanent alliance with the Bavarian majority party, the Christian Social Union (CSU), and the Social Democratic Party (SPD). Each usually receives about 40 percent of the vote, and the balance of power is held by the liberal Free Democratic Party (FDP), which has formed part

A new political voice The Green Party – *Die Grünen* – emerged in the early 1980s to argue for environmentalist and peace policies. It was the first such party in Europe to win national representation.

A controversial president Kurt Waldheim·is invested as Austria's new president, having created an international furore when he admitted to a lack of candidness about his wartime record.

of all but one of the federal governments elected since 1949.

The system of proportional representation used in federal elections allows electors two votes, one to elect a representative at the constituency level and one counted at *Länder* level where proportionality between parties is ensured. The two main parties normally win all the constituency seats, but smaller parties can win in the *Länder* count. The second vote is used strategically by voters to support the coalition partner of their preferred party; this allows the Free Democrats, who win no constituency seats from the first vote but succeed in clearing the 5 percent hurdle, to enter government as the coalition partner. In 1983 they switched their support from the SPD, with whom they had previously been in government, to

the winning CDU. In this same election that the enviromental leftist Green Party first cleared the 5 percent hurdle to enter the federal parliament. It has been the most successful new party in Germany since 1949.

Governing the cantons

History and an independent spirit have lent some rare features to Switzerland's form of democracy. A greater degree of power remains with the cantons, which have considerable freedom to decide their own form of government; the two German-speaking, predominantly Roman Catholic half-cantons of the Appenzeller mountain district still deny women the right to vote in their elections. The men meet *en masse* once a year to decide cantonal legislation.

Federal government is in the hands of a federal council. Each of its seven members, elected for a four-year term by joint session of the two-chamber federal parliament, acts as minister of state for one of the administrative departments of government. The president, who is federal head of state, is chosen for one year from among them. No two members can be from the same canton, and an informal power-sharing agreement means that the three major language groups (French, German and Italian) are represented in turn on the council.

As recently as 1984 a referendum in Liechtenstein extended voting rights to women; they are still unable to vote in local elections. Government is shared between its hereditary prince and a 15-member elected assembly.

DIRECT DEMOCRACY IN SWITZERLAND

A distinctive feature of the Swiss constitution is the right it extends to citizens to take a direct part in government through what are known as "citizen initiatives" and referendums. These are used to decide controversial issues of policy. Some 100,000 signatures (representing just over 1.5 percent of the population) are needed to initiate federal legislation; 50,000 are sufficient to call a referendum on an existing piece of legislation. Between 5 and 10 national referendums are held each year. Since the same rights exist at both cantonal and communal levels, Swiss citizens are consequently faced with a myriad of political choices, in addition to participating in regular elections to all three levels of government. Remarkably, voter participation remains among the highest in the Western world.

The right of Swiss citizenship derives from residence in a local commune (an administrative district below the level of canton) in the middle and early years of the 19th century, when censuses were first held on a regular basis. There are more than 3,000 communes, whose population today may range from 20 to 35,000. As the result of migration from rural areas to the cities over the last 150 years, most Swiss do not actually reside in the commune of their citizenship. New citizens, of whom there are few because of very stringent qualification rules, must be adopted by a commune. This is particularly difficult to achieve in the small mountain cantons, where the right of citizenship continues to be jealously guarded.

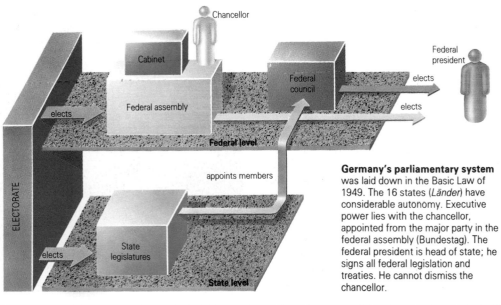

Germany's parliamentary system was laid down in the Basic Law of 1949. The 16 states (*Länder*) have considerable autonomy. Executive power lies with the chancellor, appointed from the major party in the federal assembly (Bundestag). The federal president is head of state; he signs all federal legislation and treaties. He cannot dismiss the chancellor.

LOOKING EAST AND WEST

In 1815, in the resettlement of Europe at the close of the Napoleonic wars, the Congress of Vienna guaranteed the permanent neutrality of Switzerland. It has vigorously defended this position ever since, avoiding international involvement of any kind: in 1986 a referendum overwhelmingly rejected membership of the United Nations. When the neutrality of Austria was proclaimed in its new constitution in 1955, a sizeable neutral zone was created between the Eastern bloc of communist states and Western Europe.

Discussions over the future of Austria coincided with the negotiations that preceded West Germany's entry, after initial exclusion, into the North Atlantic Treaty Organization (NATO). This had been formed in 1949 under United States leadership as a defensive alliance in response to perceived Soviet expansionism in central Europe. The Soviet Union may have thought, initially, that by offering Austrian neutrality in return for the withdrawal of its troops, it would provide an incentive to West Germans to decide to remain outside NATO, as this might establish a model for the possible reunification of Germany. However, West Germany joined in 1955, and in response the Warsaw Pact, a military alliance between the Soviet Union and the communist states of Eastern Europe, was established with East Germany as a signatory member.

By the late 1980s West Germany possessed not only the largest, but possibly the best trained and equipped, NATO force in Europe. It had renounced the possession and use of nuclear weapons, but allowed battlefield nuclear weapons on its soil under United States' control.

The path to reunification
After World War II the world's major geopolitical frontier divided Germany (the inner German frontier). The fact of partition preoccupied most West Germans. The Basic Law enshrined in the West German constitution the hope that one day Germany would be reunified. After the 1970s, West German foreign policy asserted that this would most readily happen through the gradual opening up of political and economic relations with Eastern Europe. The "eastern policy" (Östpolitik) was originated by Willy

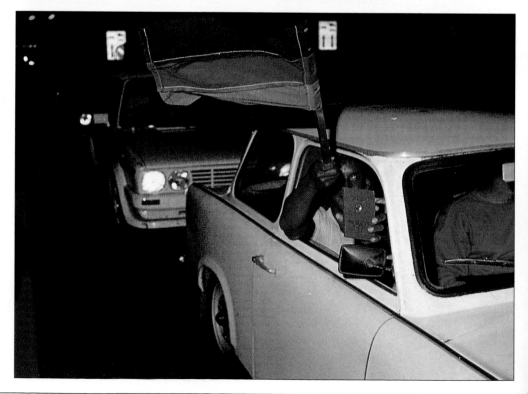

AUSTRIA – A MEETING-PLACE BETWEEN EAST AND WEST

The Soviet Union's agreement to withdraw its troops from Austria at the same time as the other Allied forces, which was a necessary prerequisite to the restoration of Austrian sovereignty in 1955, was a major surprise at the time. With hindsight it can be seen as the first thaw in the ideological freeze that descended over central Europe after World War II. The bargaining counter for the Soviet withdrawal was the guarantee of Austria's permanent neutrality, which was then written into its constitution. Thereafter, Austria's leaders acted scrupulously to avoid antagonizing the Soviet Union. From the point of view of the West, the loss of Austria as a possible NATO partner was compensated for by its adherence to liberal democratic ideals and its membership of the capitalist bloc.

Austria's capital, Vienna, was the historic heart of the Austro-Hungarian empire, which extended into present-day Czech Republic, Slovakia, Hungary, Poland, Romania, Slovenia and Croatia. It was, and is, a multinational city, which retains strong links with these countries and was promoted by the Austrian government as an international rendezvous between East and West. It was made the permanent site of the conventional arms reduction talks, and in 1979 the agreement to limit strategic nuclear weapons (known as SALT II) was signed in Vienna by

The vast edifice of UNO-City in Vienna. The rapid opening up of East–West relations in the late 1980s should give neutral Austria new opportunities to act as a meeting place in the decade ahead.

United States President Jimmy Carter and Soviet leader, President Leonid Brezhnev. A large complex of United Nations offices (UNO-City) was built there in the 1980s. The headquarters of the influential Organization of Petroleum Exporting Countries (OPEC) are also in Vienna. In 1989 Austria applied for membership of the EC.

Supporting tolerance German President Richard von Weizsacker attends a gathering of more than 300,000 people in Berlin in 1992 which called for an end to outbreaks of violence against minority groups.

Voting with their feet In summer and fall 1989 thousands of East Germans streamed into the West to be welcomed as fellow citizens. Their departure sparked off a series of dramatic political events, opening the way to reunification a year later.

Brandt (1913–92; chancellor 1969–74). He had been mayor of West Berlin in 1961 when the Berlin Wall had cruelly divided the city. Nothing better demonstrated the near total breakdown of East–West relations at the height of the Cold War. In 1972 the signing of the Basic Treaty between East and West Germany led to an exchange of permanent missions (but not ambassadors) between the two, and West Germans were allowed to travel east. West Germany's trade with the countries of Eastern Europe developed rapidly. East Germany was recognized as a member of the United Nations, but the freedom of

movement of its citizens remained severely circumscribed.

It was the fall of Erich Honecker, the leader of East Germany from 1973 and the chief architect and defender of the Berlin Wall, that led to the speedy dismantling of the barriers between East and West Germany at the end of 1989 and precipitated the sudden progress toward political reunification.

This required the assent of the four occupying powers in Berlin – the United States, the Soviet Union, Britain and France – and in February 1990 the representatives of these four countries and the two Germanies, meeting in Ottawa, Canada, reached agreement on how to proceed with reunification. Proposals agreed at a meeting of NATO in May 1990 to allow a united Germany to become a full member of the organization were acceded to by the Soviet Union in July 1990. The Treaty on the Final Settlement with Respect to Germany was signed by

all the concerned parties two months later, allowing German reunification to become a reality on 3 October 1991.

New strengths and problems

Reunification made Germany by far the largest state in Western Europe in terms of population (91 million); it was already the richest in terms of gross national product. One of the founder members of the European Community (EC), the strength of its economy gave it an increasingly important role within the Community, and in the late 1980s it became one of the most enthusiastic advocates for closer European integration, leading some of its smaller neighbors such as Denmark to fear that it would overpower them. Germany was well-placed to reap the advantages of new markets in Eastern Europe, but the breaking down of the trade barriers also brought a flood of economic refugees and immigrants from the east that have created some tensions.

A new order in Europe

For more than forty years following World War II, the front line in Europe, in the event of a war between East and West, was Germany. In 1989 West Germany was host to 760,000 troops from eight NATO member states. It provided 345,000 troops of its own to NATO central command. After the agreement reached by the superpowers in 1987 to remove all their middle-range weapons from Europe, it still housed a nuclear arsenal of about 2,000 short-range (less than 500 km/300 mi) nuclear weapons and 2,500 nuclear artillery shells.

The original signatories of the NATO agreement in 1949 included 10 Western European states, as well as the United States and Canada. They agreed that "an armed attack against one or more of them in Europe or North America shall be considered an attack against them all". Four more states joined later, including West Germany in 1955. It was in response to this that the states of Eastern Europe, led by the Soviet Union, formed a similar defensive alliance, the Warsaw Pact.

NATO coordinates the military forces of its member states under a unified central command based in Brussels. By far the largest contributor to NATO in terms of men, weapons and costs is the United States. NATO's central command is usually headed by a United States general. At the start, the Warsaw Pact had a considerable numerical advantage in troops and equipment. NATO consequently adopted the principle of "flexible response", which included the use of both battlefield and strategic intercontinental nuclear weapons to counter an attack. Through the extension of the United States' nuclear umbrella to Europe, NATO would inflict such unacceptable losses on the Warsaw Pact countries that attack across the inner German frontier would be foolhardy.

A basic imbalance in the strategies of the two blocs was caused by the United States' distance from the European conflict zone. This meant that the United States' intermediate-range (500–2,000 km/ 300–1,200 mi) nuclear weapons could hit Soviet targets from West Germany, while similar Soviet weapons could not reach the United States. When President Ronald Reagan in 1983 spoke of a "limited nuclear war in Europe", traditional West European fears of abandonment by the United States were heightened. Would the United States be willing to sacrifice New York or Los Angeles in order to save Hamburg or Frankfurt?

Helmut Schmidt (West German chancellor 1974–83) in 1977 first promoted the installation of the United States' Pershing 2 and Cruise intermediate-range missiles in Western Europe as a way of binding together the fates of the NATO partners. This prompted massive peace demonstrations throughout the NATO countries, particularly in West Germany.

The end of division in Europe

Mikhail Gorbachev's accession to power in the Soviet Union during the 1980s allowed a new stage to develop in arms limitation discussions between East and

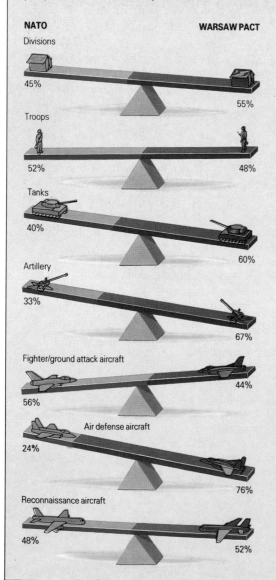

The balance of conventional forces in central Europe at the close of the 1980s was narrowly tilted toward the Warsaw Pact. But the superior quality and training of NATO forces would have favored it in a conventional attack by the other side. Mutual recognition of the near parity of forces was a first step toward reduction talks.

NATO **WARSAW PACT**

Divisions
45% 55%

Troops
52% 48%

Tanks
40% 60%

Artillery
33% 67%

Fighter/ground attack aircraft
56% 44%

Air defense aircraft
24% 76%

Reconnaissance aircraft
48% 52%

The Soviet tanks withdraw Under the terms of the Warsaw Pact, more than 300,000 Soviet troops were permanently stationed in East Germany. Their staged withdrawal began once German reunification became assured in 1990. The withdrawal of Soviet troops from Eastern Europe on the dissolution of the Warsaw Pact took longer: the last were not expected to leave Poland until 1994.

Schoolchildren in the former West Germany take little notice as NATO tanks and armored vehicles on a fullscale military exercise burst through the streets of their village. During the Cold War disturbances like this were commonplace, scarcely rippling the surface of the daily lives of these people.

West. In 1987 the Intermediate Nuclear Forces (INF) Treaty brought about the removal of all intermediate-range weapons from Europe. However, NATO still retained the option of "first use" of nuclear weapons in the event of attack from the Warsaw Pact.

The INF treaty was soon overtaken by events in Eastern Europe. The mostly peaceful revolutions of 1989 removed the communist governments from power, and the Conference on Security and Co-operation in Europe (CSCE), held in Paris in November 1990 and attended by all the governments of Europe (except Albania) together with those of the United States and Canada, formally marked the ending of the Cold War polarization of Europe. It

agreed the Treaty on Conventional Armed Forces in Europe, which severely reduced the numbers of tanks, artillery and combat aircraft that were to be sited in Europe.

In July 1991 the Warsaw Pact was dissolved, depriving NATO of its primary purpose in existing. Since then it has been searching for a new role amid growing demands in the United States that European governments should be prepared to shoulder a greater share of Europe's defense burden. The terms of the Maastricht treaty seek to develop a common defense policy by the European Community, and it is in implementing this that a role for a reshaped and redefined NATO could lie.

The writing on the Wall

West Berlin was left as a Western enclave within East German territory when the Cold War set in after 1945. By 1961 it had become the major recipient of refugees from the communist regime in the East: the flow had reached 2,000 a day, and the total number that had left was close on 3 million. This enormous exodus posed a crucial problem for the communist regime, robbing it of its vitally needed skilled, young workforce. On 13 August 1961 the Berlin Wall was built, cruelly dividing the two sectors of the city, and the border guards were given orders to shoot anybody attempting to cross. The communist regime had conceded a massive propaganda advantage to the Western powers in order to retain its labor force.

For more than a quarter of a century the Wall, with its armed guards and watchtowers, symbolized the political divide of the Cold War. After United States' President John F. Kennedy visited the Wall to declare "I am a Berliner", this monument to communist failure has attracted all major Western leaders to stand beside it to promote their image as the champion of freedom. The decision of the East German government to pull down the Wall on 9 November 1989, at the height of the Eastern European revolutions against communist rule, was therefore of the greatest political significance, and set the seal on the momentous political changes that took place during the year.

In its 28 years' existence the Wall claimed nearly two hundred people killed or injured trying to cross it; another 3,000 were captured by the East German guards.

Graffiti and slogans decorate the Wall on the West Berlin side. The smooth coping on top was to prevent people climbing over from the East. When the Wall came down in November 1989, entrepreneurs rushed to sell off pieces of it to souvenir hunters.

CHANGING DIRECTION

REDRAWING THE MAP · COMMUNISM AND AFTER · BEYOND THE WARSAW PACT

The mosaic of states that make up the political map of Eastern Europe was created out of the struggle for power that took place as the 500-year-old Ottoman empire in the Balkans began to break up in the 19th century, and again as the territories of the defeated Austro-Hungarian and German empires were carved up after World War I. The political boundaries of the newly created states rarely corresponded to the region's ethnic and linguistic divisions. For 40 years after World War II the imposition of communist rule throughout the region suppressed ethnic rivalries, but after the revolutions of 1989–90 fierce nationalisms began to reemerge. Ethnic separation led to the peaceful partition of Czechoslovakia but civil wars tore apart the pre-1991 state of Yugoslavia, and United Nations troops became involved in the conflict.

COUNTRIES IN THE REGION

Albania, *Bosnia and Hercegovina, Bulgaria, *Croatia, Czech Republic, Hungary, *Macedonia, Poland, Romania, Slovakia, *Slovenia, Yugoslavia

Disputed borders Bulgaria/Macedonia; Macedonia/Greece; Romania/Bulgaria; Romania/Moldavia

STYLES OF GOVERNMENT

Republics All countries in the region

Multi-party state All countries in the region

One-chamber assembly Albania, Bulgaria, Hungary, Macedonia, Slovakia, Slovenia

Two-chamber assembly Bosnia, Croatia, Czech Republic, Poland, Romania, Yugoslavia

CONFLICTS (since 1945)

Nationalist movements Bulgaria (Turks); former Czechoslovakia (Hungarians); Poland (Belorussians, Germans); Romania (Germans, Hungarians); former Yugoslavia (Albanians, Croatians, Serbs, Slovenes)

Internal unrest Czechoslovakia 1968, 1977; Bulgaria 1989; Hungary 1956; Poland 1970, 1976, 1980–89; Romania 1987, 1989; former Yugoslavia 1968, 1970, 1988, 1989

Coups Czechoslovakia 1948, Hungary 1946, 1956

Revolutions Albania 1946; Bulgaria 1946; Czechoslovakia 1989; Poland 1947; Romania 1946, 1989; Yugoslavia 1945

Civil wars Slovenia/Serbia 1991; Croatia/Serbia 1991; Bosnia and Hercegovina/Serbia, Croatia 1992–

Interstate conflicts Hungary/USSR 1956; Poland/USSR 1956; Czechoslovakia/USSR 1968

* former constituent of the republic of Yugoslavia, pre-1991

REDRAWING THE MAP

In the 19th century, as Ottoman power began to weaken in southeast Europe, political domination of the region was contested between the empires of Austro-Hungary, Germany and Russia. At the Congress of Berlin (1878) Bulgaria, Romania, Montenegro and Serbia, all part of the former Ottoman empire in the Balkans, were recognized as independent states, and in 1912–13 Albania freed itself of Turkish rule and established its independence of Bulgaria and Serbia, which both laid claim to it. It is the only country in the region to have retained its boundaries without change throughout the 20th century.

World War I completely altered the European map. In 1918 the defeated Austro-Hungarian and German empires were carved up to create the states of Poland, Czechoslovakia, Hungary and Yugoslavia. The last of these, known as the Kingdom of the Serbs, Croats and Slovenes until 1929, brought together Serbia and its territories of Bosnia and Hercegovina, Macedonia and Kosovo with Montenegro and the former Austro-Hungarian territories of Croatia, Slovenia and Vojvodena.

World War II had a devastating effect throughout Eastern Europe. The presence of 3 million *Sudeten* Germans in the north and northwest of Czechoslovakia provided Adolf Hitler (1889–1945) with an excuse to annex the Sudetenland in 1938 prior to invading the whole country six months later. In September 1939 Poland was invaded. A subsequent pact between Hitler and the Soviet dictator Joseph Stalin (1879–1953) led to the country being partitioned, with the Soviet Union occupying the east. Yugoslavia was overrun by Germany in 1941, and Romania, Bulgaria and Hungary were all drawn into the fighting, first as allies of the Germans and then of the Soviet Union, which emerged from the five years of fighting as the dominant power in the region.

Its territorial expansion was achieved at the expense of Poland, Czechoslovakia and Romania; its political dominance was guaranteed by introducing and supporting communist rule and centrally planned economies in Bulgaria, Hungary, Czechoslovakia, Poland and Romania. Albania and Yugoslavia, though communist states after 1945, remained independent of direct Soviet influence.

The Yalta and Potsdam agreements of 1945, made by Britain, the Soviet Union and the United States, the three major Allies against Nazi Germany, drastically shifted Polish boundaries eastward and northward; smaller territorial changes took place in Bulgaria, Czechoslovakia,

States of Eastern Europe

□ formed since 1991

The political map of Eastern Europe was mostly created early in this century. Renascent nationalism in the 1990s led to the partition of Czechoslovakia and the fragmentation of Yugoslavia.

The downfall of communism is celebrated in Bucharest on Christmas day 1989. The communist symbol has been cut out of the center of the Romanian flag. The Romanian revolution started among the country's large Hungarian minority, who had been subjected to the worst excesses of Ceausescu's repressive rule, including the destruction of their villages and enforced resettlement.

Deciding the fate of postwar Europe The three major wartime Allied leaders – British prime minister Winston Churchill, US president Franklin Roosevelt, and Soviet leader Joseph Stalin – meet at Yalta in February 1945. As a result of their deliberations, the Soviet Union became the dominant power in Eastern Europe for 40 years.

Ethnic rivalries

These postwar territorial settlements led to the movement and relocation of large groups of people, creating new problems among the ethnic minorities in the region. Two-thirds of the region's population is of Slavic origin, but Germans, Jews and Romanians form large non-Slavic groups. Elsewhere, particularly in Bulgaria, Romania and Yugoslavia, larger groups have repeatedly protested against their treatment and suppression by the state, though until the end of the 1980s such protest was effectively silenced.

Measures against minorities were severe, ranging from educational discrimination to compulsory denationalization, forced migration and direct repression. Several long-standing conflicts exploded into violence in 1989. In Bulgaria, Turks were subjected to increasing restriction of their ethnic and civil rights, resulting in a massive exodus of 300,000 people to Turkey. In December 1989 tension among the Hungarian minority in Romania erupted into violence in the city of Timisoara and culminated in the overthrow of Nicolae Ceausescu's regime.

In Yugoslavia, a federation of six republics and two autonomous regions, increasing resentment of Serbian domination of the federal government split the country apart in 1991–92. When a proposal to convert the federation into a loose union of sovereign states failed to win approval in 1991, Croatia and Slovenia declared themselves independent. Macedonia and Bosnia followed suit, reducing Yugoslavia to two federal units, Serbia and Montenegro.

The resulting rise in ethnic tensions precipitated wars first in Croatia and then in Bosnia. In the former, ethnic Serbs took control of Croatian territory, in the latter ethnic Serbs and ethnic Croatians effectively divided Bosnia, with the Muslim population caught in the crossfighting. The increasingly bitter civil war produced the greatest refugee crisis in Europe since 1945 as over 2 million people were forced to flee.

The European Community recognized the independence of Croatia and Slovenia in January 1992, and Macedonia in 1993 after a prolonged dispute with Greece. In 1992, as the war escalated, the UN imposed sanctions on the Serbian rump state of Yugoslavia and later expelled it from membership, but the war continued despite repeated UN peace initiatives.

Hungary, Romania and Yugoslavia. The most significant change in postwar Europe was the division of Germany in 1949 into two. The Federal Republic of Germany (West Germany), was set up in the western zones formerly occupied by Britain, France and the United States, and the German Democratic Republic (East Germany) in the Soviet zone. Berlin was similarly divided.

COMMUNISM AND AFTER

The common dissatisfaction that brought hundreds of thousands of demonstrators out on to the streets of Eastern Europe's major cities in 1989 is partly explained by the high degree of similarity between the structure and functions of their postwar governments. Their constitutions, mostly modeled on the Soviet Union's were designed to ensure a highly centralized state, planned economy and unlimited power for the Communist Party.

Czechoslovakia and Yugoslavia alone had federal systems. In Czechoslovakia the Slovak and the Czech regions, were theoretically given the same amount of autonomy when the constitution was established in 1948, but in practice the autonomy of the Slovak minority was drastically curtailed in the next two decades. Though new legislative and executive bodies were created in 1968 for both regions, they had little real power and most decisions were still made by central government. The federal system in Yugoslavia gave each of the six republics limited power, which gradually declined as that of the Serbian-led Communist Party hardened.

The communist political system in Eastern Europe was based on the Soviet Leninist model, with "democratic central-ism" as its guiding principle and parallel hierarchical party and government structures. Most were administratively reorganized during the 1970s in a move that was supposed to bring government closer to the people, but which in effect gave the party leadership an even greater concentration of power by distancing local government from the decision-making process. Subsequently legislation was passed in several states to try to reestablish independent local self-government.

Within central government, the legislatures of Eastern Europe served to legitimize decisions proposed by the party. The executive branches ran the daily affairs of the country and encompassed the different ministries, the army and security police. Headed officially by the state council, or presidium, of the national assembly, government was actually under the control of the council of ministers, chaired by the prime minister. The executive branch was responsible for the establishment and coordination at every level of the national economic plan.

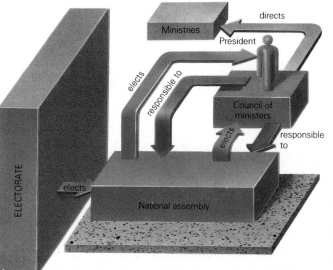

War in the Balkans A Croatian flag hangs from a battle-scarred roadside Calvary – an emblem of the bitter national and religious rivalries that erupted into violence in 1991, tearing apart the former state of Yugoslavia.

The constitution of Hungary, amended in October 1989, lays down that the highest organ of state is the one-chamber national assembly, elected every four years. It elects the president, the head of state, and the council of ministers.

Moment of decision A woman casts her vote in the fateful national elections in Czechoslovakia in June 1992, which paved the way to the eventual separation of the Czech Republic and Slovakia into two independent states.

AN UNEASY TRANSITION

Events in Hungary and Poland caused the rapid collapse of Eastern Europe's communist regimes within the space of a few months in 1989. Multiparty government and constitutional reform followed quickly on the Hungarian communist party's decision to disband and reform as a socialist party in spring 1989. In Poland the Solidarity trade union, competing in a minority of seats in elections in June 1989, was victorious in every single one, propelling it into government by August 1989. By the end of the year the communist regimes in Czechoslovakia, Bulgaria and Romania had fallen as the pace of change escalated throughout Eastern Europe. Only Albania stood out against the trend, but by March 1991 one-party rule had ended here as well.

Throughout 1990 a series of what were termed "founding elections" took place in the former communist states. With a few exceptions, the Communist Party's successors were defeated by anticommunist coalitions such as Civic Forum in Czechoslovakia, led by the dissident writer Vaclav Havel. Similaly in Bulgaria, the Union of Democratic Forces beat the Bulgarian Socialist Party (the former Bulgarian Communist Party) by a narrow margin in the first democratic elections held in 1991. But it was hard for such umbrella groups to turn themselves into broad-based competitive parties. Fragmentation and regionalism was the result. In Czechoslovakia's national elections in June 1992 the Slovak national party headed the poll in Slovakia. The decision was taken to dissolve the federation peacefully rather than risk ethnic violence, and in January 1993 the states of the Czech Republic and Slovakia came into being.

As in the Soviet Union, the Eastern European communist parties, until their collapse, had virtual control of every aspect of political and economic life.

The road to multi-party government
Membership of the region's communist parties grew rapidly after World War II but it nevertheless rarely exceeded 20 percent in any Eastern European state. In 1988 membership in Albania, Hungary, Poland and Yugoslavia was below 10 percent, and in Czechoslovakia and Bulgaria it was only a little more than that. Romania, with a membership of 16 percent, had the highest level of support.

Party membership before 1989 was predominantly male; women's membership nowhere was greater than 28 percent. Throughout the region, at the highest levels of the party organizations, women were noticeable by their absence, with only a handful of women reaching the level of politburo.

During the communist years, trades unions and workers' councils posed the greatest challenge to party rule, none more so than Solidarity, which threatened to topple the Polish government in 1980–81.

The political crises of 1989 forced the communist parties to assume greater flexibility as they struggled to retain a hold on power, and in one country after another the party's name was changed to "socialist party" to reflect this new openness. As communism was dismantled, numerous other political parties, representing all shades of opinion, sprang up, many of them remnants of pre-war political parties.

Coalition government became the norm throughout the region. In Poland, the adoption of a complicated form of proportional representation meant that as many as 27 parties were represented in the Sejm, the national assembly, after elections in 1992. In Romania, Bulgaria and Albania, especially, the reformed communist parties still wielded influence, and every country was faced with having to absorb hundreds of former Communist Party officials, secret police and military personnel.

BEYOND THE WARSAW PACT

The immense influence that the Soviet Union wielded over the postwar Eastern bloc states was amply demonstrated in 1956 and 1968 when Soviet troops invaded two member states of its own mutual defense organization, the Warsaw Pact. The Hungarian Revolution of October–November 1956 and the "Prague Spring" in Czechoslovakia in 1968 were both attempts to introduce liberal reforms and achieve political independence from the Soviet Union. They were both brutally repressed by Soviet troops and tanks - a course of action inspired by Soviet fears that the presence of communist unorthodoxy within the Warsaw Pact countries would threaten the security of the whole bloc, later justified by the so-called Brezhnev doctrine, by which Moscow claimed the right to defend socialism wherever it was under threat.

Challenging Soviet domination

With the exception of Albania and Yugoslavia, Soviet domination of the region remained strong until Mikhail Gorbachev's assumption of the Soviet leadership in 1985 led to the abandonment of the Brezhnev doctrine. The new thinking in Soviet foreign policy radically altered its relations with its satellite states in Eastern Europe, giving them greater freedom to determine their own destinies. Whether or not the change came

The Prague Spring of 1968 Cartoons mocking the Soviet leadership of the day are displayed on a shop window. Top left is a photo of Alexander Dubcek (1921–92), leader of the reforms. Later disgraced, he returned to the political stage in 1990.

Breaking the mold Gorbachev's picture above a crowd in Romania during a visit by the Soviet leader in 1987. His new thinking in foreign policy allowed the Eastern bloc satellites to go their own way in 1989–90, leading ultimately to the dissolution of the Warsaw Pact.

about as a response to the grave economic crisis affecting the whole region, rather than as a genuine ideological gesture, the structure of Eastern European politics has been irreversibly altered.

Albania and Yugoslavia had long asserted their independence of the Soviet Union. Under strong Soviet influence until 1958, Albania later became closely aligned with China until 1976. Its continued adherence to Stalinist policies left it politically isolated and it was unaffected by the dramatic political changes that swept though the rest of the region in 1989, though succumbing a year later.

Yugoslavia refused to accept Soviet domination in 1945, and under the leadership of President Tito (Josip Broz,

1892–1980) it played a prominent role in forming the nonaligned movement in the 1950s. This new movement was created to provide an alternative to domination by either of the superpower blocs of East and West in the Cold War period. Pledged to fight colonialism, its membership, with the exception of Yugoslavia, was drawn from the newly independent states of Africa and Asia.

Seeds of revolution

Even before the Soviet Union's policy toward its political satellites changed, the growth of a number of independent peace movements in Eastern Europe was posing a challenge to communist governments during the late 1970s and 1980s. Their

THE WARSAW PACT

The Warsaw Pact came into being in May 1955 in response to West Germany being admitted to the North Atlantic Treaty Organization (NATO). It was a defensive agreement signed by eight countries: Albania, Bulgaria, Czechoslovakia, East Germany, Hungary, Romania, Poland and the Soviet Union. (Yugoslavia never joined.) The pact provides for a unified military command, based in Moscow, the maintenance of Soviet troops on member states' territory, and mutual assistance.

Hungary briefly withdrew from the pact during the revolution there in 1956, but returned after a Soviet-dominated government had been installed. Albania permanently withdrew in September 1968 in protest against the invasion of Czechoslovakia; Romania effectively terminated its membership by refusing to take part in the invasion or in the pact's military exercises.

By the late 1980s, the Soviet Union's new thinking in foreign policy was beginning to affect the power and influence of the Warsaw Pact, just as Western attitudes to NATO were beginning to alter as a result of growing detente between the superpowers. The collapse of Eastern Europe's communist regimes in 1989 called into question the continued existence of the Pact, though initially the new democratic governments took a cautious attitude toward ending their membership. However, the Soviet Union's repression of Lithuanian protest in January 1991 aroused fears that Soviet force could be used again in Eastern Europe under the terms of the treaty, and Poland, Hungary and Czechoslovakia called for the dismantling of the Pact's military structures by the end of the year. A month later President Gorbachev proposed speeding up the timetable to do so to 1 April, and three months later, on 1 July 1991 at a summit meeting in Prague, a treaty was signed by all six member states that formally dissolved the Pact after 35 years; to many Western eyes it marked the end of the Cold War.

numbers grew dramatically during 1988 and 1989 as dissident political activity increased throughout the region, but they were originally fostered by international developments such as the deployment of intermediate-range nuclear weapons in Europe, the intensification of military activity by the Warsaw Pact, and concern over environmental issues following the cataclysmic nuclear disaster that took place at Chernobyl in 1986.

Charter 77 was a pioneering group in Czechoslovakia that merged campaigning for peace with human rights issues. The Roman Catholic church in Poland also played a crucial role in mediating the antigovernment protests of Solidarity and other opposition groups. Many of the new political parties that emerged after 1989 were based on these former peace and freedom movements.

After the collapse of the region's communist regimes, relations between the Soviet Union and its former client states altered very quickly. Accompanying the moves to end the Warsaw Pact were negotiations for the withdrawal of Soviet troops from Eastern Europe. All five Soviet divisions (73,500 men) had been withdrawn from Czechoslovakia by May 1991, and in October a friendship treaty with the Soviet Union included a clause condemning the 1968 Warsaw Pact invasion. A similar clause describing the 1956 invasion as unlawful was incorporated in the friendship treaty that was signed between Hungary and the Soviet Union in December 1991 after the withdrawal of the 49,500-strong Soviet garrison (completed by 19 June). But although the first convoys had begun to leave Poland in April 1991, negotiations there were more prolonged and fraught with difficulties, delaying the departure of the last troops until the end of 1993.

As the new governments of Eastern Europe sought to sever their ties with the Soviet Union and its successor states, they saw their future security lying in the forging of closer links with the rest of Europe, arguing for a redefined NATO to include Eastern Europe. Hungary, Poland and Czechoslovakia have also initiated moves to join the European Community (EC) at some time in the future.

Rolling up the iron curtain Troops in Hungary dismantle the frontier fence along the Austrian border in September 1989. The decision to open up the frontier with the West was an early indication of the new political forces at work in Eastern Europe.

Solidarity: Gdansk to government

During 1980 and 1981 the attention of the world was seized by the heroic challenge that the Polish independent trade union Solidarnosc (Solidarity) laid down to the monolith of communist rule. Polish workers had expressed their dissatisfaction before, with widespread strikes taking place in 1956, 1968, 1970 and 1976, but the unique feature of the protests of the 1980s was the way they cut across class and occupational divisions. The explosion of antigovernment feeling started in August 1980 in the Gdansk, Gydnia and Szczecin shipyards and was led by Lech Walesa, a charismatic union activist and electrician from Gdansk.

Discontent with the rigidity of communist rule and the persistent limitations on personal and civil freedoms was fueled by a rapidly deteriorating economy, food shortages and high prices. Walesa and his fellow union activists, who had originally thought only of creating a loose confederation of independent regional trade unions, soon found themselves at the head of a single national trade union - Solidarity. The new union quickly developed a national organization and structure, and by the spring of 1981 its membership had soared to 9.5 million, including over 900,000 former members of the Communist Party.

A devoutly Roman Catholic people, Poles have always sought the guidance of the church during turbulent times. It was therefore only natural that the church came to play an extremely important role in the next months and years by acting as mediator in Solidarity's struggle against the government. After martial law was declared in 1981 in response to the political crisis, the church gave refuge to hunger strikers and was prominent in offering medical, economic and moral support to Solidarity members.

The government's decision to impose martial law caught the organization unprepared. Its leaders were imprisoned, many for over a year; over 50,000 people were arrested, and more than 50 killed. Without leadership, Solidarity was dismantled and officially silenced. However, many of its activities moved underground and continued clandestinely until the union's resurgence in 1989.

The historic compromise

Poland entered a period of change after a fresh outbreak of strikes in August 1988. The country's leader, Wojciech Jaruzelski,

Commemorating the Gdansk strike Solidarity supporters tie flowers and other tokens to the shipyard's railings a year after the 1980 strike there. The pope's portrait emphasizes the supportive role played by the Roman Catholic church.

Tadeusz Mazowiecki, Poland's first noncommunist prime minister, with Lech Walesa (elected president in 1990) on his left, shortly after taking office in August 1989.

A Solidarity march shortly before the movement was forced underground after martial law was imposed in December 1981 by General Jaruzelski, Poland's new hardline prime minister. Most of the union's leadership was imprisoned.

realized that necessary economic reforms could not be carried out without popular support. The way was open for the "historic compromise", made between Lech Walesa's Citizen Committee and the government, which relaxed the political system to allow limited opposition to the Communist Party.

The agreements of 5 April 1989 legalized Solidarity, provided for a two-chamber national parliament, and established the new office of president. The Communist Party guaranteed itself a parliamentary majority by reserving two-thirds of the seats in the lower house for itself, but Solidarity was allowed to compete for the remaining seats and for the newly created 100 seats in the senate.

Elections were set for June. Solidarity's landslide victory was even larger than expected. It won 99 of the seats in the senate and all 161 seats available to it in the lower house. This prevented the Communist Party from forming a cabinet during the summer of 1989, but it was not until Walesa broke the old power base by establishing a coalition with two puppet parties, the Democratic Party and the Peasant Party, that a noncommunist government was finally installed in Poland – the first in Eastern Europe for nearly forty years.

COLLAPSE OF AN EMPIRE

COLLECTIVIZATION TO REFORM · THE MACHINERY OF COMMUNIST RULE · RETHINKING SUPERPOWER STRATEGIES

For nearly 70 years, from 1923 to 1991, the Union of Soviet Socialist Republics (USSR) was the dominant power in the region. It came into being after the 1917 communist revolution in Russia. In the ensuing civil war the communists extended their power over the Ukraine and reannexed the non-Russian territories of the former tsarist empire east of the Black Sea and in Central Asia to create the Soviet Union in 1923. Estonia, Latvia and Lithuania were annexed in 1940. In 1990 these Baltic states withdrew from the Soviet Union and were recognized internationally as independent states. The Soviet Union was dissolved the following year, to be replaced by the Commonwealth of Independent States. Mongolia, which declared itself a People's Republic in 1924, was governed on Soviet lines until 1990.

COUNTRIES IN THE REGION

Armenia, Azerbaijan, Belorussia, Estonia, Georgia, Kazakhstan, Kirghizia, Latvia, Lithuania, Moldavia, Mongolia, Russia, Tajikistan, Turkmenistan, Ukraine, Uzbekistan

Disputed border Russia/China (Ussuri river)

Disputed territory Armenia/Azerbaijan (Nagorno-Karabakh), Moldavia/Romania (Bessarabia and Bukovina), Russia/Japan (Kuril Islands and south Sakhalin island)

STYLES OF GOVERNMENT

Republics All the countries in the region

Multi-party states All the countries in the region

One-chamber assembly Mongolia

Two-chamber assembly (in 1990) All the constituent Soviet republics

CONFLICTS (since 1945)

Nationalist movements Azerbaijan (Armenians); Georgia (Ossetians); Kirghizia (Uzbeks); Moldavia (Russians in Dniestr region); Russia (Tatars, Karachai, Ingush, Volga Germans); Uzbekistan (Meskhetian Turks)

Internal conflicts Armenia 1988–89; Azerbaijan 1988–90; Latvia 1991; Georgia 1989; Lithuania 1990, 1991; Moldavia 1989–90

Civil wars Georgia 1991–; Moldavia 1992

Interstate conflicts Armenia/Azerbaijan 1991–92; USSR/Hungary 1956; USSR/Czechoslovakia 1968; USSR/China 1969; USSR/Afghanistan 1979–89

MEMBERSHIP OF INTERNATIONAL ORGANIZATIONS

Commonwealth of Independent States (CIS) Armenia, Azerbaijan, Belorussia, Kazakhstan, Kirghizia, Moldavia, Russia, Tajikistan, Turkmenistan, Ukraine, Uzbekistan

COLLECTIVIZATION TO REFORM

The head of the new Soviet government that came to power in Russia in 1917, V. I. Lenin (1870–1924), laid down the framework of communist rule in the constitution of 1918. Land was redistributed and property and banks nationalized. Three years of civil war, however, had a devastating economic effect, and the New Economic Policy (NEP) partially restored private enterprise. After Lenin's death a struggle for political power took place in the government, from which Joseph Stalin (1879–1953) emerged as leader.

For nearly 30 years of unchallenged rule, Stalin's autocratic policies were to transform the Soviet Union. Peasant agriculture was drastically reorganized into "collective" state-owned farms, and a program of rapid industrialization was undertaken. It is estimated that some 6 million peasants died as a result of collectivization. Countless more citizens were to perish in the penal labor camps and in the purges of 1936–38 that were the twin pillars, with the secret police, of Stalin's brutal regime. The legacy of the Stalin years was a socialist economy geared toward heavy industry and defense and a highly centralized one-party state dominated by a large and increasingly cumbersome bureaucracy.

Nevertheless, Stalin's total control of Soviet life may have enabled the Soviet people to withstand the German invasion of 1941. Some 20 million Soviet citizens died in the fighting and devastation of World War II, hundreds of cities lay in ruins and the economy was destroyed.

States of Northern Eurasia

- conflict
- civil unrest

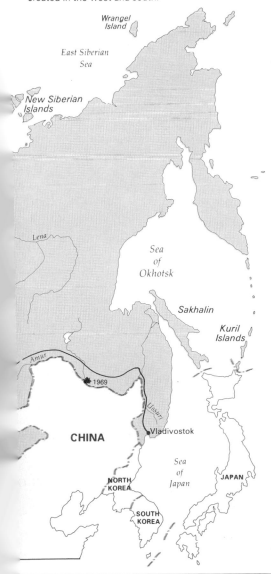

Wrangel Island

East Siberian Sea

New Siberian Islands

Lena

Sea of Okhotsk

Sakhalin

Kuril Islands

Amur

1969

Ussuri

CHINA

Vladivostok

Sea of Japan

NORTH KOREA

JAPAN

SOUTH KOREA

However, the Soviet Union emerged as the victor and dominant power in Eastern Europe, with political hegemony over a series of satellite states. Its territorial gains included the three Baltic states of Estonia, Latvia and Lithuania, part of Poland's eastern territories and Moldavia. Although it had declared war on Japan only three weeks before Japan's surrender, it also acquired south Sakhalin island and the Kuril Islands in the Sea of Okhotsk.

The Soviet Union after Stalin

In the years following Stalin's death in 1953, Nikita Khrushchev (1894–1971) emerged as first secretary of the CPSU. In a historic speech at the 20th Party Congress in 1956 he denounced Stalin's abuses and introduced important economic and political reforms, including the devolution of economic power to the regions (the *sovnarkhoz* experiment) and the extension of cultural freedoms. However, these changes were shortlived. The more conservative leadership of

Leonid Brezhnev (1906–82), first secretary for nearly 20 years from 1964, resulted in economic stagnation.

It was in response to the growing economic crisis that Mikhail Gorbachev, who became first secretary in March 1985, set in motion the twin policies of *perestroika* ("reconstruction") and *glasnost* ("openness"). *Perestroika* referred to a program of measures designed to reform the economic system, including the expansion of private enterprise and greater economic self-management. *Glasnost* represented a determination to democratize the social and political system.

The opening up of Soviet society to self-examination unleashed a storm of nationalist discontent. Once the system of rigid central control was relaxed, the much vaunted "friendship of the Soviet peoples" was revealed to have been little more than a policy enforced from above, and long-buried grievances emerged with frightening force. The Tatars, for example, transported from the Crimea in

1944 and resettled in Russia, staged a public demonstration in 1987 for the return of their homeland; in many of the non-Russian republics resentment of the influence wielded by Russian incomers was more and more openly expressed. In the Baltic republics (annexed by the Soviet Union in 1940) the demand for national self-determination grew rapidly in the late 1980s, and there were nationalist protests in Armenia, Belorussia, Moldavia, Georgia and the Ukraine. Ethnic violence was especially strong in Nagorno-Karabakh, an Armenian enclave within Azerbaijan.

The "leading role" of the CPSU came more and more into question, and in February 1990 official recognition was given to other parties. Following elections to many republic and local soviets later that year, the CPSU lost power in the Baltic republics, Armenia, Moldavia and Georgia, and in Moscow and Leningrad. Mikhail Gorbachev, elected to the new role of president in March, attempted to steer a course between reformers and conservative elements in the CPSU, but satisfied neither. In August 1991 hardline communists launched an unsuccessful coup. In the immediate aftermath of this turmoil the activities of the CPSU were suppressed, and by the end of the year the Soviet Union had passed into the history books.

THE MACHINERY OF COMMUNIST RULE

As the country's official name suggested, power in the Soviet Union was supposed to operate through the soviets (or people's councils), which were the formal organs of state power. In practice, however, real power always lay with the Communist Party through its control of all political activity.

Parallel systems of government

The soviets, elected by the citizenry, were intended by Lenin to form a pivotal role in the process described in his notion of "democratic centralism". This holds that democratization prevails through the election of all state bodies of government, which are accountable to the people. In turn, power is centralized through control from above, lower bodies being obliged to obey the directives of higher ones.

At the lowest level of the state apparatus were the 50,000 local soviets. Above them were the regional soviets. Each of the 15 republics of the Union had its own supreme soviet, and above them was the supreme soviet of the Soviet Union, based in Moscow.

Over the decades democratic centralism resulted in the CPSU becoming the only dominant political force in the Soviet

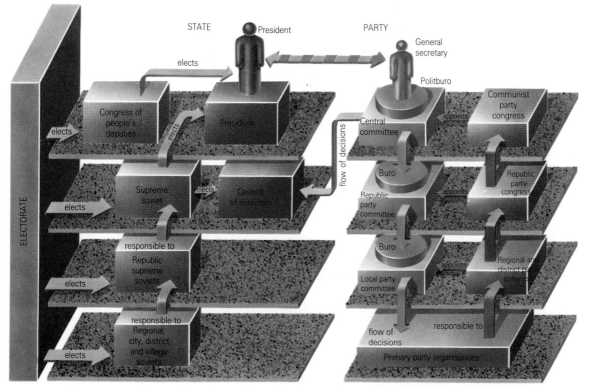

The statue of Lenin dominates the proceedings of the 27th Party Congress in 1986, which was the launching pad for the *perestroika* reforms. The party congresses, held every five years, were the forum for the CPSU to announce new directives and rubber stamp decisions. Day to day power lay with the politburo, the inner cabinet of the CPSU's ruling central committee.

The highest state legislative body (*left*) in the Soviet Union was the two-chamber supreme soviet, which met only for two or three days each year – between sessions the presidium acted as the supreme state authority. The council of ministers was the highest executive body, but it was the CPSU that dictated policy decisions. The Congress of People's Deputies was established in 1989 as a working legislature. In 1990 it elected Mikhail Gorbachev, general secretary of the CPSU, as the first and last president of the Soviet Union. The congress's power was to be shortlived – the political structure of the Soviet Union collapsed in 1991.

THE COUP THAT FAILED

The overthrow of Eastern Europe's communist regimes at the end of 1989 threw the Soviet Union's policy-makers into confusion. Many welcomed the changes as signaling the need for greater democratic reform and argued that the country's worsening economic situation called for a speeding up of moves toward a market economy. The loss of a "buffer zone" of satellite states on the Soviet Union's western flank, however, led hardliners within the CPSU to call for controls to be tightened rather than loosened.

Gorbachev was forced to follow a delicate path between these opposing camps. After his failure to condemn the revolutions in Eastern Europe was criticized at the 28th Party Congress in July 1990 he took steps to weaken the power of the CPSU, and pushed through a series of measures including recognition of the unification of Germany and the abolition of press censorship. But by late 1990 he appeared to be bowing to conservative pressure when he failed to implement promised economic change.

By the summer of 1991 there was a swing back toward reform. Steps were taken to allow the sale of state-owned enterprises, and on 24 July agreement was reached with 9 of the 15 republics on a draft Union Treaty decentralizing power in the Soviet Union.

All of this was too much for the old guard. Two days before the Union Treaty was due to be signed on 20 August an ultimatum was delivered to Gorbachev, on holiday in the Crimea, demanding that he resign and declare a state of emergency. On refusing to do so he was placed under house arrest. In Moscow a self-proclaimed State Committee for the State of Emergency (SCSE) assumed power.

The overthrow of the coup came through the efforts of Boris Yeltsin, president of the Russian Federation since 1990. From his position in the Russian government building ("the White House") he organized opposition on the streets and as columns of tanks entered Moscow on 20 August tens of thousands of people gathered to defend the building. Three people were killed, but the coup quickly collapsed. The leaders were arrested and Gorbachev returned to Moscow.

It was, however, Yeltsin who emerged as the clear winner. In the days after the coup the Russian government took over the functions of the all-Union government. In a showdown on television on 23 August Yeltsin signed a decree suspending the activities of the Russian Communist Party, which took Gorbachev completely by surprise. The ultimate result of the failed coup was the collapse of the Soviet Union.

Union. Proclaiming itself the "leading and guiding force in Soviet society", it was charged both with making policies and ensuring that decisions were carried out promptly and efficiently.

At the apex of the CPSU was the central committee and its "inner cabinet", the politburo. Made up of between 11 and 15 members and 6 and 8 non-voting (candidate) members, the politburo, headed by the general secretary, exercised real power over day-to-day decisions. It had great influence in policy making.

The hierarchical nature of the CPSU's organization ensured that its wishes were carried out. Party organizations existed at all levels of Soviet society – Union republic, region, city, town and district – and each was accountable to the tier above it. At the lowest level were the primary party organizations. These existed in every farm, factory and enterprise, and the CPSU relied on this grassroots organization to report on problems of inefficiency and mismanagement.

The power of the CPSU was not confined to government. It took responsibility for the political education of all Soviet citizens. Through its control over the "party list" system it had the power to make appointments to all positions in the country's economic and social life.

The end of Soviet rule

The political superstructure of the Soviet Union collapsed like a deck of cards after the failure of the conservative coup in August 1991. Following Boris Yeltsin's suspension of the Russian Communist Party, Gorbachev resigned as general secretary of the CPSU and nationalized all its assets. More than 70 years of communist government were at an end.

The State Council (President Gorbachev and the heads of the participating republics) assumed executive control and recognized the independence of Latvia, Lithuania and Estonia. A treaty of economic community was signed by most of the remaining republics, but Russia,

Ukraine and Belorussia led the opposition to a proposed Union of Sovereign States, with centralized control of the armed forces, and its own government.

In December 1991 the leaders of these three republics met to set up a Commonwealth of Independent States to be held together by voluntary ties. It aimed to provide stability to the former Soviet republics as they moved toward independent statehood. On 28 December all the other republics except Georgia agreed to join and the Soviet Union was dissolved.

By 1994 many of the republics were still struggling to establish new political systems. In Russia, stalemate led to President Yeltsin's dramatic suspension of parliament in October 1993. This provoked armed rebellion that was defeated by forces loyal to the President. Parliamentary elections held in December 1993 resulted in defeat for radical reformers but presidental powers were endorsed and extended in a referendum held at the same time.

RETHINKING SUPERPOWER STRATEGIES

The Soviet Union rapidly emerged as a global superpower after World War II, rivaled only by the United States in its military and political influence. The Soviet troops remaining in Eastern Europe after its liberation in 1945 helped to establish socialist states under Soviet control in Bulgaria, Czechoslovakia, Hungary, Poland and Romania, as well as the eastern half of Germany (the German Democratic Republic).

Formal treaties – the Council for Mutual Economic Assistance (COMECON) and the Warsaw Pact (a defensive agreement to counter the North Atlantic Treaty Organization, NATO) – bound these countries economically and militarily to Moscow, and completed the Soviet Union's domination of the region. The Soviet Union also assumed political leadership of a numerically powerful bloc of socialist states in Asia, including Mongolia, North Korea (1948), China (1949) and North Vietnam (1954). In 1961 Cuba joined the Soviet bloc.

By the mid-1950s official Soviet foreign policy emphasized coexistence between capitalist and socialist states, yet in practice continued to give active support to anticolonial liberation movements in the Third World. During the "Cold War" years of extreme political hostility – which reached its height with the Cuban missile crisis of 1962 – the Soviet Union and the United States embarked upon a nuclear arms race that came to dominate global politics for the next two decades.

The Brezhnev doctrine

By the early 1960s the Soviet Union was finding it increasingly difficult to maintain control over the geographically fragmented bloc of socialist states. Even within Eastern Europe tensions erupted. On two notable occasions – Hungary (1956) and Czechoslovakia (1968) – Soviet troops were sent in to restore order. Explicit justification for such intervention was spelt out in 1968 in the so-called Brezhnev doctrine, whereby Moscow claimed a right to intervene in the internal affairs of socialist countries wherever socialism was deemed to be threatened.

Relations between China and the Soviet Union had become very strained. Under the leadership of Mao Zedong

A grim display of military might Rockets are paraded through Red Square at the height of the Cold War. Next to Khrushchev on the viewing platform, in the soft hat, is the bearded figure of Cuba's Fidel Castro.

Meeting in Moscow US President George Bush with President Gorbachev at a Kremlin summit in July 1991. On the left is Boris Yeltsin, then the newly elected president of the Russian republic.

End of a superpower

By the mid-1980s the Soviet Union was spending 15 percent of its GNP on defense (compared with the United States' 6 percent). This became an increasingly unacceptable burden on its ailing domestic economy. The election of Mikhail Gorbachev as general secretary in 1985 heralded a new era in East–West relations, summed up in the Kremlin's so-called "new thinking". This placed emphasis squarely on the need to reduce military and nuclear commitments, and called for all countries to cooperate in resolving the world's problems.

The signing of the historic Intermediate Range Nuclear Force (INF) treaty in December 1987 was the first step in reversing the postwar arms race. It was followed by the withdrawal of Soviet troops from Afghanistan, the scaling down of the Soviet military presence in Mongolia and the reduction of aid to socialist movements in the Third World. As the communist regimes of Eastern Europe fell in 1989 the Soviet Union chose not to assert the Brezhnev doctrine. In November 1990 Gorbachev attended a meeting of the Conference of Security and Cooperation in Europe (CSCE) in Paris ending the 40-year ideological division in Europe, and that same year was awarded the Nobel peace prize.

The collapse of the Soviet Union brought new security problems. At first it was hoped that the CIS would set up a common military policy, but by spring 1992 all the republics had decided to establish national armies. Russia and Ukraine disagreed over the ownership of the Black Sea fleet, but cooperation continued over the command of nuclear weapons under Russian control. Belorussia, Kazakhstan and Ukraine agreed to return all tactical warheads to Russia for destruction and to decommission their strategic systems by 1994.

Within the region itself ethnic conflict continued, especially in the Caucasus where the problem of Nagorno-Karabakh, which declared its own independence in 1991, took on the aspect of an interstate dispute between Armenia and Azerbaijan. Civil war in Georgia threatened to spill over into Russia, and in Moldavia differences between the indigenous Romanian population and Russian-speakers east of the Dniestr led to an escalation of the civil war and to tension with both Romania and Russia.

(1893–1976), China chose to follow an increasingly independent socialist path, and by 1960 all the Soviet Union's advisors and technicians in China had been withdrawn. The Soviet invasion of Czechoslovakia aroused fears that the Brezhnev doctrine might be used as a pretext to bring China back into the Soviet fold. A sharp war of words led to armed border clashes on the Ussuri River in March 1969.

The signing of the 1972 Strategic Arms Limitation Treaty (SALT I) marked a thaw in relations between the Soviet Union and the United States, improved further by the withdrawal of United States troops from Vietnam after 1973. But detente did not prevent the Soviet Union from increasing its supply of arms to Third World countries, or from sending military advisors to wars in Angola and the Horn of Africa. The Soviet invasion of Afghanistan in December 1979 brought a revival of Cold War attitudes.

SOVIET INTERVENTION IN AFGHANISTAN

The Soviet Union's 10-year embroilment in Afghanistan in a war it could not win has been likened to the United States' costly involvement in Vietnam between 1964 and 1973. In December 1979 over 100,000 troops were sent to support a pro-Soviet government that was engaged in a civil war with Afghan tribesmen, who called themselves Mujahidin (holy warriors). They saw the government's socialist reforms as a threat to their traditional Islamic culture and values. The Soviet Union feared that socialism in Afghanistan was in danger of being replaced by a fundamentalist Islamic regime of the sort that had seized power in neighboring Iran.

The Brezhnev doctrine was used to justify this occupation of a socialist neighbor state. Other motives have been cited. The possession of military bases in Afghanistan would bring Soviet aircraft within striking distance of the Strait of Hormuz at the southern end of the Gulf, through which vital oil imports pass to the West. An additional factor may have been the long-held desire (dating back to tsarist days) to acquire a warm-water port – through Pakistan, which was considered vulnerable to revolution.

The immediate concerns, however, were undoubtedly the threat posed to the Soviet Union's national security by the possibility of the Mujahidin allying themselves with Iran, and fear that Islamic fundamentalism could find support among the 40 million Muslims living in Soviet Central Asia.

The occupation of Afghanistan, condemned by the Western powers, China and throughout the Third World, achieved nothing politically in the longterm and proved highly damaging to Soviet morale. By the time it withdrew its troops in 1989 the Cold War was virtually over and it had voluntarily abdicated its role as a superpower.

The Baltic States: return to sovereignty

Incident in Lithuania Thousands of mourners gather for the funerals of 15 civilians killed when Soviet special troops (the Black Berets) stormed the TV tower in Vilnius in January 1991 and fired on a crowd of peaceful demonstrators.

State of emergency Following the Soviet assault, Lithuanian nationalists form an armed blockade to defend the parliament building against possible attack, while a line of Red Cross volunteers stand by to attend the wounded.

The Baltic states of Estonia, Latvia and Lithuania were part of the tsarist Russian empire until its collapse in 1917. On 3 March 1918 the new revolutionary Bolshevik government in Moscow concluded the Treaty of Brest-Litovsk with Germany and relinquished control of the old empire's Baltic provinces. With the end of World War I in November, the provisions of this treaty lapsed, leaving the Baltic provinces with an uncertain future. Revolutionary forces in Estonia set up a workers' commune on German withdrawal, but this was overthrown in the ensuing civil war with the assistance of British naval forces operating in the Baltic. A democratic republic was proclaimed in May 1919. Democratic republics were set up in the same year in Latvia and Lithuania. But these regimes did not survive the difficult interwar years. The first regime to be toppled by a coup was in Lithuania in December 1926, followed by Estonia in March 1934 and Latvia in May 1934.

World War II signalled the end of independence for all three states. In a secret protocol of the Soviet-German agreement signed on 23 August 1939, the Baltic republics were assigned to the Soviet sphere of interest. In June 1940 a Soviet ultimatum forced the governments of all three to resign, to be replaced by those acceptable to Stalin. A month later these puppet governments declared themselves Soviet Socialist Republics and applied to join the Soviet Union, which they duly did in August. Less than a year later all three were invaded by German forces, but in the withdrawal of the occupying armies in 1944 they returned to the Soviet Union.

From Soviet republic to independence

During the Cold War, the Baltic Soviet republics, which have direct access to the North Atlantic, were of strategic importance to the Soviet Union, and were firmly entrenched as an integral part of its war machine. An important naval base was built at Liepaja (Latvia). In the redrawing of the political boundaries after World War II the former German Baltic port of Königsburg, renamed Kaliningrad, was made an enclave of the Russian republic and crucial communication lines across Lithuania linked this high security zone with the rest of Russia.

Glasnost opened the way to a new assertion of national identity among the Baltic peoples. In the Soviet Union's first open elections to the Congress of People's Deputies in March 1989, "people power" in the Baltic region was expressed by throwing out the CPSU's nominees, including the prime ministers of both Latvia and Lithuania, and electing a new national leadership. From this point on the movement toward regaining independence was swift but by no means smooth.

In February and March 1990 multiparty elections to the republics' supreme soviets resulted in total victory for the nationalists in Lithuania. Sajudis, the nationalist movement, won by a large majority and on 11 March it became the first union republic to declare its independence from the Soviet Union. On 30 March the Estonian soviet voted 70 to 0 (3 abstentions) to proclaim the 1940 Soviet occupation of their country illegal. In May it renamed itself simply the "Republic of Estonia" and restored the pre-1940 flag and national anthem. The Latvians too voted the 1940 occupation illegal and resolved to return to their 1922 constitution.

International recognition of the Baltic states was not forthcoming, and the concentration of Soviet armed forces in the

Baltic states was a barrier to progress toward independence. On 10 January 1991 President Gorbachev told Lithuania that it must accept Soviet central control. This precipitated popular demonstrations in the capital Vilnius, and on 13 January Soviet paratroopers fired on an unarmed crowd killing 15 people and injuring 140. As international opinion turned against Gorbachev, Lithuania responded by voting to secede from the Soviet Union and organized a plebiscite on the question. Gorbachev declared the plebiscite illegal but it went ahead nonetheless, with over 90 percent voting for independence. Similar ballots in Latvia and Estonia in March recorded large majorities for independence. Stalemate had been reached.

The Soviet coup of August 1991 totally transformed the situation. The international discrediting of the communist regime gave the Baltic republics the chance they had been looking for. Within days of Gorbachev's return to Moscow, the European Community had taken the lead in recognizing the independence of the three republics on 27 August. Two weeks later the Soviet State Council accepted the inevitable and finally recognized the independence of Estonia, Latvia and Lithuania on 6 September. All three countries were granted membership of the United Nations in the same year.

The founding fathers of Marxist-Leninism

Marxism as a modern political movement largely developed in Europe during the last quarter of the 19th and the early years of the 20th century. In terms of political ideas and subsequent propaganda, the key architects were Karl Marx (1818–83) and Friedrich Engels (1820–95), whose ideas were taken and adapted by Lenin to fit the particular circumstances of tsarist Russia.

The partnership between Marx and Engels is one of the most influential intellectual collaborations in history. Their great project was to produce a new political and economic understanding of society in order to create a more humane and equal world. Their famous *Communist Manifesto* was published in 1848, though it was not until the 1870s and the formation of the German Social Democratic party that their ideas began to have important political influence.

Lenin's particular contribution to their political legacy was to promote the role of the Communist Party in making and consolidating the revolution. It is this interpretation of Marxism that is the Marxist-Leninism propagated by the communist regimes of the 20th century.

The influence of the ideas of these three men can be gauged from the fact that for most of the second half of the 20th century about one-third of the world lived under Marxist-Leninist regimes. In Eastern Europe they were the victims of the revolutions of 1989, and two years later Marxist-Leninism had been discredited as the official doctrine of the Soviet Union as well.

The mighty three Parading troops are dwarfed beneath the portraits of Marx, Engels and Lenin in Winter Palace Square, Leningrad.

STATES IN FERMENT

FROM NOMADISM TO SOVEREIGNTY · THE STRUGGLE FOR POWER · STATES OF WAR

The predominantly Islamic states of the Middle East, lying on the ancient trade routes that connect Asia and Africa with Europe, have been the scene of numerous wars. Before World War I the dominant rulers in the region were the Turkish Ottomans. After entering the war on the German side in 1914, the vast Ottoman empire was divided up by the victorious allies in the Treaties of Sèvres (1919) and of Lausanne (1923). From these territories, together with Persia (now Iran), Afghanistan and Israel (created in 1948), today's states have emerged. The region's geopolitical location and enormous oil wealth meant that influence was keenly contested between the United States and the Soviet Union during the Cold War. In 1991 it was the scene of the first post-Cold War international confrontation.

COUNTRIES IN THE REGION

Afghanistan, Bahrain, Iran, Iraq, Israel, Jordan, Kuwait, Lebanon, Oman, Qatar, Saudi Arabia, Syria, Turkey, United Arab Emirates, Yemen

STYLES OF GOVERNMENT

Republics Afghanistan, Iran, Iraq, Israel, Lebanon, Syria, Turkey, UAE, Yemen

Monarchies Bahrain, Jordan, Kuwait, Oman, Qatar, Saudi Arabia

Federal state UAE

Multi-party states Afghanistan, Israel, Lebanon, Turkey

One-party states Iran, Iraq, Syria

States without parties Bahrain, Jordan, Kuwait, Oman, Qatar, Saudi Arabia, UAE, Yemen

CONFLICTS (since 1945)

Coups Iran 1953; Iraq 1963, 1968; Syria 1970; Turkey 1960, 1971, 1980; Yemen 1976–78

Revolutions Afghanistan 1978; Iran 1979; Yemen 1962

Civil wars Afghanistan 1978–92 (with USSR involvement 1979–89); Iraq 1961–75; Jordan 1970–71; Lebanon 1976–92; Oman 1969–75; Yemen 1967–69; 1979

Internal unrest Israel 1987– (West Bank); Saudi Arabia 1979; Syria 1982

Independence war Aden (S. Yemen)/UK 1962–67

Interstate conflicts Israel/Arab states 1948–49, 1956, 1967, 1973, 1978, 1982–; Oman/S. Yemen 1965–75; S. Yemen/Saudi Arabia 1969; Iraq/Kuwait 1973; Turkey/Cyprus 1974; Iran/Iraq 1980–88; Iraq/Kuwait (with UN involvement) 1990–91

MEMBERSHIP OF INTERNATIONAL ORGANIZATIONS

Arab League Bahrain, Iraq, Jordan, Kuwait, Lebanon, Oman, Qatar, Saudi Arabia, Syria, UAE, Yemen

Colombo Plan Afghanistan, Iran

North Atlantic Treaty Organization (NATO) Turkey

Organization of Petroleum Exporting Countries (OPEC) Iran, Iraq, Kuwait, Qatar, Saudi Arabia, UAE

FROM NOMADISM TO SOVEREIGNTY

Patterns of sovereignty in the Middle East have changed greatly over the centuries, and the present territorial arrangements cannot be regarded as permanent. The region contains 16 states of equal status, each of which claims absolute sovereignty over its land and the airspace above, and a measure of sovereignty over adjacent maritime zones as well. The indigenous peoples of the Middle East, especially where nomadism was dominant, were unfamiliar with this rigid division of space, instead recognizing ill-defined tribal territories.

The map of the Middle East today is largely the product of European activity in the 19th century, and of the peace treaties that concluded World War I. Britain and France, stepping in to fill the vacuum left by the decline of Turkish Ottoman influence in the region, played leading roles in deciding most of the international land boundaries of the Middle East. Several boundaries were drawn as straight lines to satisfy the ambitions of European imperialism, without reference to the aspirations of the local people or to the geographical realities.

A region that suffers from acute aridity needs rational management of scarce water resources, but several river basins were divided between rival states. Routes that had been followed by nomadic migrants for centuries were now impeded by international boundaries. Some of the world's longest unfixed boundaries lie in the desert between Saudi Arabia and

Guns and oil wells Enormous wealth from oil has flowed into the revenues of the desert kingdoms of the Middle East this century, strengthening the power of their traditional rulers.

An area of constant conflict Since 1945 the Middle East has been in almost continuous ferment. The Arab–Israeli conflict continues to disrupt the lives of thousands of people. The Iranian revolution of fundamentalist Shi'ia Muslims (1979) threatened to destabilize its Sunni Muslim neighbors. The Iran–Iraq war (1980 to 1988) was the 20th century's longest conflict. The 1991 Gulf war polarized the region politically. The US-led international force to eject Iraq from Kuwait was based in Saudi Arabia. Syria and Turkey supplied troops, but Jordan and Yemen refused to join the alliance.

Oman, and Saudi Arabia and Yemen.

Middle Eastern states were among the first in the world formally to agree maritime boundaries. The process began with the need of the states sharing the Gulf coastline to allocate oil concession areas: the earliest continental shelf agreement was between Bahrain and Saudi Arabia in 1958. Seven further boundaries have subsequently been agreed in the Gulf, but a number of serious disputes remain between states; in particular, several islands in the Gulf and the Red Sea are disputed.

Nations without states

A continuing cause of bitter conflict in the region was the failure of the European diplomats to fulfil the ambitions of the large Kurdish and Armenian minorities in Turkey and Iraq, who after World War I had hoped for states of their own. The Armenians suffered badly at the hands of the Turks, particularly in 1915 when massacres and forced deportations resulted in appalling loss of life. Turkish estimates suggest that between 500,000 and 600,000 Armenians died; the Armenians put the number at 1.5 million.

The 10-million or so Kurds, living in the mountains of southeastern Turkey, western Iran and northern Iraq, are fighting for a state of their own. By 1985

they had secured a large tract of territory along the Iraqi border with Iran. With the ending of the Iraq–Iran war (1988) Iraqi troops launched a fierce onslaught against them, in which chemical weapons were used. A similar campaign after the Gulf war (1991) was thwarted by the establishment of safe havens for the Kurds supervised by the United Nations.

Palestinian Arabs, displaced after the creation of the Jewish state of Israel in 1948, have consistently demanded a return to a Palestinian homeland. These demands have been spearheaded by many rival groups and factions, but the Palestine Liberation Organization (PLO) under the leadership of Yasser Arafat increasingly claimed the right to speak for the Palestinian people.

Inbuilt potential for conflict

States such as Iraq, Jordan, Lebanon and Syria, created at the end of World War I as artificial political units, came into being with inbuilt potential for conflict, which was enhanced after 1948 by the arrival of Palestinian refugees, especially in Jordan, Lebanon and Syria. These states are still having to face almost impossible tasks of national unification. Power-sharing between Christians and Muslims in Lebanon, fragmented by years of civil war, broke down completely in the 1980s.

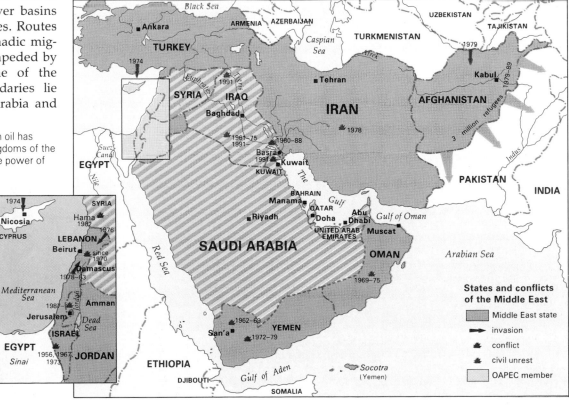

States and conflicts of the Middle East
- Middle East state
- invasion
- conflict
- civil unrest
- OAPEC member

THE STRUGGLE FOR POWER

The states of the Middle East embrace many styles of government, including autocratic monarchy (Oman, Saudi Arabia), constitutional monarchy (Jordan), parliamentary democracy (Israel, Turkey), a religious republic (Iran) and revolutionary socialism (Iraq, Syria). In 1990 the separate governments of the Yemen Arab Republic and the People's Democratic Republic of Yemen (Southern Yemen) amalgamated to become one country. In one state, Lebanon, there has been an almost total breakdown of effective government. The region contains one federation – the United Arab Emirates (UAE) – made up of seven tiny traditional sheikhdoms. Federal authority is exercised by a supreme council of rulers.

Almost nowhere in the region are the forms of democratic government fully observed. Israel's political structures closely follow the Western model. Its 120-member parliament (the Knesset) is elected by secret universal suffrage every four years. The president is elected by the Knesset. There are a large number of political parties; these range from communists on the far left to influential right-wing orthodox religious groups.

Arabs in the West Bank ceased to vote in elections to Jordan's national assembly after Israel's occupation in 1967.

Severe economic problems brought Turkey to political crisis in 1980 and the army took control. Elections were restored in 1983, and martial law lifted in 1987. However, Turkey's record on human rights (particularly its repression of its Kurdish population) led to the rejection of its request to join the European Community.

Repression and revolution

In nearly every Middle Eastern state, of whatever political color, there is a regrettably high degree of repression and denial of civil rights, with heavy spending on internal security. Most have experienced difficulties in maintaining internal order in recent years. Seven states have witnessed revolutions that captured world headlines: Afghanistan (1973,1978), Iran (1921,1979), Iraq (1958), Yemen (1962), Southern Yemen (1978), Syria (1963) and Turkey (1923, 1980). In six of these upheavals traditional rulers were replaced by radical one-party regimes, and in seven of them military governments were brought to power.

Several states have experienced civil wars, some of them costly and bitter struggles prolonged by the interference of outside powers: Iraq (1961–75), Southern Yemen (1962–69), Oman (1969–75), Jordan (1970), Lebanon (1976–1992) and Afghanistan (1978–1992) are notable examples. There have been numerous other instances of violent internal conflict, including the Syrian army's ruthless crushing of the fundamentalist Islamic movement, the Muslim Brotherhood, in

An Arab monarchy Bahrain's traditional ruler possesses virtually absolute power, and several ministries are headed by members of the emir's family.

Resurgent Islam Thousands of Iranian women, heads traditionally covered, took to the streets of Tehran to welcome the return of Ayatollah Khomeini.

REVOLUTION IN IRAN

Few Middle Eastern revolutions have been as far-reaching as that of Iran in 1979. This popular uprising, led by Islamic religious leaders, brought about the downfall of a seemingly strong monarchy that claimed descent from the ancient rulers of Persia. Under Shah Muhammad Reza Pahlevi (1919–80), who had succeeded to the throne in 1941, Iran – then the world's second largest oil exporter – was prosperous and undergoing rapid modernization. Its powerful armed forces were equipped with sophisticated weapons provided by the United States. But many of Iran's Shi'ia Muslims (90 percent of the population) resented the excessive westernization and materialism encouraged by the shah's authoritarian regime. There was widespread opposition, centered on the figure of Ayatollah Ruhollah Khomeini (1901–89), a religious leader who had been exiled in 1964 for denouncing the shah's policies.

Protests grew more violent throughout 1978, forcing the shah to flee Iran on 16 January 1979 as his support finally crumbled. Ayatollah Khomeini returned to Iran on 1 February to an ecstatic welcome. A new constitution made him religious and political leader for life and placed supreme power in the hands of the Muslim clergy.

The Islamic revolution transformed life in Iran. Islamic laws were introduced governing such things as women's dress and banning popular music and alcohol. Iran became nonaligned, but gave encouragement to Shi'ia groups in neighboring states. The revolution enjoyed popular support, but its many opponents, found particularly among the middle classes, minority nationalist groups, and left-wing and Marxist supporters were harshly treated. The revolution had major international repercussions, notably by increasing oil prices, and led to a costly war with Iraq (1980–88), which strengthened its hold. In 1989 Ayatollah Khomeini called on Muslims to execute the British writer Salman Rushdie. His own death a few months later was greeted with mass demonstrations of mourning but Iran's new leaders adopted more conciliatory policies to the West.

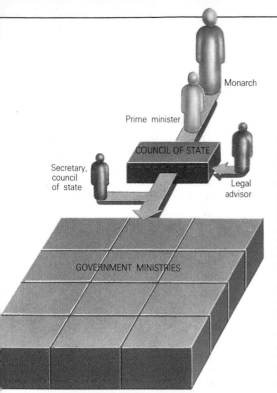

Monarch

Prime minister

COUNCIL OF STATE

Secretary, council of state

Legal advisor

GOVERNMENT MINISTRIES

the city of Hama in 1982; the seizure of the great mosque in Mecca by religious extremists in 1979 and their bloody encounter with Saudi security forces; the Palestinian uprising (Intifada) against Israel's occupation of the West Bank and Gaza since 1987; and the repression of the Kurds and Shi'ia Muslims by Saddam Hussein's regime in Iraq (since 1991).

Desert autocracies

Problems of internal control are not new to the region, however. Nomadic populations inhabiting vast desert areas often lay beyond the effective reach of government. In the inaccessible mountainous regions of Afghanistan, Turkey and Yemen, independent-minded tribesmen have proved resistant to integration into a modern state. Ancient feuds between tribes have sometimes boiled over into fierce local fighting. The desert states of the Arabian peninsula are in effect coalitions of former nomadic tribes. In these traditional monarchies government is still largely without parliaments, elections or political parties, but the rulers do seek to maintain consensus with the people. The process of modernization has put strain on these political systems, but rising living standards, thanks to oil revenues, have kept most citizens content. After the Gulf war there was pressure on the ruling Sabah regime to democratize Kuwait.

STATES OF WAR

The Middle East comprises two geo-politically distinct groups of states. The "northern tier" of Afghanistan, Iran and Turkey are large, populous, extensively mountainous states, which – particularly in the case of Iran and Afghanistan – embrace considerable tribal and ethnic diversity. To the south lies a block of a dozen Arab-speaking states, sharing a common religion (Islam) and all belonging to the Arab League, the organization of Arab states established in 1945. Embedded in the western flank of these states is the Jewish state of Israel. Apart from Egypt, which signed a peace treaty in 1978, no Arab state recognizes the state of Israel.

Since World War II every state in the region except Bahrain and Qatar has had an armed confrontation with its neighbors. Israel fought wars with its Arab neighbors in 1948, 1956, 1967, 1973 and 1982–84, while the Iran–Iraq war (1980–88) was one of the bloodiest encounters of this century. The Turkish invasion of Cyprus in 1974 left the island in effect partitioned, though the Turkish Republic of North Cyprus (declared in 1983) has failed to gain international recognition. On the other hand, cooperation between states in the region has sometimes been at a remarkably high level, particularly in the Arab world.

Obstacles to Arab unity

Arab unity remains the ideal of many Arabs, though it has little prospect of becoming reality. The Arab League, and the powerful Baathist parties in Iraq and Syria, are deeply committed to the idea, while most Arab leaders pay lip-service to the concept.

The obstacles to Arab unity are very great. States created after World War I have generated their own nationalisms, vested interests and divergent styles of government. Ideologically there is little in common between the traditional monarchies and the more radical socialist regimes. In some regions the division of Islam between the Sunni and the Shi'ia is another impediment to close cooperation. The economic incentives for integration are limited, given current patterns of trade. Geographical factors – vast distances, fragmentation of settled population, and inadequate communications –

have always hindered Arab unity. For the people of the Arabian peninsula, loyalty to clan, tribe or region is usually more important than loyalty to state.

Attempts to unite Arab states have proved ineffectual. The merger of Egypt and Syria as the United Arab Republic in 1958 was shortlived. Southern Yemen and Yemen took nearly 20 years to unite, yet in 1993, political dispute flared into civil war. Possibly the most successful practical venture among Arab states is the Gulf Cooperation Council, founded in 1981 by the Gulf states (Bahrain, Kuwait, Oman, Qatar, Saudi Arabia and the UAE) primarily for defense, though it has led to fruitful collaboration in several spheres.

Territorial questions have caused a number of quarrels between Middle East states. Armed conflict has taken place over international boundaries between

Armed for victory A triumphal arch in Baghdad commemorates Iraq's role in the war against Iran (1980–88). The arms holding the swords have been modeled on those of Iraq's leader Saddam Hussein himself, and piled below are the helmets of hundreds of Iranian soldiers.

Renouncing terrorism Yasser Arafat, leader of the PLO, at the UN in 1988. He conceded the state of Israel's right to exist – the first time a Palestinian spokesman had done so – and renounced violence. In 1993, Israel and the PLO signed an accord.

Yemen and Southern Yemen (1956, 1979), Southern Yemen and Saudi Arabia (1969), Syria and Jordan (1970), Iraq and Kuwait (1973), Southern Yemen and Oman (1975), and Yemen and Saudi Arabia (1980). In September 1980 Iraq invaded Iran, ostensibly to protect the Shatt-al Arab waterway, giving access from the Euphrates to the Gulf, ownership of which was disputed with Iran. The war quickly esca-

THE GULF WAR

The event that most sharply exposed the fragility of pan-Arab unity was the Gulf war of 1991, which was fought in response to Iraq's annexation of the tiny oil-rich state of Kuwait. All three of Kuwait's much larger neighbors – Saudi Arabia, Iraq and Iran – had at one time or another harbored expansionist ambitions toward it, and there had been previous armed border clashes between Iraq and Kuwait in 1973. In August 1990, almost exactly two years after the end of its damaging eight-year war with Iran, Iraq invaded Kuwait, using the excuse that Kuwait had been sabotaging the Iraqi economy.

There is every reason to suppose that Saddam Hussein, Iraq's leader, may not have expected the degree of international outrage the invasion aroused. During the Iran–Iraq war Western countries opposed to the spread of pro-Iranian Islamic fundamentalism, in particular the United States, had provided Iraq with weapons and support, enabling it to build up the world's fourth largest army. There were growing fears that Iraq was covertly acquiring the technology to give it nuclear weapon capability.

Immediately after the invasion the UN imposed mandatory sanctions on Iraq. A multinational military force, under joint United States' and Saudi command, was established in Saudi Arabia to counter any further Iraqi aggression. More than 500,000 United States' troops, 42,000 British, 20,000 Egyptian and 15,000 French joined the force, with smaller contingents from several other nations. Alone of Middle Eastern Arab states, Jordan and Yemen refused to join the anti-Iraqi alliance.

Saddam Hussein ignored all deadlines to withdraw from Kuwait, and in January 1991 a massive six-week air offensive was launched against Iraq. Allied bombs destroyed Iraqi equipment and aircraft, infrastructure and inflicted heavy casualties. Iraq launched retaliatory missile raids against Israel and Saudi Arabia, but little damage was done. A short ground war forced the evacuation of Iraqi troops from Kuwait, who first sabotaged oil storage tanks and oil drilling stations, causing extensive economic and environmental damage.

The Iraqi defeat encouraged uprisings among the Kurds in northern Iraq and the Shi'ia population in the south, which were harshly suppressed. More than a million Kurdish refugees fled to Iran, and thousands more were trapped in the northern mountains until safe havens could be established. Harassment of the southern rebels continued, despite international protests.

lated as Iraq extended its territorial claims in Iran. By the time it ended, in August 1988, more than a million people had been killed and there had been massive bombing of cities on both sides. Nothing was altered territorially by the war.

Territorial disputes do not always end in armed conflict. Many have been amicably resolved; in 1965, for example, Saudi Arabia and Jordan exchanged territory to give Jordan a longer coastline on the Gulf of Aqaba. Boundaries are often used as tools in international politics; borders are often closed at times of political tension, sometimes interrupting communications for years.

Superpower rivalry
The Middle East has long been the sphere of superpower rivalry. Before World War I Britain, France and Russia were rivals in the region, while during the Cold War it was the United States and the Soviet Union who sought clients there. Thus Turkey became a member of NATO, United States' military bases were estab-

lished in Bahrain and Oman, and Israel's policy toward its Arab neighbors was strongly supported by the United States. Afghanistan was under Soviet occupation from 1979 to 1988, Southern Yemen had a radical pro-Soviet regime, and the Soviet Union also enjoyed close relations with Syria and Iraq.

In the 1970s Lebanon was dragged into the conflict between Israel and the PLO, exacerbating the longstanding tensions between its Christian and Muslim communities. In 1976 the Syrian army entered Lebanon to restore order, while Israel's forces invaded in 1978 and from 1982 to 1984. The long civil war in Lebanon has destroyed the former prosperity of its capital, Beirut, and left the country fragmented between rival militia groups.

Hostage taking, bombing and international terrorism have all been a feature of conflict and violence in the Middle East. The readiness of outside powers to sell arms to Middle East states has led to the region accquiring one of the world's most formidable concentrations of weapons.

The emergence of Israel

The scattered Jewish people had long dreamed of returning to their historic homeland in Palestine. During the 1880s this dream became reality as small groups of ardent Zionists, driven out of Russia and Eastern Europe by persecution, began to settle in Palestine, then part of the declining Ottoman empire. Approximately 500,000 Arabs were living in Palestine at that time; by 1918 there were some 644,000 Arabs and 56,000 Jews. Jewish immigration was boosted when Britain was given the mandate to govern Palestine by the League of Nations in 1920. Undertakings made by Britain in the Balfour Declaration of 1917 promised to establish a Jewish national home in Palestine. The Palestinian Arabs deeply resented this policy, and as Jewish immigration grew in the 1920s and 1930s their anger and frustration deepened.

Proposals to partition Palestine that clearly deprived the Arabs of the best parts of their country did not help: a UN partition plan made in 1947 allocated 56 percent of Palestine to the Jewish state. With the ending of the British mandate in May 1948 war broke out between the Jews and Arab armies from neighboring states, and left the new state of Israel in control of 77 percent of the land. During the fighting approximately 600,000 Arabs fled from Israel, and large-scale Jewish immigration began. Many of the first arrivals were refugees from the horrors of Nazi persecution in Europe. By the end of 1948 Israel's population of 915,000 comprised 83 percent Jews and 17 percent Arabs.

Since 1948 the settlement pattern of Israel has been transformed, with over

The formation of Israel
- Israel 1947
- land gained 1948–49

LEBANON · SYRIA · Mediterranean Sea · Tel Aviv · JORDAN · Jerusalem · Gaza · Dead Sea · Jordan · ISRAEL · EGYPT · Negev

Uprising on the West Bank In a gesture of defiance, a Palestinian youth waves the outlawed flag of Palestine. The Intifada – or uprising – against Israeli occupation of the West Bank began in 1987. Violent clashes between unarmed Palestinians and Israeli troops, shown on television throughout the world, aroused much sympathy for the Palestinian cause. Israelis argued that retention of the West Bank was necessary to safeguard the security of the Jewish state.

Where are the Palestinians?
Israel's wars with its Arab neighbors in 1948 and 1967 created large numbers of Palestinian refugees who settled in Jordan, Lebanon, Syria and other Arab states. Jordan has absorbed more than a million Palestinians in recent years. Together with Palestinians already living there, they now comprise more than half the national population, and three-quarters of the population of the capital, Amman.

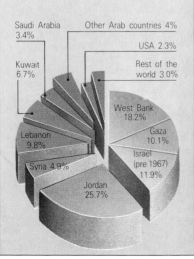

Saudi Arabia 3.4% · Other Arab countries 4% · USA 2.3% · Kuwait 6.7% · Rest of the world 3.0% · West Bank 18.2% · Gaza 10.1% · Lebanon 9.8% · Israel (pre 1967) 11.9% · Syria 4.9% · Jordan 25.7%

500 new rural settlements and some 40 new towns, planned to accommodate over 1.8 million Jewish immigrants. About one-third of the newcomers were Jewish refugees from North Africa and the Middle East. Jewish immigration subsequently fell markedly (though there was a large influx of Russian Jews in the early 1990s), while Israel's Arab population has maintained a very high rate of natural increase.

Israel's quest for security

Although Israel is well established in the international community, the Arab states

The state of Israel
- Israel 1967
- land occupied by Israel 1967
- land returned to Egypt 1975–82

LEBANON

Golan Heights

SYRIA

Mediterranean Sea

Tel Aviv

WEST BANK

Jerusalem

Gaza

Dead Sea

ISRAEL

Negev

JORDAN

Suez Canal

Sinai

Gulf of Suez

Gulf of Aqaba

SAUDI ARABIA

EGYPT

as a whole have continued their strong opposition to the creation of a Jewish state in their midst, at the expense of the rights of the Palestinians. As a result Israel's per capita defense spending has been among the highest in the world. Israel's quest for security has led to major armed intervention against neighboring states on several occasions.

As a result of the Six-Day war of 1967 Israel gained control of Jerusalem, the West Bank area of Jordan, the Sinai peninsula of Egypt (since relinquished) and the Golan Heights in Syria. There were subsequent invasions of Lebanon in 1978 and 1982, and Iraq's nuclear reactor at Tammouz was bombed in 1981.

The Palestinian uprising in the West Bank and Gaza after 1987 was severely repressed by Israeli troops, winning considerable world sympathy for the Palestinian cause. This was partially reversed when the Palestinians supported Iraq's Scud missile attacks on Israel in the Gulf war. In August 1991, Israel and the Arabs began US–Soviet sponsored peace talks in Madrid. A peace accord drawn up in September 1993 came into effect in April 1994 and Israeli troops withdrew from Gaza and Jericho.

The expansion of the Israeli state The 1947 UN plan that divided Palestine into a Jewish and an Arab state was rejected by the Arabs. At the end of the first Arab–Israeli war in 1948 the greater part (77 percent) of Palestine was left in Israeli hands. Most of the remainder was taken into Jordan; Egypt occupied the Gaza strip. The Six-Day war brought further substantial territorial gains. The Gaza strip, West Bank and Golan Heights remained under Israeli occupation until 1994; the Sinai peninsula was returned to Egypt in 1981.

Praying at the Wailing Wall All that remains of the temple in Jerusalem destroyed by the Romans in AD70, the Wailing – or Western – Wall is the Jews' most sacred shrine. Partition in 1948 placed it in the part of the city controlled by Jordan, and it was prohibited to Jews. On 7 June 1967 – the third day of the Six-Day war – it was taken by Israeli troops. It is a powerful symbol of national survival, and a center of pilgrimage.

Conflict in Afghanistan

During the 1950s the mountain kingdom of Afghanistan, engaging in its traditional rivalry with neighboring Pakistan, an ally of the United States, became dependent on the supply of arms from the Soviet Union, bordering it to the north. In 1973 an army coup toppled the monarchy, and five years later the government became dominated by local communists. Quarrels among them raised the possibility of political turmoil on its border, precipitating the Soviet occupation of the country in December 1979.

The regime in the capital, Kabul, was generally regarded as a Soviet puppet, and the West supported the Muslim resistance fighters, the Mujahidin. They conducted a classic guerrilla war campaign, using the mountainous territory for hit and run actions that kept a Soviet army of over 100,000 men bogged down in an unpopular war it could not win. Many commentators pointed out the similarities with the United States' involvement in Vietnam: the common lesson was that, for all their military strength, superpowers can be denied victory and their armies forced to retreat.

By the Geneva Accords of 1988, the Soviet Union agreed to pull its troops out of Afghanistan, and this was completed by February 1989. The Kabul regime immediately proclaimed a state of emergency in response to the intensification of the civil war by the Mujahidin, who set up an interim government in exile in Pakistan. They rejected a UN peace plan and in 1992 went on to overthrow the government in Kabul, replacing it with an Islamic fundamentalist regime. The civil war had damaging effects on the remote, impoverished country. Between 3 and 4 million of its population fled as refugees to Pakistan after 1979. The countryside remains littered with thousands of unexploded land mines, which are the daily cause of injury or death.

A Mujahidin guerrilla lays down his arms to pray. Afghanistan's bleak mountainous terrain gave the rebel army a tactical advantage despite the Soviet Union's superior strength and firepower.

ARAB STATES IN AFRICA

A POSTCOLONIAL MOSAIC · PROBLEMS OF CONTROL · NEIGHBORS AT ARM'S LENGTH

North Africa was subject from classical times to domination by successive external powers – Roman, Arab and Ottoman – whose empires lasted for centuries. Modern European rule was, by contrast, shortlived, though Spain and Portugal had gained footholds in northwest Africa as early as 1415. France began its conquest of Algeria in 1830, and almost all the rest of the region was partitioned by contending European powers – Britain, France, Italy and Spain – between 1870 and 1914. Only Ethiopia (known then as Abyssinia) remained independent, largely because its mountainous terrain repelled invaders. (It was briefly occupied by Italy, from 1935 to 1941.) Two of the region's states – Algeria and Sudan – are among the ten largest in the world, while Djibouti, on the Gulf of Aden, is among the smallest.

COUNTRIES IN THE REGION

Algeria, Chad, Djibouti, Egypt, Eritrea, Ethiopia, Libya, Mali, Mauritania, Morocco, Niger, Somalia, Sudan, Tunisia

Disputed borders Chad/Libya, Egypt/Libya, Mali/Burkina, Somalia/Ethiopia, Somalia/Kenya

Disputed territory Morocco/Polisario Front (Western Sahara)

STYLES OF GOVERNMENT

Republics All countries of the region except Morocco

Monarchy Morocco

Federal state (since 1991) Sudan

Multi-party states Chad, Egypt, Ethiopia, Mali, Mauritania, Morocco, Sudan, Tunisia

One-party states Algeria, Djibouti, Libya, Somalia

States without parties Niger

Military influence Algeria, Libya, Mauritania, Niger, Sudan

State without effective government (since 1991) Somalia

CONFLICTS (since 1945)

Coups Algeria 1965, 1992, Chad 1990, Egypt 1952, Ethiopia 1977, 1991, Mali 1968, 1991, Mauritania 1978, 1984, Niger 1974, Somalia 1969, Sudan 1969, 1985, 1989

Revolutions Ethiopia 1974–5, Libya 1969, Sudan 1969

Civil wars Chad 1965–87 (with French and USSR involvement), Ethiopia (with USSR intervention) 1962–91; Sudan 1956–72, 1987–, Somalia 1988–91

Independence wars Algeria/France 1952–62; Morocco/Polisario Front (with Algerian and Libyan involvement) 1975–

Interstate conflicts Egypt/Israel 1956, 1967, 1973; Egypt/UK, France 1956, Egypt (with UN force)/Iraq 1991; Ethiopia/Somalia (with Cuban and USSR intervention) 1977–78; Libya-Chad 1980–81; Libya/USA 1989; Mauritania/Senegal 1989

MEMBERSHIP OF INTERNATIONAL ORGANIZATIONS

Arab League Algeria, Djibouti, Egypt, Libya, Mauritania, Morocco, Somalia, Sudan, Tunisia

Organization for African Unity (OAU) All countries except Morocco

Organization of Petroleum Exporting Countries (OPEC) Algeria, Libya

A POSTCOLONIAL MOSAIC

Independence came to most of the states of Northern Africa in the two decades following World War II. Libya, formerly an Italian colony, came under British and French control in 1942. In 1951 it achieved independent status as the United Kingdom of Libya. Nationalist movements in the other Islamic states of Northern Africa were greatly encouraged by the overthrow of the pro-British Egyptian monarchy in 1953 and the subsequent rise to power of Gamal Abdel Nasser (1918–70). This resulted in 1956 in the final withdrawal of British troops from the Suez Canal zone, and brought forward the granting of independence to Sudan.

France divided its vast territories of French West Africa and French Equatorial Africa into 11 sovereign states in 1960. Four of these – Chad, Mali, Mauritania and Niger – extend across the Sahara. France held on much more determinedly to rule in Algeria. The temperate coastal zone had attracted nearly a million French and other European settlers, who owned about 40 percent of the cultivated land. The bitter fight for independence lasted from 1954 to 1962, and resulted in perhaps 1 million Algerian dead. 500,000 people were wounded, and some 2 million others uprooted from their homes.

The European imperial powers carved out their territories in Northern Africa without considering either the wishes of the people or the underlying political and geographical realities. In several parts of the region sovereignty still remains unresolved, and territorial and boundary disputes abound. In 1975 Spain withdrew abruptly from the Spanish Sahara, the last colony in Northern Africa. As Western Sahara, it was divided between Morocco and Mauritania. When the latter withdrew in 1979 the whole area was annexed by Morocco. Polisario guerrillas, opposing Moroccan rule, received backing from Algeria and Libya; Morocco constructed a defensive wall of rock and sand across the desert to repel guerrilla invasions. Attempts to agree a ceasefire have failed.

In Chad there has been almost continuous civil war since independence, with Muslim northerners struggling against domination by the central government. In 1973 Libya occupied 90,000 sq km (30,000 sq mi) of northern Chad, an area known as the Aouzou strip, believed to contain iron ore and uranium. This huge territorial gain was made on the basis of a treaty agreed between France and Italy in 1935 that was never implemented.

In Sudan animists and Christians in the south are fighting the Muslim central government for greater autonomy. In 1972 the three southern provinces (later split into six) were granted their own regional assemblies, but fighting resumed in 1983, with appalling consequences for the civilian population. In 1991 a federal system was introduced, dividing the country into nine states. In Djibouti a delicate political balance has to be maintained between the

Morocco's Green March In 1975 more than 350,000 unarmed Moroccans crossed the border to occupy the Western (Spanish) Sahara. The subsequent partition of the colony ignored local wishes for independence, leading to guerrilla war by the Polisario Front.

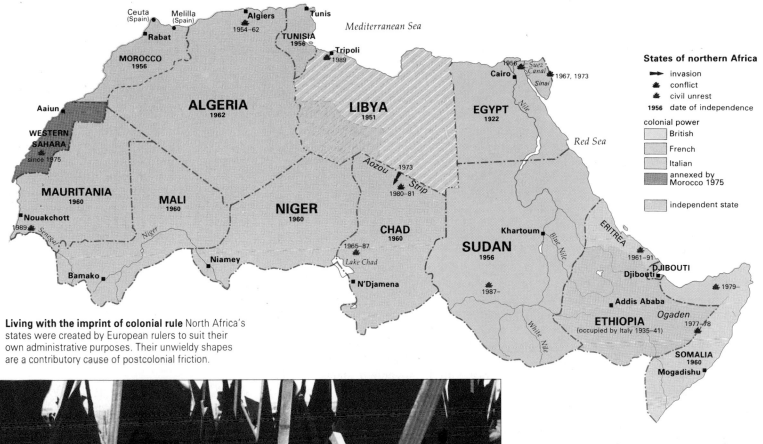

States of northern Africa

➤ invasion
♠ conflict
♠ civil unrest
1956 date of independence

colonial power
British
French
Italian
annexed by Morocco 1975
independent state

Living with the imprint of colonial rule North Africa's states were created by European rulers to suit their own administrative purposes. Their unwieldy shapes are a contributory cause of postcolonial friction.

majority Issa (Somali) tribe and the minority Afar (Danakil) nomadic tribe.

The troubled Horn of Africa

Conflicts in the Horn of Africa and adjacent areas, compounded by famines, have created millions of refugees and cost countless lives. Ethiopia, one of the oldest states in the world, as well as one of the poorest, has extraordinary ethnic diversity. Some 70 languages and more than 200 dialects are spoken by 40 ethnic groups. The northern province of Eritrea, forcibly annexed in 1962, fought a long war to secede, and in the late 1970s other groups also took up arms against the Soviet-backed communist regime, which had come into power after a military coup in 1974 deposed the emperor Haile Selassie (1892–1975). In 1991 the Eritrean People's Liberation Front captured the capital, Addis Ababa and overthrew Mengistu, the Ethiopian leader. This provided the opportunity for the largely Muslim Eritrea to seize independence. It was recognized as a separate state on 24 May 1993.

Somalia, created from British and Italian Somaliland, never accepted the delineation of the former colonial boundaries, which left large groups of Somali-speaking people in Djibouti, Kenya and Ethiopia. There was an armed border dispute with Kenya in 1963, and in 1978 Somalia took advantage of the civil war in

Ethiopia to claim the province of Ogaden, but was defeated after an eight-month war. In the late 1980s civil war broke out in the north, and in 1991 the Somali government was overthrown. The complete breakdown of political authority prevented much-needed food aid from reaching the victims of famine, and in 1992 United States and French troops were deployed to restore order.

The conflicts in the Horn of Africa and adjacent areas have created as many as 6 million refugees and have cost countless thousands of lives.

PROBLEMS OF CONTROL

Government of the states of Northern Africa has proved difficult in the years since independence. Ten states have experienced at least one military coup, and nine have suffered civil wars. Monarchies have been overthrown in Egypt (1953), Tunisia (1957), Libya (1969) and Ethiopia (1974). Written constitutions have frequently been abrogated or suspended, usually following military coups. In many cases, new constitutions

have subsequently been adopted; other states, such as Mauritania and Niger, are governed by military councils.

Most states have experimented with a variety of constitutional arrangements. Nearly all have some form of people's assembly or national assembly, with varying degrees of legislative power. The collapse of international communism and pressure from Western aid donors led many one-party states to recognize opposition parties in the early 1990s, but democracy is often very fragile. When the fundamentalist Islamic Salvation Front

THE POLITICS OF FAMINE

The series of disastrous famines that have caused great loss of life in parts of Northern Africa since the 1970s have been made even worse in their effects by political conflict. Drought over a period of years led to crop failure and deterioration of grazing in the Sahel region, which stretches through southern Mali, Niger, Chad and Sudan. In Ethiopia and Somalia the problems caused by population pressure as well as soil erosion compounded by drought, resulted in particularly severe famines.

International efforts to bring aid to the victims of famine are sometimes impeded by the actions of governments or separatist groups. In Somalia in 1992 shipments of aid were unable to leave the docks because of fighting between local warlords. In other cases aid is made conditional on certain political undertakings being given. For example, an Ethiopian scheme to compel 1.5 million people to move from northern Ethiopia to the south was suspended in 1986 after several Western agencies had cut famine relief in protest. National governments may restrict the amount of relief to particular groups. The post-1983 civil war in southern Sudan between the Sudan People's Liberation Army (SPLA) and government forces created desperate food shortages, but the people were often deprived of relief supplies to weaken SPLA support. Food aid was withheld from the rebellious Tigre and Eritrean provinces of Ethiopia in 1987 and 1988.

Armed conflict has a more direct effect on the victims of famine by destroying crops and interfering with the transport of supplies. International efforts to find longterm solutions to the problem of famine, for example by locust control, have been impeded by continuing civil wars in the region.

The plight of Africa's homeless A refugee camp in Sudan. Scenes like this have raised world concern for the millions who have been displaced by war or famine, but finding solutions is not easy.

Egypt's presidential system of government The 1971 constitution places great power in the hands of the president, who is nominated by the people's assembly and confirmed by plebiscite for a six-year term. He is supreme commander of the armed forces, and may appoint one or more vice presidents; he also appoints both the prime minister and the cabinet.

(FIS) won the first round of Algeria's first multi-party elections in 1991, the second round was canceled and the FIS ordered to disband.

Contrasting forms of government

At opposite ends of the political spectrum, and with strongly contrasting forms of government, are Morocco and Libya. Morocco is the last of North Africa's traditional monarchies – the present king, Hassan II, succeeded his father in 1961 as heir to a thousand-year-old throne. Some limited forms of constitutional government have been introduced in recent years, and legislative power is held by an elected chamber of representatives. Executive political power, however, remains heavily concentrated in the hands of the king, who appoints the prime minister and cabinet, as well as the governors of Morocco's 49 provinces.

Since 1977 Libya has been governed as a "socialist" state. It has a constitution that is designed to allow every adult to share in policy-making through some 2,000 People's Congresses, which appoint popular committees to execute policy. The General People's Congress, at the center of the structure, decides policy at national level. Executive power is exercised by the General People's Committee. This consists of 18 secretaries who are at the head of 18 departments. In practice, however, a considerable degree of power remains with Colonel Muammar Qaddafi, who was the leader of the revolution that deposed King Idris in 1969.

Strong regional administration based on provinces is a common feature of

ELECTORATE

confirmed by plebisite

elects ⅔

elects

President

Shura council

appoints ⅓

nominates

People's assembly

appoints

appoints

Vice president

appoints

Prime minister

Council of ministers

internal government in North Africa. In the past, provincial governors exercised considerable local authority, and many still have an extensive administrative role. Provinces are subdivided, sometimes into an elaborate hierarchy of administrative units, particularly in the most densely populated regions.

Internal communications

In some states provincial governors are appointed by the state president, reinforcing direct central control. Nonetheless, internal cohesion remains the greatest problem confronting all Northern African governments. The European administrators who carved up huge tracts of land without regard for existing populations created awkwardly shaped states (Mali and Somalia, for example), which has made effective central administration even more difficult.

Every state, except Djibouti and Tunisia, contains extensive areas of empty or sparsely populated land, which adds to problems of communication and transport. This is particularly true in the Saharan areas of Algeria, Chad, Mali, Mauritania and Niger. Libya has invested heavily to improve internal communications, including air, road, radio and telecommunications, in an effort to foster national unity.

Many of the region's crowded capital cities owe their importance to the past, when they were the centers of colonial administration and (in the case of coastal states) major ports providing links with imperial Europe. Today they may not be geographically well placed to serve as centers of government, but have acquired overwhelming dominance in population size and in economic and political power. Libya's capital city has switched twice between Benghazi and Tripoli, the capitals of the colonial provinces of Cyrenaica and Tripolitania. There are now plans to build a new national capital at Sirte, halfway between the two.

Egypt is a notable exception to these problems of government. Areas of settlement have for a long time been linked by a network of roads and railways as well as by the river Nile. The capital, Cairo, is located almost at the population center of gravity. Throughout the centuries Egypt has generally enjoyed effective rule and a genuine sense of national identity, advantages that cannot be found so markedly in any other state in Northern Africa.

NEIGHBORS AT ARM'S LENGTH

It might be expected that a fair measure of cooperation and political harmony would exist between the states of Northern Africa. All but Chad, Ethiopia, Mali and Niger are members of the Arab League, the organization of Arab states established in 1945. European colonialism (British, French, Italian) is a shared experience. They all face development problems associated with postcolonial dependency and problems of aridity. Nonetheless, there are marked differences of interest and outlook that work against the establishment of closer regional ties.

The landlocked states of Chad, Mali and Niger look to west Africa for their routes to the sea; Mali has signed an agreement for eventual economic and political integration with neighboring Guinea. Along the Mediterranean coast Algeria, Morocco and Tunisia, which have all been profoundly affected by European history and culture, have cooperation agreements with the European Community (1976). Egypt has been a major contender for leadership of the predominantly Middle Eastern Arab world. Ethiopia's interests lie both within the turbulent Horn of Africa and toward the Red Sea region.

Barriers to unification

The Maghreb – a term meaning "west" in Arabic, and usually applied to Algeria, Morocco and Tunisia – is often regarded as a distinct region of Northern Africa. Geographical coherence, a common history and political maturity seem to provide a good basis for cooperation, even for some kind of federal unification. However, regional differences exist within the states themselves, as well as contrasting political regimes. In spite of much talk – and some action – the states remain at arm's length.

The most energetic attempts to promote unification in North Africa have been initiated by Libya. There have been plans at various times to unite the country with Algeria, Chad, Egypt, Morocco, Sudan and Tunisia, but these have all come to nothing. Political differences apart, sheer distance across sparsely populated or empty desert is a major geographical barrier to unification.

There are relatively few incentives for regional trade and technical cooperation, except between neighbors. At times rela-

EGYPT IN THE ARAB WORLD

Under Gamal Nasser's charismatic leadership (prime minister 1954–56, president 1956–70), Egypt assumed political direction of the Arab world – a role that its central position between the Islamic states of the Middle East and those of North Africa enabled it to play. Nationalization of the Suez Canal in 1956 was to bring Nasser widespread admiration, and his reputation was further enhanced by the completion of the Aswan High Dam project, funded by the Soviet Union after the withdrawal of Western aid.

Nasser became a symbol of Arab nationalist aspirations. But his efforts to create political unity within the Arab world were largely unsuccessful. The United Arab Republic of Egypt and Syria lasted only from 1958 to 1961. Even defeat in the 1967 war with Israel scarcely dented Nasser's reputation. Outside the Arab world he achieved international status as a spokesman for the Third World. With Prime Minister Jawaharlal Nehru of India and President Tito of Yugoslavia he was responsible for setting up the nonaligned movement in 1961.

Nasser's successor, Anwar Sadat (1918–81), sought friendship with the West. He expelled Nasser's Soviet advisors, and launched an attack on Israeli forces along the Suez Canal in an attempt to restore Egypt's military prestige. Spiraling military costs eventually led him to seek better relations with Israel. The peace agreement signed in 1979 with Israel's prime minister, Menachim Begin caused anger throughout the Arab world. Egypt was suspended from the Arab League between 1979 and 1988, and became increasingly isolated from its neighbors. Sadat was assassinated by Muslim fundamentalists in October 1981. Under his successor, Hosni Mubarek, Egypt's influence began to be felt once again in the Arab world. In 1991 Egypt – which sent a force to fight in the Gulf war against Iraq – played a major role in convening Middle East peace talks.

Extreme passions A poster bitterly condemns Anwar Sadat's pro-Israeli and pro-United States policies after the 1973 Arab–Israeli war. The 1979 Egyptian–Israeli peace, which arranged for Israel's phased withdrawal from the Sinai peninsula, held since 1967, was welcomed throughout the world. President Sadat and Israel's premier Menachim Begin shared the 1978 Nobel Peace Prize for negotiating its terms. But the peace was denounced by nearly every Arab government. As a protest, the Arab League headquarters were moved from Cairo to Tunis.

Striking a blow for Arab nationalism President Nasser in 1956 announces his intention to nationalize the Suez Canal, owned by an Anglo-French company. The United States refused to lend support to an abortive invasion plan by Britain, France and Israel to regain control of the canal. Nasser's bold action not only won him enormous standing among Arabs, but also gave inspiration to nationalist groups throughout the world.

tions between neighboring states have been strained, resulting in closed frontiers, expulsion of nationals, and acts of sabotage and subversion. Disputed boundaries and territorial sovereignty are a frequent source of trouble within the region, and flashpoints are many.

Algeria and Libya temporarily supported Polisario claims to an independent Western Sahara. Libyan forces have operated within Chad to support the Muslim northerners in the long-running civil war. In 1989 ancient resentments between the Muslims of Mauritania and

black Africans of Senegal exploded into outbreaks of violence. Offshore boundaries have also caused international disputes. Rival Libyan and Tunisian sea-bed claims were settled only after prolonged wrangling and an International Court ruling in 1982. Libya's controversial claim to the waters of the Gulf of Sirte led to armed clashes with the United States navy at least twice, in 1981 and 1989.

Superpower involvement

Proximity to some of the world's vital strategic waterways – the Bab-el-Mandeb at the entrance to the Red Sea (Djibouti, Ethiopia), the Cape route (Somalia), the Strait of Gibraltar (Morocco) and the Suez Canal (Egypt) – draws the region into the arena of global politics and confrontation. Regional rivalries were frequently exploited during the Cold War, particularly in the Horn of Africa, where Cuban and Soviet forces and equipment helped Ethiopia to reclaim the Ogaden region from Somalia in 1978. Until the late 1980s, Soviet aid was forthcoming to help Ethiopia to retain control over the province of Eritrea, where Muslim secessionist groups were keenly supported by several of the oil-rich Gulf states.

Egypt's stand against Israel has four times involved it in armed conflict (1948, 1956, 1967 and 1973) and loss of territory, though the Sinai, occupied by Israeli forces in 1973, was peacefully restored in 1982. Libya's erratic foreign policy aims have heightened international tension. In a region of such potential and actual conflict, where superpowers competed for patronage, arms procurement was easy. Defense spending in relation to income was high: in the late 1980s Northern African states accounted for about a tenth of the world's arms imports.

The regimes of Ethiopia and Libya both had close ties with the Eastern bloc. Other states in the region are linked both economically and politically with the West; France retains a military presence in Chad and Djibouti. However, Egypt, Ethiopia, Mali, Somalia and Sudan all switched alliances in the past. Algeria pursued a policy of nonalignment. Its experience in one of the most bitter independence wars in recent colonial history fundamentally affected its political outlook, creating a determination to avoid external domination again, and it achieved high international standing, particularly in the Third World.

The Libyan factor

Since Colonel Muammar Qaddafi seized power in 1969, Libya's foreign policy has appeared erratic and contradictory. Despite lack of manpower, in relation to its size it has exerted a disproportionate influence on international affairs. This has been achieved by a program of high military spending and lavish financial support to various political causes and groups abroad. Libyan spending on weapons in the last two decades has been among the highest in the world, roughly equaling that of all the other states of Northern Africa put together.

Certain themes run through Qaddafi's seemingly gadfly behavior in foreign affairs. It was a commitment to nonalignment that resulted in the closing of United States and British bases in Libya in 1970; however, following a huge arms deal in 1975, relations with the Soviet Union became very close. Passionate support for Arab unity led to several abortive proposals for federation with other Arab states – Egypt, Morocco, Sudan (twice), Syria (five times) and Tunisia. Qaddafi's determination to promote Arab unity has sometimes led to the provision of Libyan financial and military assistance in some unexpected places, as diverse as the Philippines and Uganda.

As the leader of a state dominated by foreign rule for many centuries, Qaddafi is deeply committed to bringing about the end of colonial rule everywhere. Libya has allegedly supplied funds, weapons and military advice to liberation movements throughout Africa: in Burkina, Chad, Liberia, Sudan and Uganda. Until 1982 significant support was given to the Polisario Front in Western Sahara in its struggle with Morocco.

Farther from home, such support extends to extremist groups whose activities are regarded as terrorist by those to whom they are opposed. These include the Irish Republican Army (IRA), which acquired arms from Libya for campaigns against Britain in Ulster, and the Palestine Liberation Organization (PLO). The plight of the Palestinians is of particular concern to Qaddafi, who advocates the ultimate military defeat of Israel. The Egyptian–Israeli peace agreement of 1979 left him bitterly disillusioned.

Qaddafi has come to see Libya as the champion of all the world's oppressed peoples. Libyan interference has been reported in the internal affairs of states as far away as New Caledonia and Vanuatu

(in the Pacific), Nicaragua and Surinam (in Central and South America). In the mid-1980s Libyan agents were suspected of involvement in a series of terror bomb attacks in Europe. Although not all the activities attributed to Libyans can be proved, Libya's behavior is widely distrusted in the West. The implication of two Libyans in the midair bombing of a transatlantic flight in December 1988, in which 270 people died, brought mounting international pressure.

A number of governments have ceased diplomatic relations with Libya, and its

activities have twice provoked fierce raids on its coastal towns, the first from Egypt in 1977. The second took place in 1986, when the United States bombed Qaddafi's headquarters in Tripoli in retaliation for his alleged complicity in terrorist activities against United States' personnel in Europe. The United States' navy also shot down two Libyan jets over international waters in 1989, when feelings ran high over a factory said to be producing chemicals for military use at Rabta, 60 km (37 mi) from Tripoli.

Libya's policies have undoubtedly

created much instability and confusion, particularly in Africa where its activities have been widespread and persistent. Their effect may not be so significant. From a Libyan viewpoint few of its foreign ventures, particularly those with its neighbors, appear to have been successful but in a regime where dissent is not tolerated, and opponents living in exile in several countries have been hunted down and murdered, opposition cannot be openly expressed. The most unpredictable factor in Libya's affairs remains Qaddafi himself.

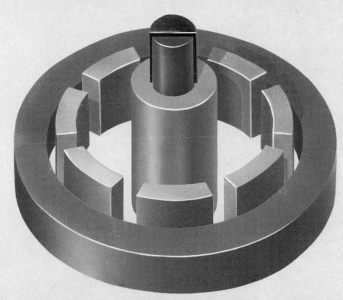

Components of government

- ■ General Secretariat of General People's Congress
- ■ General People's Committee
- ☐ General People's Congress
- ■ people's committees
- ■ basic popular congress

Libya's system of government extends through interlocking rings of popular congresses and people's committees. It claims to find its inspiration in the extended family organization of the nomadic Bedouin people of the desert, from whom Qaddafi comes.

Leader of the Revolution Muammar Qaddafi in thoughtful mood. In 1969 he led the military coup that overthrew King Idris and established the Libyan Arab republic. His world vision of pan-Arab union has been the guide for many of his policies.

An impressive show of military strength Libya's lavish spending on weapons, and its support of other Muslim states and groups abroad, are made possible by its extensive oil wealth. Prosperity undoubtedly stifles criticism of Qaddafi's regime at home.

POSTCOLONIAL QUANDARIES

The territorial divisions of this huge region go back barely a century, when Africa was carved up by Europe's competing empire builders. In the centuries before this, control of the trade routes across the Sahara to the Mediterranean and beyond created a succession of powerful states in the region. Their economic and political stability was increasingly disrupted by the growth of Europe's transatlantic trade in slaves from west Africa, which reached a peak in the 18th century. European colonization remained confined to a number of small trading stations along the coasts of both west and east Africa until, in a short period between about 1880 and 1914, the entire region was seized and rapidly partitioned in a race for possession between Belgium, Britain, France, Germany, Portugal and Spain.

COUNTRIES IN THE REGION

Benin, Burkina, Burundi, Cameroon, Cape Verde, Central African Republic, Congo, Equatorial Guinea, Gabon, Gambia, Ghana, Guinea, Guinea-Bissau, Ivory Coast, Kenya, Liberia, Nigeria, Rwanda, São Tomé and Príncipe, Senegal, Seychelles, Sierra Leone, Tanzania, Togo, Uganda, Zaire

Dependencies of other states   British Indian Ocean Territory (UK)

STYLES OF GOVERNMENT

Republics   All countries in the region

Federal state   Nigeria

Multi-party states   Benin, Burkina, Cameroon, Cape Verde, Congo, Gambia, Liberia, Nigeria, São Tomé and Príncipe, Senegal, Sierra Leone, Togo, Uganda, Zaire

One-party states   Burundi, Central African Republic, Equatorial Guinea, Gabon, Guinea-Bissau, Ivory Coast, Kenya, Rwanda, Seychelles, Tanzania

States without parties   Ghana, Guinea

Military influence   Burundi, Equatorial Guinea, Ghana, Guinea, Guinea-Bissau, Liberia, Nigeria, Sierra Leone, Togo

---

CONFLICTS   (since 1945)

Coups   Benin 1960–73 (6 coups); Burkina 1966, 1980–87 (4 coups); Burundi 1966, 1976, 1987; Central African Republic 1965, 1979, 1981; Congo 1968, 1977; Equatorial Guinea 1979; Ghana 1966, 1972, 1978, 1981; Guinea 1984; Guinea-Bissau 1980; Liberia 1980, 1990; Nigeria 1966, 1975, 1983, 1985; Rwanda 1973; Sierra Leone 1967, 1992; Togo 1963, 1967; Uganda 1966, 1971, 1985

Civil wars   Burundi 1972, 1988; Liberia 1990–; Nigeria 1967–70; Rwanda 1962–65; Uganda 1978–79, 1985–86; Zaire 1960–63; Rwanda 1993–

Independence wars   Kenya/UK 1952–59; Cameroon/France 1955–60; Congo (Zaire)/Belgium 1959; Guinea-Bissau/Portugal 1956–74

MEMBERSHIP OF INTERNATIONAL ORGANIZATIONS

Economic Community of West African States (ECOWAS)
Benin, Burkina, Cape Verde, Gambia, Ghana, Guinea, Guinea-Bissau, Ivory Coast, Liberia, Nigeria, Senegal, Sierra Leone, Togo

THE LEGACY OF EUROPEAN RULE

The territorial states of colonial Africa were completely artificial constructs, which reflected only the trading interests of the dominating European powers and their competition for markets and raw materials. State boundaries were drawn without any regard to the people living within them, ignoring precolonial nationalities and cutting across linguistic areas. A deliberate policy of "divide and rule", pursued to weaken potential opposition, sowed the seeds of future disunity and communal diversity.

Only Liberia was free of colonial rule. It had been settled by freed slaves from the United States after 1817, and was established as an independent republic in 1847. Its True Whig party was, until 1980, the longest continuously ruling party in the world, with 102 years in power.

The struggle for independence

The impetus for independence from colonial rule grew rapidly in central Africa after World War II, as it did in other parts of the world. African troops had fought

with the armies of Britain and France for freedom and democracy in Europe, yet they lived under authoritarian and racist colonial rule, and this irony was not lost on them. Neither the Soviet Union nor the United States, who emerged in the postwar years as the new world superpowers, had anything to gain by defending European rule in Africa. They both encouraged in their own way the rising tide of African nationalism.

Articulate Africa leaders such as Kwame Nkrumah (1910–72) in British-ruled Ghana (then the Gold Coast) and Jomo Kenyatta (1889–1978) in Kenya emerged to spearhead the nationalist movements. Ghana, which achieved its independence in 1957 following a brief period of internal self-government from 1954, provided the model for decoloniza-

tion in the remaining British colonies in west Africa (Gambia, Nigeria and Sierra Leone) between 1960 and 1965.

A complicating factor in Britain's colonies in east Africa (Kenya, Tanganyika, Uganda) was the presence of well-established white settlers (numbering 40,000 in Kenya, the largest settlement) with their own aspirations for political ascendancy. This led to revolt (the Mau Mau insurrection of 1952–56) and delayed the granting of independence in Kenya.

The flood of nationalism proved unstoppable. France soon followed Britain's lead, and in 1960 granted independence to its extensive African territories, creating six states in west Africa – Benin (then called Dahomey), Burkina (then Upper Volta), Guinea, Ivory Coast, Senegal, Togo – and four in central Africa – Cameroon, Central African Republic, Congo and Gabon. In the same year Belgium decided to decolonize and, almost overnight and virtually without preparation, gave the vast central African

Senegalese soldiers stand ceremonial guard outside the parliament building in Dakar. Senegal's first president, who led the country to independence, was the poet Léopold Senghor. His espousal of black African culture and values won wide acclaim.

The roads to independence Central Africa experienced some of the most peaceful, as well as some of the bloodiest, passages to independent statehood as the rising tide of nationalism throughout Africa resulted in the region's rapid decolonization.

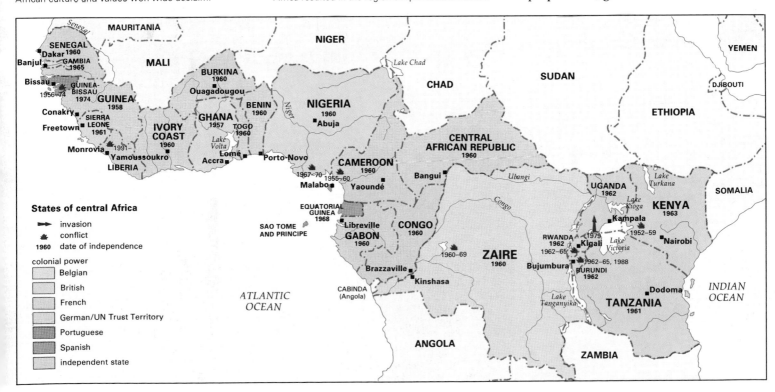

States of central Africa

→ invasion
⚔ conflict
1960 date of independence

colonial power
- Belgian
- British
- French
- German/UN Trust Territory
- Portuguese
- Spanish
- independent state

A new Napoleon Emulating his French hero, Jean-Bédel Bokassa is crowned emperor of the Central African Republic, renamed the Central African Empire, in 1977. His reign was costly and brutal; he was removed in a coup less than two years later.

Interrupted elections Voters at a polling booth in Nigeria during the 1983 elections. After an army coup later that year, a military council took over, but a long-promised return to elections was made in 1992.

colony of the Congo its independence: the result was a bloody civil war to prevent the breakaway of the province of Katanga (1960–63). Portugal alone resisted decolonization for more than a decade despite mounting guerrilla war, until revolution at home brought hard-won independence for Guinea-Bissau and for the offshore islands of Cape Verde and São Tomé and Príncipe in 1974–75.

The task of nation-building

Most of the newly independent states included peoples of different ethnic, linguistic or religious groups: Zaire (the Belgian Congo) had 150 major ethnic groups speaking over fifty languages. Opposition to European rule had been sufficient to unite such disparate elements during the independence struggle, but afterward it could not prevent a jockeying for power among members of contending political elites, who often engineered their way to the top by distributing wealth and influence to supporters coming from the same ethnic, regional or religious groups.

The first task on independence was to overcome this communal diversity by finding ways of creating national cohesion. Many in power soon felt that the

federal systems bequeathed to them in constitutions drawn up by colonial administrators would hinder, rather than help, these efforts, and feared that federalism would foster the wish among ethnically or linguistically defined groups for some form of separate national identity.

Consequently, the federal components were swiftly removed from the constitutions of Ghana and Kenya, and the federal relationship of Buganda with the rest of Uganda was forcibly ended in 1966. The loose federal arrangements that had been set up among the former colonies of French West Africa and French Equatorial Africa soon collapsed.

In the huge country of Nigeria the federal structure that was inherited from the British recognized the major ethnic divisions between the Yoruba people in the west, the majority Hausa-Fulani in the north, and the Ibo in the east. This exacerbated regional tensions, and in the late 1960s a bloody separatist war broke out when the Ibo people tried to establish a separate independent state of their own (Biafra). Nonetheless, Nigeria has not abandoned federalism, but has instead sought to diffuse the potential for conflict by creating many more smaller states within the federation.

BREAKING THE CONSTITUTIONAL MOLD

Colonial rule provided no genuine preparation for democratic parliamentary government. It bequeathed constitutions modeled on the British or French systems, but these proved unworkable without stable economic foundations. All the states of the central African region suffer from weak economies, and this has made the task of government and of nation-building more difficult. Scarce resources make it hard to create the complex political systems, as well as the health and educational services, transport and communications, that all help to build up national integration. Little wonder, therefore, that after independence multiparty government collapsed in many countries, giving way to a pattern of alternating military and one-party rule with frequent coups and counter-coups. Benin, for example, had six coups in 13 years, and Burkina four in seven.

It has been argued that authoritarian one-party rule was the only way to end factionalism and achieve national unity, and that it reflected the basic consensus of African society; others saw it as a legacy

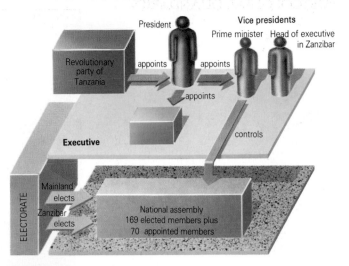

Tanzania's 1977 constitution vests sole authority in the Revolutionary Party (CCM), and executive power in a president who is also party head. The national assembly has elected and appointed members, a fixed number of whom represent Zanzibar.

Julius Nyerere, president from 1962 until 1985, devised Tanzania's distinctive form of socialism and was an advocate of African unity.

of colonial rule. Often the party was used by strong leaders as the means to attain power. It was then allowed to wither away as the leadership consolidated its hold on power through control of the machinery of state. Across the region, in Senegal, Ivory Coast, Zaire, Kenya and many other states, party rule was replaced by presidential rule.

The early 1990s saw a return to multi-party government in many central African countries, brought about by the withdrawal of Soviet aid for socialist-dominated governments and by growing financial pressure from Western countries. Even the ruling socialist Revolutionary Party of Tanzania (CCM), which had provided stable one-party rule since 1965, agreed in 1992 to recognize opposition groups outside the party.

In some cases, notably Benin, the transition to multi-party rule was peaceable, but often only the semblance of democracy was introduced; promised elections were frequently delayed, and when they did take place the re-election of the ruling party gave rise to accusations of manipulation and cheating. In a few countries, for example Togo, strong military influence continued to be exerted within the framework of multi-party rule; in

others such as Ghana and Guinea the army continued to rule without political parties.

While economic weakness perhaps made it inevitable that political instability would characterize the early independence period, there are states where the military has never intervened in government, such as the Ivory Coast and Senegal, and others – Gabon (1962), Gambia (1981), Kenya (1982) and Cameroon (1984) – where coups have been attempted, but failed. Military support from the former colonial power improves stability. Another important ingredient for stability is the skill of the political leadership. Factors that may lead to an army takeover are a blatantly corrupt regime, serious economic crisis, personal ambition and factional rivalry. This last led in 1990 to the execution of President Samuel Doe in Liberia, who had himself overthrown the government in 1980. The continuing civil war led to the intervention of a west African peacekeeping force in 1991.

The greatest test of a regime's stability arises when the incumbent leader dies. In far more cases than not the succession is orderly: between 1971 and 1988 there were 15 peaceful transitions of leadership in the region. Only once, in Guinea in

TANZANIA: A SOCIALIST PROGRAM IN ACTION

Pragmatic self-reliance became the guiding principle of Tanzania's brand of socialism in 1967. "Let others go to the moon," Tanzania's first president, Julius Nyerere, said. "We must work to feed ourselves." He advocated rural cooperatives, called *ujamaa* in Swahili ("familyhood"), as a framework for self-help in the battle against poverty, ignorance and disease. Many Tanzanians opposed the *ujamaa*, which took away their economic independence, and the government resorted to the forced resettlement of urban slum dwellers to fill their model communities. The cooperative program was eventually abandoned in 1976.

Despite this setback, Tanzania has continued to pursue its socialist ideals, both by public ownership of large enterprises and by giving priority to the improvement of social services. Tanzania's achievements in the field of education and health are a measure for other African nations – literacy rates are 30 percent higher than the African average and child mortality rates 10 percent lower.

Tanzania's approach to health care is particularly innovative. District nurses, not doctors, are the foundation of the medical system. After basic training they dispense preventive medicine, provide health education and make diagnoses using simple manuals. This ensures that Tanzanians in remote villages have access to modern medicine, despite an acute shortage of doctors.

Tanzanian socialism could not, however, withstand the economic pressures of the early 1980s. The failure of agricultural commodity prices to keep up with the domestic rate of inflation meant that there was little incentive for Tanzanians to produce more. Farming slipped back toward subsistence agriculture, and industry stagnated. In 1984 Tanzania reluctantly agreed to reduce the state's role in production and to increase opportunities for free enterprise in order to secure foreign aid from Western donors.

1984, did the army move in to take power on the death of a leader, though in a handful of other cases the succeeding leader remained in power only briefly before being removed by the army. All this suggests that the charge of political instability in central Africa is perhaps exaggerated, and that the political structures are being established that will diminish it still further.

These dramas of national politics are, in any case, generally confined to the capital cities and major provincial centers. Only occasionally do they affect the lives of the mass of the population, who live in rural areas. Idi Amin's genocidal regime in Uganda (1971–79) was a tragic example. Such instances should be set beside the many positive examples of government action reaching out to change people's lives. In Cape Verde, for example, tens of thousands of people have been involved in tree-planting schemes to tackle the problems of drought and soil erosion. Tanzania has achieved one of the best public health services in Africa, and a universal primary school system.

The shattered portrait of Idi Amin, Uganda's dictatorial ruler, lies among the debris left by Tanzanian troops who gave military backing to an internal movement to oust him from power in 1979.

Central Africa's French-speaking states retain close links with France. President Mitterrand's portrait is paraded during a visit to Burkina. French troops have sometimes intervened in postcolonial disputes.

NEW PARTNERS OR PAST MASTERS?

The colonial exploitation of Africa's mineral and agricultural resources to feed European trading and industrial needs meant that, on independence, the new states were left with fragile economies that were dependent at best on a handful of export products, at worst on a single product. Africa's leaders were pulled in two directions. Economic dependency and the ties of history continued to link them to the former colonizing power; but the awareness of a shared African past and culture looked for common areas of agreement nearer home.

Pan-Africanism, which had the political aim of a united Africa, was championed by the pioneers of African nationalism, such as Kwame Nkrumah of Ghana, as the only alternative to continuing economic and cultural subordination to Europe. Political independence without economic independence only created a new form of colonialism. The Pan-African idea lives on in the shape of the Organization of African Unity (OAU), formed in 1963; it is perhaps honored more in the word than the deed.

Links between France and its former colonies have remained particularly strong: on independence, only Guinea chose to sever all connections, after a referendum held in 1958. In part, this is the legacy of the strong centralized rule from Paris practiced by France's colonial administrators. France continues to give extensive military and other aid to its former colonies. An annual conference of Franco-African heads of state maintains these links more informally. A similar role is played by the biennial meetings of Commonwealth leaders, attended by the heads of Britain's former colonies.

Closer associations

The OAU has taken the view that any attempt to redefine the territorial boundaries imposed by colonial rule, however unsatisfactory these are, would lead to chaos. Nonetheless a number of attempts have been made within the region to reshape territories, the most successful of which were the union of Ghana and British Togoland in 1957, of the (British) Southern Cameroons and the (French) republic of Cameroon in 1961, and of Tanganyika and Zanzibar (Tanzania) in

1964. In 1981 a confederal arrangement between Senegal and Gambia (Senegambia) was agreed, but by the end of the decade had been abandoned.

There have been other failures. These include the proposed federation of Ghana, Guinea and Mali in west Africa, and the East African Community (EAC) of Kenya, Tanzania and Uganda formed in 1967 and dissolved 10 years later. An attempt was also made to unite the two former Portuguese colonies of Cape Verde and Guinea-Bissau, which finally broke down in 1980.

In the case of the EAC, breakdown was prompted by the emergence of strong political and ideological differences between each of the states involved. Relations between Tanzania's socialist government and the more capitalist policies of Kenya became strained, and after the rise to power of Idi Amin in Uganda Tanzania's President Julius Nyerere refused all contact with his bloody regime, giving military support to the internal armed opposition that overthrew him in 1979. By 1992, such political differences appeared to be resolved, and talks began to revive the EAC.

Developing strategies for the future

More successful than these efforts at political unification have been moves within the region toward technical and economic cooperation. Several groups have been formed among the French-speaking states, including the Economic Community of West Africa and the Economic Community of Central Africa. Worldwide, many former colonial states have entered into collective agreement with the European Community (EC) through the convention signed in Lomé, in Togo, in 1975. Signatories included 46 African, Caribbean and Pacific states; the number subsequently rose to 66.

It was this development that provided the impetus for the formation of the Economic Community of West African States (ECOWAS). Its 16 member states – which are both French- and English-speaking – are pledged to work toward the free movement of goods and people between them. Progress, however, has been halting, particularly after Nigeria's expulsion of almost two million non-nationals in 1983.

The Lagos Plan of Action devised by the OAU in 1980 envisaged the creation of a single Pan-African common market by the year 2000. Such a development remains a far-off dream while Africa's economies remain dependent on the economies of the industrialized nations. Effective forms of regional cooperation have to be built gradually and upon firm foundations if they are successfully to offer the states of Central Africa a long-term hope of overcoming the problems inherited from their colonial past.

TOWARD A UNITED AFRICA?

The Organization of African Unity (OAU) was established in 1963 with the aim of eradicating colonialism, and of building up economic, cultural and political cooperation throughout Africa. It has provided a forum for discussion, given limited support to liberation movements, and helped resolve a few at least of the disputes that have emerged between member states. It has also served as a useful umbrella under which multilateral agencies such as the United Nations are able to work.

By and large, however, it has not proved particularly effective in achieving its foremost aims of African unity. One reason for this is that, though founded on the principle of nonalignment, a number of member states remain militarily aligned, in particular the French-speaking states. French troops have been dispatched to a number of states in the region. Kenya has offered military facilities to the

United States, and China and the Soviet Union in the past gave support to a number of African states.

Adherence to the principle of non-interference in the internal affairs of member states means that the OAU has failed to speak out against abuse of human rights where it has occurred. For example, the massacres in Burundi by the ruling Tutsi minority in 1972, and the political murders carried out by the regimes of Jean-Bédel Bokassa in the Central African Empire (1965–79), of Francisco Macias Nguema in Equatorial Guinea (1976–77), and of Idi Amin in Uganda (1971–79) went uncondemned. Its members were severely split over what action to take during Biafra's attempt to secede from Nigeria. As a result of this and other differences, the authority and influence of the OAU has declined considerably since its inception, and the prospects for African unity seem as remote as ever.

Kenya: the path to one-party rule

The death in 1978 of Jomo Kenyatta, Kenya's first prime minister and its president from 1964, made headline news around the world; his funeral in Nairobi was attended by many heads of state. Kenyatta had brought considerable political stability to Kenya. It was one of the few countries in the region to have avoided an overthrow of its civilian government by the army, and to hold regular elections: the succession to power of Vice President Daniel arap Moi was orderly. But stability had been achieved, at the cost of an increasingly presidential style of government that allowed little room for dissent.

Kenya's independence was gained in the aftermath of a fierce armed struggle, fought mainly by the Kikuyu people against white settler rule. More than 50,000 British troops were flown in to combat the Mau Mau uprising in 1952, and by the end of the emergency 13,000 Africans had been killed and over 80,000 held in detention. Though it had been crushed, the resentment of the landless Kikuyu lived on, and the Kenya African National Union (KANU), led by Jomo Kenyatta, was swept to power at independence in 1963 by harnessing this widespread popular feeling.

Within a year Kenya's other nationalist party, the Kenya African Democratic Party (KADU), founded to protect the

Jomo Kenyatta's powerful personality dominated Kenyan politics for nearly half a century. His leadership provided stability and sustained economic growth, but the opposite side of the coin was the consolidation of one-party rule and the banning of dissent.

Huge crowds in Nairobi gather to celebrate Kenyatta Day on 20 October. Kenya's young, fast-growing population is among the largest in central Africa.

interests of the minority communities against Kikuyu domination, had amalgamated with KANU. However, divisions soon appeared within the ruling party. A radical group led by the vice president, Oginga Odinga, criticized the slow pace of land resettlement and economic reform. In 1966 Odinga broke away to form the more radical Kenya People's Union (KPU), whose support lay among the second largest ethnic group, the Luo. In 1969, however, Odinga was arrested, and the KPU was banned.

Kenya was now effectively a one-party state. Elections were held in 1969, 1974 and 1979, but they allowed no scope to replace either the president or the ruling party. In the early 1980s, after arap Moi's succession, dissent increased. There was an attempted coup in 1982, which was led by members of the air force and supported by university students; it was heavily suppressed, and later that year the national assembly declared Kenya officially a one-party state.

Elections within a one-party state

Limited democracy existed within the Kenyan system. Periodic elections to the national assembly allowed a channel for dissent by enabling voters to remove incumbent members. In the 1974 elections, for example, 737 KANU candidates contested 158 seats; 88 of the sitting members were defeated.

A secret ballot, used in general elections, was not a requirement at the preliminary polls, which determined the party candidates to go forward for election. The churches voiced their opposition to the intimidatory procedure used instead, which compelled voters to line up publicly in front of the candidate of their choice.

In reality, the national assembly had no effective power and acted simply as an advisory body to the president. Political opponents were bought off through a system of patronage. Half the elected members of the national assembly held ministerial appointments. There was rigid police control, and opposition was illegal. Dissenters were frequently detained without trial. However, antigovernment riots in the early 1990s spearheaded demand for political reform, at first resisted by Daniel arap Moi. Later a return to multiparty politics and reforms were announced by arap Moi, with elections to be held in 1992.

POLITICS IN BLACK AND WHITE

PATHS TO INDEPENDENCE · DIVERSE POLITICAL SYSTEMS · FRONTLINE STATES

The first European settlers in southern Africa were Dutch and Portuguese traders, who established coastal stations in the 16th and 17th centuries to provision ships on their way to Asia. After 1806 the British took control of the area around the Cape, and the original Dutch settlers (Afrikaners or Boers) began to move northward, most notably in the exodus known as the Great Trek (1831–38). They came into increasing conflict with the indigenous African peoples, particularly the Nguni people in what is today Natal, who were reorganizing themselves into a powerful militarized nation. The success of the Zulu clan of the Nguni pushed a number of defeated clans northward, and a disruptive and chaotic period (the Mfecane wars) preceded the Afrikaner, and later the British, advance from the Cape.

The inauguration of President Mandela in May 1994 heralded a new era of multiracial government. Addressing an international audience of millions he announced, "The time for healing of wounds has begun".

COUNTRIES IN THE REGION

COUNTRIES IN THE REGION

Angola, Botswana, Comoros, Lesotho, Madagascar, Malawi, Mauritius, Mozambique, Namibia, South Africa, Swaziland, Zambia, Zimbabwe

Dependencies of other states Ascension, St Helena, Tristan da Cunha (UK), Mayotte, Réunion (France)

STYLES OF GOVERNMENT

Republics Angola, Botswana, Comoros, Madagascar, Malawi, Mozambique, Namibia, South Africa, Zambia, Zimbabwe

Monarchies Lesotho, Swaziland

Federal state Comoros

Multi-party states Angola, Botswana, Mauritius, Mozambique, Namibia, South Africa, Zambia, Zimbabwe

One-party states Comoros, Madagascar, Malawi, Swaziland

State without parties Lesotho

Military influence Lesotho

CONFLICTS (since 1945)

Internal unrest South Africa 1960, 1976, 1984–

Coups Comoros, 1976, 1978, 1989; Lesotho 1987; Madagascar 1972

Civil wars Angola 1975–91; Mozambique 1977–; Zimbabwe 1965–80

Independence wars Madagascar/France 1947–58; Angola/Portugal 1961–74; Mozambique/Portugal 1964–74; Namibia/South Africa (with Angolan and Cuban involvement) 1966–89

Interstate conflicts South Africa/Angola (with Cuban involvement) 1975–1991; South Africa/Mozambique 1980–89

MEMBERSHIP OF INTERNATIONAL ORGANIZATIONS

Organization for African Unity (OAU) Angola, Botswana, Lesotho, Madagascar, Malawi, Mauritius, Mozambique, Namibia, Swaziland, Zambia, Zimbabwe

Southern African Development Coordination Conference (SADCC) Angola, Botswana, Lesotho, Malawi, Mozambique, Swaziland, Zambia, Zimbabwe

---

PATHS TO INDEPENDENCE

During the last quarter of the 19th century the rush by European states to acquire colonies in Africa speeded up, driven by the need for raw materials to feed domestic industries and by the determination not to be outdone by colonial rivals. In southern Africa Britain penetrated northward from the Cape to establish colonies in Nyasaland, and in Northern and Southern Rhodesia (today Malawi, Zambia and Zimbabwe). Portugal responded to the threat posed by Germany's seizure of South West Africa (Namibia) by expanding inland from its trading stations of Angola and Mozambique on the west and east coasts. The French, who had lost the island of Mauritius to the British in the Napoleonic wars, strengthened their hold on Madagascar and the Comoros islands in the Indian Ocean.

White settler power

Attracted by the good climate, rich soil, mineral wealth and potential cheap black labor of southern Africa, a stream of Europeans came to settle and exploit its resources. This was in contrast to the rest of Africa, which – except for Algeria and Kenya – was left mainly to trading and concession company employees, colonial administrators and missionaries. In the period after World War II, decolonization was taking place in many other parts of the world, while here white communities, economically and politically powerful though numerically small, strenuously resisted the growing demand for independence under black majority rule. They had before them the model of white-ruled South Africa.

In 1910, after the Boer war (1899–1902) when the British took over the Afrikaner republics of Natal, the Orange Free State and Transvaal, with its valuable gold mines, the Union of South Africa came into being as an independent British dominion. Tensions between the English- and Afrikaans-speaking communities grew until, in 1948, the Afrikaner-dominated National Party won control of government, which it has retained ever since. The policy of black and white racial segregation known as apartheid ("separateness") was given legal form.

An attempt was made to consolidate white rule in the three British colonies farther to the north by creating the Federation of Rhodesia and Nyasaland. This lasted from 1953 to 1960. White influence was strongest in Southern Rhodesia, which had never known direct colonial rule: until 1923 it had been governed by the British South Africa Company, when it opted to become a self-governing colony. African nationalist movements in Northern Rhodesia and Nyasaland brought about an end to the federation, and their independence as Zambia and Malawi was granted in 1964.

A proving ground for guerrillas Armed nationalist groups were formed in the 1960s to loosen Portugal's tenacious grip on its colonies. These FRELIMO soldiers are training in Mozambique.

Postcolonial conflict After independence guerrilla groups continued to fight among themselves, backed by external left-wing and right-wing governments. In this arena for conflicting ideologies, Namibia's independence (achieved in 1989) became a bargaining counter.

Independence followed for the British protectorates of Botswana (formerly known as Bechuanaland), Lesotho (Basutoland) and the territory of Swaziland between 1966 and 1968.

The end of decolonization

Portugal clung longer to Angola and Mozambique. As the poor man of Europe, it needed to reap the economic advantages of its colonies and to sustain the ideology of empire that underpinned its dictatorship at home. Here, as well as in Namibia, Southern Rhodesia and South Africa, nationalist movements were banned and protest turned to armed struggle. In April 1974 the dictatorship in Portugal that had kept the colonial war going was overthrown by its own troops. Democratic government was restored in Portugal, and a year later both Angola and Mozambique won independence.

In Southern Rhodesia the white ruling party rejected the terms for full independence that had been proposed by the British government, and in 1965 made a Unilateral Declaration of Independence (UDI). Two armed nationalist groups – the Zimbabwe African People's Union (ZAPU) and the Zimbabwe African National Union (ZANU) – continued to fight for majority rule, which was achieved in 1980. In Namibia the struggle against rule from South Africa lasted until the end of the 1980s.

Most of the states of the region contain a diversity of ethnic and linguistic groups – a direct legacy of the way in which artificial boundaries were imposed by the colonizing powers. For example, there are more than seventy ethnic groups, speaking a variety of languages, in Zambia. Although this can sometimes exacerbate political tensions within individual states, the existence of common languages that cross frontiers can also aid greater cooperation within the region.

Map: Southern Africa

CONGO
Cabinda (Angola)
ZAIRE
KENYA
TANZANIA
Lake Tanganyika
INDIAN OCEAN
Luanda
1961–74
ANGOLA
1975
1975–91
Lake Nyasa
MALAWI
1964
ZAMBIA
1964
Lilongwe
Lusaka
Zambezi
COMOROS
1975
MOZAMBIQUE
1975
Harare
1964–74, 1977
1979 (South African troops)
1965–80
ZIMBABWE
1980
Mozambique Channel
Antananarivo
1966–89
BOTSWANA
1966
MADAGASCAR
1960
Walvis Bay (South Africa)
Windhoek
Gaborone
NAMIBIA
1989 (to South Africa 1919)
Pretoria administrative capital
Maputo
Johannesburg
Mbabane
SWAZILAND
1968
Orange
Bloemfontein judicial capital
Maseru
LESOTHO
1966
ATLANTIC OCEAN
since 1984
SOUTH AFRICA
1910
Cape Town legislative capital

States of southern Africa

→ invasion
⚔ conflict
1950 date of independence

colonial power
▨ British
▨ French
▨ German/South African
▨ Portuguese

DIVERSE POLITICAL SYSTEMS

On independence, Angola and Mozambique became single-party socialist states. In Angola the strategy of divide and rule by which Portugal governed its colonies had exploited ethnic rivalries, helping to perpetuate deep-seated differences between nationalist guerrilla groups. These conflicts, exacerbated by external intervention, continued to prevent the development of national cohesion after independence.

The rise to power of the Marxist-led People's Movement for the Liberation of Angola (MPLA) was achieved with Soviet and Cuban support. It was opposed by the South African-backed National Union for the Total Independence of Angola (UNITA). Throughout the 1980s South Africa launched attacks within Angola, in part to destroy bases of the South West Africa People's Organization (SWAPO), who were fighting for Namibia's liberation. In Mozambique, South Africa also sought to destabilize the socialist government of the Front for the Liberation of Mozambique (FRELIMO), and discourage its support of the African National Congress (ANC), the banned South African nationalist movement.

The price of conflict

In the early years of independence, both Angola and Mozambique launched ambitious social welfare programs, with some early successes in education and primary health care. But their economies – already among the poorest in the world – were devastated by the effects of war. Between 1975 and 1983 Mozambique had doubled the number of its teachers and school students and tripled the number of health posts and centers. As a result of attacks by the South African-backed Mozambique National Resistance (MNR) guerrillas, 800 clinics and 2,600 schools were closed, its infant mortality rate rose sharply, almost half the population was dependent on food aid, and nearly 4 million people (out of a population of 15 million) were made homeless.

The collapse of the communist bloc and the withdrawal of aid from the former Soviet Union and other socialist allies in the late 1980s heightened the economic crisis, forcing FRELIMO to abandon its Marxist-Leninist stance. One-party rule was officially ended in 1990, but the failure to obtain a ceasefire in the civil war with RENAMO delayed the introduction of elections. The results of Angola's first multi-party elections, held in 1992, were violently disputed by UNITA.

In Malawi and Zambia one-party rule became centered on the person of the president. Hastings Banda, Malawi's first president, was made president for life in 1971. However, political unrest in 1992 pushed Banda toward democracy, first through referendum and then, in May 1994, to multiparty elections in which he was defeated. The United Democratic Front led by Bakili Muluzi scored a substantial victory. Although the style of government practiced by Kenneth Kaunda in Zambia was less authoritarian, responsibility for all political appointments was kept in his hands, and the ruling elite remained a small, closed circle. Elected eight times as president, most recently in 1988, in 1991 he was defeated in a contested presidential election by his long-term political rival, Frederick Chiluba.

According to Zimbabwe's multi-party constitution, hammered out in protracted independence negotiations in 1979, some 20 seats in the 100-seat house of assembly were reserved for white voters. The ZANU party, headed by Robert Mugabe, won a massive victory in the first post-independence elections. In 1987 it was united with the other major party, ZAPU. Seats reserved for whites were replaced by non-constituency seats fixed by the house of assembly sitting as an electoral college. Other changes introduced a move toward an executive presidency.

The state of Botswana is a multi-party democracy. In addition to the national assembly, elected every five years, there is a house of chiefs representing the eight principal ethnic groups. Executive power lies with the president, who is elected by the national assembly.

Reform in South Africa

Growing economic crisis in the late 1980s forced the South African government to lift some apartheid restrictions. The pace of reform was accelerated by F.W. de Klerk, who became president in 1989. The ban was lifted from the ANC in 1990 and its leader Nelson Mandela released from prison after 27 years. In 1991 negotiations began at the Convention for a Democratic Africa (CODESA).

By 1991 apartheid legislation was withdrawn, and by 1992 a new constitution was in place. Formerly South Africa had a three-chamber parliament representing white, mixed race and Asian interests, and excluding black representation. The 1994 parliament was divided into two houses: a 400-seat National Assembly, elected by the whole population; and a 90-seat senate. Under the new constitution, the president is elected by both houses. In May 1994 Nelson Mandela became South Africa's first black president.

Mothers in a refugee camp queue to have their babies vaccinated. Health care in Mozambique has been badly affected by the conflict with South Africa.

Preparing to celebrate Namibia's freedom A SWAPO rally in 1988 marks the 22nd anniversary of the fight against South Africa, a year before its end.

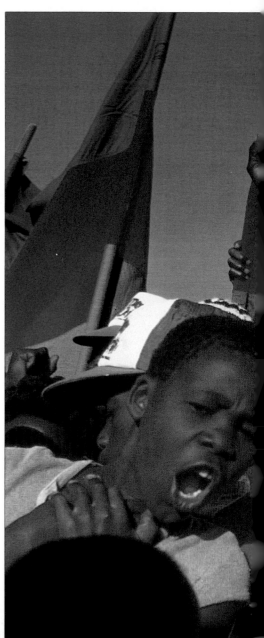

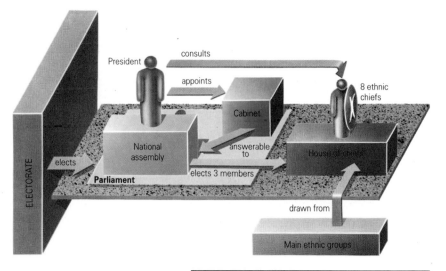

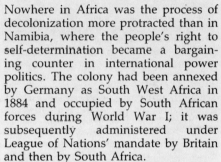

An African democracy Botswana's national assembly elects a president to head the government. He is advised by a house of chiefs, elected from eight ethnic groups. There are two main parties and several minor ones.

AFRICA'S LAST LARGE COLONY

Nowhere in Africa was the process of decolonization more protracted than in Namibia, where the people's right to self-determination became a bargaining counter in international power politics. The colony had been annexed by Germany as South West Africa in 1884 and occupied by South African forces during World War I; it was subsequently administered under League of Nations' mandate by Britain and then by South Africa.

The territory came into dispute after 1946 when South Africa proved unwilling to recognize the United Nations' responsibility for the territory as successor to the League of Nations. In 1966 it refused to comply with the United Nations' decision to terminate its mandate – a measure supported by a ruling of the International Court of Justice in 1971 – but continued to occupy and administer the territory, renamed Namibia by the United Nations in 1968. The application of the apartheid laws roused widespread unrest among the black Africans.

In 1973 the United Nations recognized the South West African People's Organization (SWAPO), whose leader Sam Nujomo had been exiled in 1960, as the sole representative of the Namibian people. A UN resolution was passed in 1976 arranging for independent elections to be held under UN supervision. Meanwhile, SWAPO guerrillas operating from bases within Angola carried on armed resistance, particularly in the north of the country. Negotiations dragged on as, at the instigation of the United States, the withdrawal of Cuban and Soviet troops and advisers from Angola was made a condition of independence. Agreement was finally reached on this and other points in 1988. In the first elections (1989) SWAPO won 57 percent of the vote and in 1990 its leader, Sam Nujoma, was elected the first president.

FRONTLINE STATES

South Africa's readiness aggressively to defend its commitment to white minority rule dominated regional politics in southern Africa in the 1970s and 1980s. The central aim of its strategy was to close off support from the black majority states of the region to the antiapartheid movement in South Africa. It provided both financial and military support to anti-government guerrilla groups in Angola, Lesotho, Mozambique and Zimbabwe, and conducted a relentless campaign to destabilize their governments, particularly those of Angola and Mozambique, in order to demonstrate that black majority rule and multiracialism could not work. The cost in both human and economic terms was enormous.

The term "frontline states" came to be used to describe those states in the forefront of the struggle against South Africa in the 1970s. At this time the leaders of Mozambique, Tanzania (to the north of the region) and Zambia – Samora Machel (subsequently killed in an aircrash near the South African border in 1986), Julius Nyerere and Kenneth Kaunda – were all working closely together to support black nationalist movements in what was to become Zimbabwe. Out of their common effort to coordinate diplomatic action to achieve majority rule came the commitment to more formalized regional cooperation in the shape of the Southern African Development Coordination Conference (SADCC), established in 1980.

Strength through cooperation
SADCC brought together all the frontline states of the region, and also included Tanzania in east Africa. Its first commitment was to reduce regional economic dependence on South Africa, and it developed sound strategies to achieve this. Each member state was entrusted with a particular ministerial portfolio: Angola, the only oil-producing state, had responsibility for energy; Mozambique, with its ports and railways, took charge of transport and communications. Policy was set each year at a conference of heads of state, and a council of ministers, which reported to the conference, was responsible for its coordination and execution. Most member states are landlocked, and following the destruction of Mozambique's rail links by MNR rebels, the transit

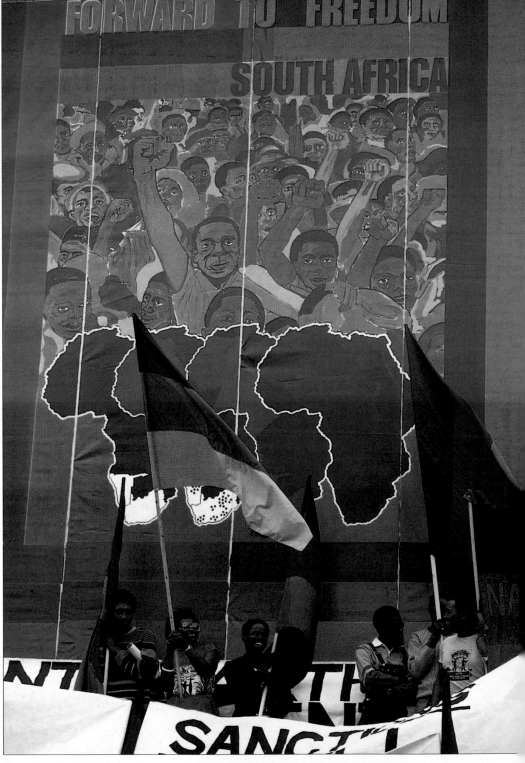

An antiapartheid rally in London's Trafalgar Square calls for sanctions against white South Africa. As a result of public pressure, some Western companies reduced the level of their investment in South Africa, but world leaders were unable to reach common agreement on applying full sanctions.

Former President F. W. de Klerk (*right*), campaigning for the National Party in March 1994, during the build up to South Africa's first multiracial elections. As early as February 1993 de Klerk and Mandela, the President of the ANC, proposed to form a government of national unity together, a decision which alienated the Zulu leader Mangosuthu Buthelezi.

INTERNATIONAL SUPPORT FOR THE ANC

Founded in 1912, the African National Congress was the first nationalist movement in the continent. During the 1950s it led a nonviolent campaign of civil disobedience in protest against the apartheid laws, but was made illegal in 1960. Its leaders were arrested. They included Nelson Mandela, who was imprisoned for life for alleged sabotage: for more than twenty-five years, while in jail, he was regarded as a symbol of the antiapartheid struggle throughout the world. The ANC, forced into exile, made its headquarters in Lusaka, Zambia. Under the leadership of Oliver Tambo it built up an extensive network of international support.

The antiapartheid movement spearheaded many campaigns around the world to focus attention on the struggle for majority rule within South Africa. It led to international boycotts both against South African goods and against banks and companies with investments in South Africa. As a result of its representations, many countries refused to have sporting links with white South African teams. The aim was to isolate the South African regime and to bring economic pressure to bear on it to change its policies.

The antiapartheid movement was unable to achieve the comprehensive support for mandatory sanctions that was urged by the UN in 1985, opponents arguing that they would bring economic ruin that would destroy the lives of black South Africans. Nevertheless, there were some successes. An international arms boycott, which helped to deprive the South African defense forces of air superiority, contributed to the victory of Angolan and Cuban forces at the decisive battle of Cuito Cuanavale in 1988, speeding up the peace process in neighboring Namibia. The start of the CODESA negotiations in 1991 allowed sports boycotts to be lifted in time for a South African team including black and white athletes to participate in the 1992 Olympic Games held in Barcelona, Spain.

of goods to the coast had to go through South Africa. For example, as much as 80 percent of Zimbabwe's trade was carried along this route in the 1980s. This forced the frontline states into still greater economic dependency on South Africa, and increased their vulnerability to aggressive countermeasures. In the period before political reform in the early 1990s, South Africa threatened to close its railroads and ports to traffic from other states in the region in retaliation for trade sanctions. Much of SADCC's work, therefore, focused on constructing the vital rail corridor to link the Mozambique port of Beira with Zimbabwe.

The independent black states of the region are members of the Organization of African Unity (OAU), which was set up to improve economic, cultural and political cooperation throughout Africa. Though it is generally regarded as having little more than symbolic influence, it does still represent a common African voice on a number of issues, and provided diplomatic and limited material support to the frontline states in their confrontation with South Africa in the 1980s.

The presence of Marxist-inspired governments in Angola and Mozambique supported by the former Soviet Union and Cuba heightened international tension after 1975, and inevitably led to growing superpower confrontation in the region. South Africa was seen as an ally by the United States to counter the alleged threat of communist domination in southern Africa. Both countries gave substantial support to UNITA rebels fighting in Angola. As a result, South Africa became closely tied in to Western defence strategies, and in spite of continuing trade sanctions, it provided anchorage facilities to the United States' fleet at its naval base of Simonstown. In the late 1980s detente, springing from the new foreign policy of Soviet leader Mikhail Gorbachev, led to the negotiated withdrawal of Cuban troops from Angola in 1988. Peace accords between government and rebels were agreed for Angola in 1990 (leading to elections in 1992) and were negotiated for Mozambique in 1991, but these soon broke down.

Throughout this entire period, all the former British colonies of the region, except South Africa, were members of the Commonwealth association of states. Although not as powerful as it was in the 1940s and 1950s, the Commonwealth still provides, particularly through its conferences of heads of state, an informal forum for consultation and it remains influential. Following the election of Nelson Mandela as president in May 1994, South Africa was readmitted to the Commonwealth, making the number of member nations up to 51.

A negotiated withdrawal In front of the world's press at Luanda airport the Cuban commander signs the agreement to withdraw his men from Angola, witnessed by his Angolan counterpart. An officer of the UN observes. Cuban troops were in Angola from the mid-1970s, when they helped the Marxist-aligned MPLA to take control of the country. The South African government accused them of aiding SWAPO guerrillas based within Angola; their withdrawal was made a condition of Namibia's independence.

Rehabilitation of a nation

Shortly after taking office in September 1989, President F. W. de Klerk announced his determination to accelerate the dismantling of apartheid – a process that had begun with Pik Botha's limited reforms in the late 1980s. The seriousness of de Klerk's intentions were further reinforced with the release of Nelson Mandela in February 1990, as the South African government finally set out on the road toward universal suffrage, a new non-racial constitution and its first democratically elected government.

In December 1991 the Convention for a Democratic South Africa was set up with a commitment to seeking a peaceful, negotiated settlement between all South Africa's peoples. In March of the following year, 70 percent of the white population approved a referendum supporting an end to white minority rule.

Talks continued throughout 1992 against a backdrop of civil unrest and political assassination. But this did not deter Nelson Mandela and President de Klerk from their commitment to continue the gradual process of rapprochement between some 26 disparate political groups. By September the ruling national executive had accepted a plan approved by 23 of South Africa's political parties to set up a multiracial Transitional Executive Council. This organization was intended to oversee preparations for South Africa's first one-person-one-vote elections to be held on 27 April 1994. The inauguration of the Executive Council in December 1993 finally put an end to three centuries of white minority rule in South Africa.

In the meantime, South Africa's international rehabilitation gathered momentum. Many trade and cultural sanctions had been lifted in 1992 and that same summer, South Africa was readmitted to the Olympic Games, to which it sent its first integrated team of athletes. Toward the end of a difficult year, Nelson Mandela and President de Klerk were jointly awarded the Nobel Peace Prize in recognition of their efforts to restore peace and prosperity to a divided nation.

Birth of a New Era

During the run up to polling day, the ten homelands (where black people were obliged to live by law) were formally abolished, and the whole nation was divided into nine provincial legislatures, each one democratically represented.

Despite numerous acts of political terrorism committed by white extremists as a prelude to election day, nothing could deter the overwhelming desire of South Africa's black population to cast their votes for the first time. Many waited for hours, if not days, with queues at some polling stations stretching for more than a kilometre. Lack of adequate preparations and a serious miscalculation of the huge numbers that would turn up to vote resulted in many stations running out of ballot papers. The unexpected volume of voters nearly triggered a fiasco, and the voting deadline had to be extended into a third day. Many black voters in rural areas were illiterate and found it difficult to deal with an enormously long ballot paper bearing the names and logos of 18 parties. Nevertheless, the outcome was as predicted, with the African National Congress (ANC) achieving over 60 percent of the vote.

On 6 May 1994 F. W. de Klerk handed over leadership to Nelson Mandela – South Africa's first black president. Having made an emotional commitment to achieving his goal of "hope, reconciliation and nation building", Mandela initiated a massive development plan that would bring improved living conditions to millions of impoverished blacks. The task facing the new federal government is a daunting one, with half of its black citizens illiterate and half of its available workforce without jobs. But the country looks toward the future with optimism: the elections, coupled with the lifting of remaining economic sanctions in 1994 brought an end to 46 years of apartheid and over 30 years of economic isolation. In June 1994 South Africa rejoined the British Commonwealth and readmission into the United Nations is also expected.

Much of the future economic success of the region depends on South Africa's reemergence into the international arena. With its vast natural resources and huge potential for trade and industrial growth, the prospects for Africa as a whole are very bright. If the new, democratic South Africa can achieve the stability needed for growth and to attract investment, then it may be able to arrest the downward economic spiral in the African continent.

Election day in Soweto, May 1994. South Africa's black citizens had waited all their lives to vote, and few were deterred by the enormous queues outside local polling stations. In the final count the ANC polled over 12 million votes, followed by de Klerk's National Party with almost 4 million.

THE POLITICS OF DIVERSITY

NEW STATES IN SOUTH ASIA · DILEMMAS OF DEMOCRACY · BETWEEN EAST AND WEST

The fertile plains of the Indian subcontinent have attracted successive waves of conquerors since Neolithic times. In the 18th century the 200-year-old Muslim (Mogul) empire, which at its height extended eastward to Bengal and southward to the Deccan, fragmented at the same time as European trading companies began to expand into the region. By the mid-19th century nearly all the subcontinent, except the kingdoms of Bhutan and Nepal, was under British rule. The nationalist Congress movement led the struggle for India's independence, which was won in 1947, when the state of Pakistan was also created. In 1971 East Pakistan seceded as the new state of Bangladesh after a civil war. The islands of Sri Lanka and the Maldives, both former British colonies, achieved their independence in 1948 and 1965.

NEW STATES IN SOUTH ASIA

The steps by which Britain granted independence to India and Pakistan were taken in the aftermath of World War II; few states can have been created amid such political turmoil. Despite the claim of the nationalist Congress Party, led by Mohandas Gandhi (1869–1948) and Jawaharlal Nehru (1889–1961), to represent all Indians, it failed to unite Hindus and Muslims. The Muslim League under the leadership of Muhammad Ali Jinnah (1876–1948) became the main representative of India's minority Muslim population, heading demands for the creation of a separate Muslim state of Pakistan.

"Two nations" in India

Elections were held in 1946 as the first step toward self-government. These confirmed the existence of "two nations" in India: the Muslim League made a clean sweep of seats in all the Muslim areas, and Congress won most of the remainder. British proposals for an independent India as a loose federation of states floundered, as riots and massacres in Calcutta in August 1946 led to intense communal strife. The result was partition as the two independent states of India and Pakistan.

The areas chiefly inhabited by the Muslim population lay in the northwest of the subcontinent and in eastern Bengal. These became the provinces of West and East Pakistan, separated by 1,600 km (1,000 mi) of territory lying in India. The movement of refugees was enormous as over 12 million people, both Hindu and Muslim, found themselves on the wrong side of the new boundary; the partitioning of Bengal and Punjab claimed half a million lives in each. The new states went to war over Kashmir; the 1949 ceasefire line still remains the northern boundary between them.

The new state of India – which declared

New states in the making Border disputes were at first frequent between the newly independent states of the subcontinent, continuing the religious and ethnic divisions that had brought about their partition. More recently, cooperation within the region has grown.

States of the Indian subcontinent

- ╍╍╍ state boundary
- ⚒ conflict
- ⚒ civil unrest
- ✈ airlift
- ▬▬ ceasefire line 1949
- ▒ under Chinese military control
- ▓ annexed from Portugal 1961

A tragic dynasty Rajiv Gandhi performs the funeral rites for his mother, Indira Gandhi, killed by Sikh militants in 1984. Following her as India's prime minister, he was himself assassinated in 1991.

The trauma of partition Refugees shelter in a mosque from Hindu violence. In order to end the massacres Gandhi – himself a Hindu – began a protest fast in January 1948: he was murdered by a Hindu extremist.

itself a federal republic in 1950 – was made up of a chaotic mosaic of political territories inherited from the British administration. Some 14 major languages are spoken by 90 percent of the population (in India there are 1,652 "mother tongues"). It was feared that the creation of states based on linguistic divisions would endanger national unity, and in 1950 a federal constitution came into being that instead divided India into 27 "administrative" states.

These arrangements did not work. In 1952 the Telegu-speaking state of Andhra Pradesh was formed, and in 1956 the political map was redrawn to produce 14 language-based states. The government was forced to abandon its plans to make Hindi the sole language used at national level. Further states were created in the next few years. Goa was annexed from Portugal in 1961 following a short war. Sikkim was persuaded to join the federation in 1975, the last of the princely states to do so. There are now 25 states in the republic, as well as seven union territories under federal administration.

Challenges to national cohesion

In the early 1990s, increasing violence between Hindus and Muslims throughout India appeared to threaten the ideal of nonsectarianism on which the modern state of India was founded. There were other challenges to national cohesion. The Sikhs of the Punjab have never ceased to agitate for a separate homeland (Khalistan). In Jammu and Kashmir mounting violence between Muslim separatists and the Indian army led to the imposition of central rule in 1990. Tamil nationalists in the south found common cause with the Tamils in Sri Lanka.

In Pakistan the death of Jinnah in 1948, only a year after independence, removed from the political scene possibly the only figure capable of giving national cohesion to the two separate territories of West and East Pakistan. Increasing demands for self-government from East Pakistan led to the imposition of military rule in 1958. The country's first direct popular elections were held in 1970. The Awami League, campaigning for increased regional self-government, won 167 of the 169 seats in East Pakistan, giving it an overall majority in the national assembly. The upshot was civil war: East Pakistan seceded as the state of Bangladesh.

Separatist demands in Sri Lanka from the Hindu Tamils (about 20 percent of the population, many of Indian origin) took increasingly violent form in the early 1980s, leading to a civil war that cost thousands of lives. A peace pact signed by the Sri Lankan president Junius Jayawardene and India's prime minister Rajiv Gandhi in 1987 proposed the creation of a Tamil-run province in the northeast of the island, but the Tamil Tigers, the strongest and most violent of the guerrilla groups, continued to press for complete independence.

DILEMMAS OF DEMOCRACY

The states of South Asia are governed in a variety of ways. The Maldives are without political parties. Executive power is held by an elected president, and there is a 48-member parliament (*majlis*). Political parties are also unrecognized in the Himalayan kingdom of Bhutan, where the king rules as an absolute monarch. Its neighbor Nepal has been a constitutional monarchy since 1990. On independence, the states of the subcontinent formerly under British rule all adopted constitutions that embody some form of parliamentary democracy. They have been sustained ever since with varying degrees of success.

Vulnerable democracies

Both Bangladesh and Pakistan have experienced long episodes of military rule. Pakistan's first coup, in 1958, was led by an army general, Mohammad Ayub Khan (1907–74), who became president in 1960. He was deposed in 1969 by General Yahya Khan, whose decision to allow popular elections in 1970 led to the civil war that resulted in East Pakistan's secession as Bangladesh.

Civilian rule returned in 1971 under Zulfiqar Ali Bhutto (1928–79), but the victory of his Pakistan People's Party in the 1977 elections was disputed amid widespread accusations of malpractice. Riots ensued, and martial law was then imposed under General Mohammad Zia ul-Haq (1924–88). Bhutto was executed in 1979. In 1986 political parties were allowed to operate again, and Bhutto's daughter, Benazir Bhutto, returned from exile to head the People's Party. Zia was killed in an air crash in 1988, and in elections later in the year Benazir Bhutto was elected prime minister but in 1990 was removed by the president in a "constitutional coup". She lost the ensuing elections.

In Bangladesh, martial law was imposed between 1975 and 1979, and again between 1982 and 1986. The military, with General Hussain Mohammad Ershad as president, remained in control, but in 1990 Ershad was forced to resign after antigovernment riots. Parliamentary government was restored in 1991 with the holding of national elections.

Sri Lanka's style of government is based on the Western competitive multiparty model. Electoral support has been mostly divided between two main parties – the United Nationalist Party (UNP) and the Sri Lanka Freedom Party (SLFP), which is socialist and more narrowly Sinhalese in appeal. In 1972 the SLFP,

A woman in a man's world Benazir Bhutto became the first woman prime minister of an Islamic state in 1988, but was unable to withstand the strength of traditionalist opposition and was later dismissed.

Voting in an Indian election India has a large number of political parties. Because of widespread illiteracy among voters a clearly recognizable symbol is allocated to each one, which appears next to the candidate's name on the ballot paper. These symbols have great potency: Indira Gandhi's Congress Party was a cow and calf, implying care and protection.

India's federal constitution India is a federal republic: its 25 states are self-governing. Power is exercised at national level by the prime minister, who controls the Lok Sabha (House of the People) in the two-chamber federal parliament, elected every five years. The president is chosen by the federal and state parliaments: in practice the prime minister's nominee is always elected.

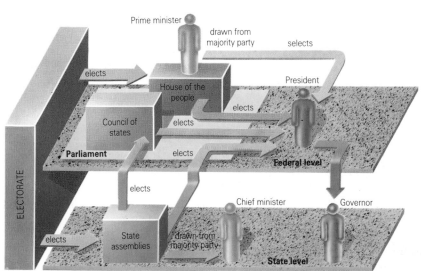

led by Mrs Sirimavo Bandaranaike, introduced a new constitution: the name of the country was changed from Ceylon to Sri Lanka, and extended the life of parliament to six years. The UNP, which came to power in 1977, adopted a presidential form of government a year later. It enjoyed uninterrupted power throughout the 1980s and early 1990s, while the issue of Tamil separatism, which came to dominate Sri Lankan politics, led in 1983 to the declaration of a state of emergency.

The largest democracy in the world

Perhaps the most notable thing about India's democratic system is its sheer size: in a country that has the second largest population in the world, it takes some 2 million officials to run a national election to the federal parliament. The Congress Party, heir to the nationalist independence movement, dominated national politics; no major party challenged

COMMUNAL STRIFE IN SRI LANKA

The dispute in Sri Lanka between the minority Tamils and the majority Sinhalese first surfaced in political conflict over language policy in the newly independent state. The decision of the SLFP government led by Solomon Bandaranaike to establish Sinhalese rather than English as the official language precipitated Tamil riots, and in 1959 the prime minister was assassinated by a Buddhist monk.

The SLFP government that was led by his widow, Sirimavo Bandaranaike, between 1970 and 1977 introduced pro-Sinhalese policies in education and employment. These exacerbated Tamil complaints of discrimination, and in 1976 the Tamil United Liberation Front (TULF) was formed to campaign for a separate Tamil state (Eelam) in the north and east of Sri Lanka. It emerged as the main opposition party to the

victorious UNP in the 1977 elections, the same year that renewed communal violence broke out.

The existence of a Tamil state (Tamil Nadu) in southern India complicated matters. The Indian government became increasingly worried that Tamil unrest would spread there from Sri Lanka. In 1987, the Sri Lankan government agreed to allow Indian troops into the Jaffna peninsula, the stronghold of the Tamil Tiger guerrillas, to disarm the rebels in return for substantial concessions, including the creation of a Tamil province. The Tamil Tigers, however, refused to observe a ceasefire. The negotiations for India's withdrawal were protracted, and the last troops did not leave until 1991. The violence continued, and in March 1991 more than 2,500 Tamil Tigers were killed by Sri Lankan troops.

its rule for 20 years, though numerous small and regionally based political parties contested power in state elections.

The period of one-party dominance ended abruptly in 1967 when Congress lost control of eight state governments, initiating a period of instability and confusion. The landslide victory of the Congress Party of Indira Gandhi (1917–84) in 1971 appeared to bring a return to normality. However, severe economic problems created a sense of national crisis, and it was in this atmosphere that Mrs Gandhi was found guilty of "corrupt electoral practices". Rather than face disqualification from politics, she declared a state of emergency: political opponents were imprisoned and the press censored.

After restrictions had been lifted, four opposition parties united to form the Janata Party to contest elections in 1977. Congress was heavily defeated, but returned to power in 1980. In 1984, in the aftermath of the assassination of Indira Gandhi, it won overwhelmingly, and her son Rajiv (1944–91) succeeded as prime minister. However, the party lost its ruling majority in 1989. The election campaign of 1991 was thrown into disarray by the assassination of Rajiv Gandhi by Tamil nationalists. The rise of the sectarian Hindu party Bharatiya Janata Party (BJP) provoked rioting throughout the campaign, but Congress was able to form the government, led by P. V. Narasimba Rao, on a minority vote.

BETWEEN EAST AND WEST

The independent states of the Indian subcontinent were among the first to be created in the breakup of Europe's colonial empires after World War II, which brought far-reaching changes in world politics. India found itself much the largest independent state in what would later be known as the Third World, and was strongly placed to exercise influence in the worldwide movement against colonial power.

Independence coincided with the growth of the Cold War, a period of heightened hostility between the new global superpowers of the Soviet Union and the United States. By choosing not to place itself in one camp or the other, India played a significant part in bringing a new force, the nonaligned movement, to bear in world politics.

Nehru's foreign policy
The architect of these achievements was India's first prime minister, Jahawarlal Nehru. He won wide respect for his moral approach to international relations, which

rejected the setting up of aggressive "defense" pacts as the means of solving world problems, and sought to find positive ways to achieve peace between nations. He was hailed as a peacemaker for his part in bringing an end to the Korean war (1953). Nehru undertook numerous major foreign policy trips during his premiership, including visits to China, the Soviet Union and the United States; the number of visits that world leaders paid to New Delhi in these years

A dominant power in South Asia
Military hardware is paraded through New Delhi on Republic Day (26 January). Estimates suggest that up to 70 percent of India's military equipment in the 1980s was supplied by the Soviet Union.

Nehru on the world stage Soviet leader Nikita Khrushchev greets Nehru on one of his visits to Moscow. In 1955 he was the first noncommunist world leader to address a Soviet audience: 100,000 people heard him speak in the Dynamo Stadium. In the Cold War atmosphere of the 1950s Nehru refused to side with either superpower, or let India be dominated by them.

is an indication of his unique position as international statesman.

Nonetheless, the realities of India's foreign policy were often at variance with its stated intentions. Disputes over boundaries and other issues of national security were many in a region in which states had been founded in conditions of civil war to preserve religious and ethnic differences. In the case of India and Pakistan there was armed conflict on three occasions: 1947, 1965 and 1971, when Indian

NEHRU AND NONALIGNMENT

The nonaligned movement began in the 1950s and 1960s as a collaborative effort by newly independent Asian and African states to create an alternative voice in world politics that was free of dependence on the dominant Western and Eastern blocs. The term was originally coined by Nehru. A conference of African and Asian countries held in Bandung, Indonesia in 1955 laid down the principles later adopted by the nonaligned movement: the condemnation of colonialism throughout the world, and the promotion of peaceful coexistence and of international cooperation through the strengthening of the United Nations.

It was the work of three men – Nehru, President Josip Tito of Yugoslavia (1892–1980) and President Gamal Abdel Nasser of Egypt (1918–70) – that led to the first conference of nonaligned states at Belgrade in Yugoslavia in 1961. Attended by representatives from 27 states, its concluding declaration made clear the movement's opposition to blocs of any kind, economic or political. It condemned action that inhibited the independence of individual states, including the maintenance of military bases in another state's territory, and the pursuit of racialist policies.

The next conference was held in Cairo, Egypt in 1964 and was attended by 47 states, with 11 more coming as observers. As the movement grew in size, regional concerns tended to cut across global policies; increasing attention was given to economic conditions in the Third World as one of the few areas in which common agreement for action could be found.

troops invaded East Pakistan in support of the separatist movement there and then remained to oversee the creation of the independent state of Bangladesh. Tension between India and Pakistan over Kashmir was renewed in 1989. Despite the signing of a nonaggression agreement between India and China in 1954, border disputes along the border with Tibet (under Chinese control since 1951) erupted into fullscale war in 1962.

The three states that separate India and China along the Himalayan border lie in what India considers to be its sphere of influence. The largest, Nepal, has been able to assert most independence. The formation of closer ties with China (including the purchase of arms and the completion of a Chinese-built road link) led India in 1989 to impose a trade blockade ostensibly to control smuggling. India maintains an advisory role over Bhutan's foreign policy. Sikkim, the smallest of the states, was incorporated into the Indian state in 1975.

The quest for regional parity

In the early years after independence Pakistan, much the weaker in terms of population and resources, responded to India's growing dominance in the region with a pro-Western foreign policy. In sharp contrast to India's nonalignment position, it joined the series of defensive pacts set up under United States' leadership to combat communism in Asia: the South East Asia Treaty Organization (SEATO, 1954), the Baghdad Pact (1955) and also its successor, the Central Asia Treaty Organization (CENTO, 1959).

The loss of East Pakistan in 1971 left India the undisputed leading power in South Asia. Despite nonalignment it had close ties with the Soviet Union, which supplied military aid. Its dominant role in the region grew still further in the 1980s, following its intervention in Sri Lanka and the landing of a small force on the Maldives to restore order after an attempted coup. Both India and Pakistan are believed to have limited nuclear capability, though in Pakistan's case this has always been denied.

Under Zia ul-Haq, Pakistan in the late 1970s saw its role lying toward the Islamic world. At the same time it received massive aid from the United States, which aimed to build up its strategic position in the Gulf and Indian Ocean. After 1979 this aid was channeled to the Islamic Mujahidin guerrillas fighting the Soviet occupation of Afghanistan. In 1989 Pakistan rejoined the Commonwealth, which it had left in 1972 following the decision of other Commonwealth members to recognize the independent state of Bangladesh.

Since independence, Bangladesh has had minor border disputes with India, as well as continuing differences over the sharing of the waters of the River Ganges; the annual influx of Bangladeshi refugees into Assam and West Bengal continues to damage relations. Nonetheless, Bangladesh's major contribution to the future development of South Asia was the establishment of the South Asian Regional Cooperation organization (SARC), which was proposed by General Zia ur-Rahman in 1980. The agreement signed by all seven states in the region at its inaugural meeting at Dhaka in 1985 pledged cooperation in combating terrorism and drug trafficking, and launched ten cooperative programs in agriculture, telecommunications and transport.

Food is distributed to new arrivals at an Afghanistan refugee camp at Nasirbagh in Pakistan. More than 3 million people fled across the border after the Soviet invasion of Afghanistan in 1979.

Bangladesh: the birth of a nation

The partitioning of Bengal, in the northeast of the subcontinent, in 1947 left the western part of the province in India; the eastern, predominantly Muslim part joined the new state of Pakistan. Pakistan's unique arrangement of two constituent but geographically separate provinces set a severe challenge to national integration, which the government failed to meet.

The relationship between the two parts was always unbalanced. East Pakistan had the greater population, but the center of civil and military government was in West Pakistan. Although most of Pakistan's export earnings came from jute produced in the East, investment was concentrated in the West.

These economic grievances were compounded by the second Indo-Pakistan war in 1965. East Pakistan was cut off from the West within an hour of the war starting, and none of Pakistan's forces were assigned to defend it. In the event India did not fight on the eastern front, but the lesson was clear for all to see: at a time of national emergency East Pakistan, and therefore the majority of Pakistan's population, was expendable.

Given this background of resentment, it was not surprising that in Pakistan's first popular elections in 1970 the Awami League, which fought its campaign on the need to renegotiate the federal constitution and curtail the power of central government, won an overwhelming victory in East Pakistan. As the more populous province, it had a majority of seats in the national assembly. However, talks on rewriting the constitution soon broke down.

For a short period in March 1971 Sheikh Mujib ur-Rahman, as head of the Awami League, was in effect leader of a separate state in East Pakistan, to which the Pakistani president was invited as a "guest" for discussions on the crisis. Then the Pakistan army struck. Mujib was arrested and sentenced to death, and East Pakistan placed under military control.

Secession followed. Declaring itself the state of Bangladesh ("Bengal nation"), a government in exile was set up in Calcutta, from which resistance to Pakistani rule was organized. Administration broke down completely. The flight of 10 million refugees to India resulted in the Indian army's intervention in the nine-month civil war. In two weeks the Pakistan army was defeated, and on 16 December 1971 Indian troops entered Dhaka, the provin-

Taking to the streets A protest group in Dhaka in December 1971 demands an end to the Indian government's intervention in Pakistan's affairs. Within days the Indian army had launched its blitzkrieg attack against East Pakistan, to enter Dhaka on 16 December.

Concrete pipe dreams Thousands of the 10 million refugees who fled to India from East Pakistan find shelter in a temporary city of concrete pipes outside Calcutta. Their plight forced the Indian government to take action to end Pakistan's civil war.

cial capital. The new state of Bangladesh was proclaimed and rapidly gained international recognition.

Building the new state

Despite the devastation of the civil war, and with the world's eighth largest population, Bangladesh embarked upon its independent life with high hopes. Mujib became prime minister in a parliamentary government, and the task of reconstruction began. The optimism did not last. Within three years a state of emergency was declared, and Mujib became president of a one-party state. Months later he was assassinated. Severe economic problems prevented any subsequent government from maintaining any degree of popular support. One other president (General Zia ur-Rahman) was assassinated in 1981, and martial law has been imposed on more than one occasion.

Bangladesh has enjoyed considerable success in foreign affairs. In 1978 it was elected to the Security Council of the United Nations, in preference to Japan, and succeeded to the presidency of the council one year later. It is a member of the Commonwealth, the Islamic Conference, and the nonaligned movement.

However, while participating fully in the world community of states, in the face of frequent flood and famine Bangladesh has not been able to overcome the enormous problem of providing a decent minimum standard of living to the majority of its people. It is a major recipient of international aid. In these circumstances the army acted as the guarantor of law and order. However, the resignation of General Ershad as president in 1990, and the restoration of elections, gave the hope that Bangladesh was moving toward a more democratic future.

Assault on the Golden Temple

The Golden Temple at Amritsar is the center of the Sikh religion and its most revered shrine. In 1984, after 298 people had been killed during a period of communal violence between Sikhs and Hindus, the Sikh fundamentalist Jarnail Singh Bhindranwale turned his headquarters in the Golden Temple into an armory for Sikh militants.

On the night of 5–6 June, in Operation Bluestar, the Indian army carried out an assault on the Golden Temple with simultaneous actions on 45 other Sikh shrines. A bloody three-day siege ensued: it ended with 576 dead, including Bhindranwale. In the aftermath of the battle the call for an independent Sikh state, Khalistan, gained ground. In July and August Sikh gunmen hijacked two Indian airliners. The army returned control of the temple to five head priests in September, but this did not overcome the sense of sacrilege that was felt by many Sikhs.

On 31 October two of prime minister Indira Gandhi's Sikh bodyguards shot her down as she walked from her home to an adjacent office. Twenty-one bullets were found in her body. New Delhi experienced its greatest violence since partition, with some 2,700 killed in communal clashes. Rajiv Gandhi succeeded his mother immediately, but no solution to the Punjab problem was found. The Sikh community continued to be disrupted by the terrorist activities of Sikh extremists, which the Indian government was unable to suppress.

Armed Sikhs within the Golden Temple, or Harimandir, the holiest shrine of the Sikh religion. An amalgam of the Muslim faith and Hinduism, it was founded in the 15th century.

THE LARGEST NATION ON EARTH

China has had its share of tremendous political turmoil this century. In 1912 a republic was proclaimed by the Nationalist (Guomindang) Party, ending the centuries-old empire. The country soon disintegrated into mini-states, under the rule of local warlords. In 1926 the Guomindang entered into an alliance with the newly formed Chinese Communist Party (CCP), but civil war soon broke out between them. The Guomindang and the communists formed a common front against the Japanese, who invaded China in force in 1937–38, but the civil war resumed after the Japanese surrender in 1945. The victorious communists established the People's Republic of China in 1949. The Guomindang leader Chiang Kai-shek (1897–1975) fled to Taiwan, where his nationalist government continued to claim sovereignty over mainland China.

COUNTRIES IN THE REGION

The People's Republic of China, Taiwan

Island territories Hainan (China)

Dependencies of other states Hong Kong (UK); Macao (Portugal)

Disputed borders China/India, China/USSR, China/Vietnam

STYLES OF GOVERNMENT

Republics China, Taiwan

Federal state China

Multi-party state Taiwan

One-party state China

One-chamber assembly China, Taiwan

CONFLICTS (since 1945)

Internal unrest China 1966–76, 1986, 1989

Revolution China 1949

Civil wars China 1945–49, 1950–59 (Tibet)

Interstate conflicts China/Tibet 1950; China/South Korea 1950–53; China/Taiwan 1954, 1958; China/India 1962; China/USSR 1969; China/Vietnam 1979

Notes: Hong Kong is due to be returned to China in 1997, Macao in 1999.

The People's Republic took over China's seat in the United Nations from Taiwan in 1971.

CHANGES IN DIRECTION

In 1949 the 4,500,000-strong Communist Party faced the task of governing 540 million people (a number that was to double within 30 years). The immediate method was to impose central control on the Soviet model – Party members were placed at every level of power. Many of them were People's Liberation Army (PLA) fighters, and it was the PLA that formed the link between the districts, the regional bureaus and the CCP's central committee in Beijing. In 1954 a Soviet-style constitution was adopted. Industries were nationalized, and the Soviet Union provided economic and technical aid for China's first Five-year Plan (1953–57) of heavy industrialization.

Mao's Great Leap Forward

In 1958 Mao Zedong (1893–1976), chairman of the CCP since 1935, embarked on the Great Leap Forward. This reorganized the countryside into about 26,000 agricultural and small-scale industrial communes. The initiative failed disastrously, resulting in famine. In 1960 relations with the Soviet Union collapsed, and the Soviet Union pulled out of all its aid projects, bringing home its advisors.

Mao Zedong's influence declined in the early 1960s, and a "recovery program" was begun under Liu Shaoqi (1898–1969). Communes were reduced in size, and private plots reintroduced. Mao's reply was to launch the Cultural Revolution against what he regarded as a return to capitalism. Its campaigns lasted from 1966 to 1976 and during its first three years were increasingly anarchic, causing large-scale economic and social disruption.

By 1970 Mao had sided with Prime Minister Zhou Enlai (1898–1976) to restore order. In the early 1970s many disgraced Party officials were allowed to return to public life, including Deng Xiaoping, head of the CCP secretariat. A policy of detente towards the United States began, following China's admittance to the United Nations in 1971.

The deaths of Zhou Enlai and Mao Zedong in 1976 unleashed a succession struggle between the ultra-left "Gang of Four" (which included Mao's widow Jiang Qing) and Deng Xiaoping. By 1979 Deng had gained effective control of the government, and had introduced a series of major economic reforms which led eventually to the disbanding of the communes and the reduction of state intervention in favor of a more market-oriented economy. Since the trial and imprisonment of the Gang of Four, a new constitution was adopted and the new leadership introduced an "open door" policy to encourage foreign trade and investment. Special Economic Zones were set up in coastal regions, aimed at introducing advanced technology to modernize China. Such changes were highly significant in a state that once prided itself on having no foreign investment and debt, and show how far things had swung toward capitalism since the Cultural Revolution.

National minorities

China is the home of the Han people, who make up 93 percent of the population; the rest are non-Han (Mongolians, Tibetans, Uighurs and others), whose homelands account for two-thirds of the land area. These enormous regions – Xinjiang alone is as large as Western Europe – provide the Han with buffer zones against potentially hostile neighbors. Within months of setting up the People's Republic of China, the CCP had sent teams into Inner Mongolia and Xinjiang to establish a firm hold over these outlying areas, and in 1951 the

Chinese seized control in Tibet. A rebellion in 1959 was ruthlessly crushed by the Chinese; Tibet's religious ruler, the Dalai Lama, fled into exile.

In 1965 Tibet became an autonomous region of China. It is useful to China as a tourist center earning foreign currency. Electrification and other improvement schemes have brought no benefits to the Tibetans themselves, who bitterly resent the eradication of their culture, language and religion. In 1989, the 30th anniversary of the Tibetan rebellion, renewed fighting broke out.

The CCP's policy toward its non-Han peoples continues the Han's belief in their superiority over neighbors who had long been treated as sources of tribute to maintain the empire. There is antagonism toward China's other minority nationalities. In Xinjiang, in the northwest, there have been protests over the use of grazing grounds belonging to the Uighur nomadic herdsmen for farming and nuclear tests, which until recently were carried out above ground.

Chairman Mao visits a tea plantation Mao Zedong, through his immense personal appeal and through his writings, dominated the CCP for over fifty years.

Riots in Tibet A monk defies Chinese troops. The presence of the Dalai Lama in exile in the West draws world media attention to Tibet's plight.

China's administrative divisions The shape of China today remains remarkably similar to the empire established by the Qing dynasty in the 3rd century BC, when the Great Wall of China was built to keep out invaders from the north. The five autonomous regions, inhabited by China's national minorities, still act as buffer zones. Administratively China is divided into provinces. In 1984, 15 coastal cities were added to the Special Economic Zones to develop foreign trade.

THE DILEMMAS OF PARTY CONTROL

The highest organ of state power in China is the National People's Congress (NPC). It meets each year (its assemblies were suspended for 11 years from 1964 to 1975) and appoints the state council of ministers, headed by the prime minister. Effective control, however, has been wielded through the CCP, with its parallel hierarchy of local, regional and central committees.

In the early 1980s China's new leadership attributed the country's serious economic problems to excessive Party interference. Party domination of government was seen to be a constraint on the economy, stultifying initiative. Intervening at every level, the Party would insist that political criteria be applied to all situations. Most people were keen to toe the Party line, no matter what that line was, as being seen to oppose the Party could have serious consequences.

This mentality led to the falsifying of production figures to please superiors. There was evidence that even after reform statistics should be treated with care – grain output figures might be exaggerated, or the number of births be underreported to give the semblance of success to China's policy to ensure that couples have only one child, for example.

Reducing the Party's dominance

The CCP leadership tried hard to reduce its domination of both government and economy. This was an immensely difficult task because a position in the Party hierarchy brought with it lifelong security, status and other advantages. There were signs, however, that the NPC had acquired some real influence to check the raw power of the Party, and the power shift was felt, too, at lower levels. People such as factory managers and local administrators, who had been dominated by the power of CCP committees to interfere at all levels, were given greater autonomy in decision-making.

In the rural commune system the CCP previously determined what, where, and how much was produced: the setting of inadequate financial rewards to motivate peasants and workers during the ultra-political Cultural Revolution had made productivity levels unacceptably low. The economic reforms of the 1980s were designed to link people's rewards more closely to individual and family effort. In the villages families were given control of their own land and were made responsible for agricultural output; they were allowed to keep or sell any surplus. In industry, managers became answerable to their balance sheets, and workers were rewarded by bonuses.

By the late 1980s private businesses had become widespread, particularly in the larger cities and nearby rural areas, and state-owned enterprises had more freedom to control their supplies, market their products, and invest profits.

The widespread initial enthusiasm for economic reforms evident in the early 1980s had gone very sour by the end of the decade, especially among inhabitants of the towns and cities. With an annual inflation rate of at least 20 percent, the threat of more "price reforms" (ending state subsidies and letting goods and services such as grain, housing, health care and transport sell at market rates) led to serious social unrest. The more relaxed economy meant that corruption and self-seeking were rampant. Ironically, the shift in economic policy to reward entrepreneurs meant that some local CCP leaders were able to make use of their existing power to set themselves up in business ahead of others.

The chief architect of the reforms was Deng Xiaoping. His status – he was neither Party chairman, prime minister nor president – showed some of the illogicalities of China's political system. Though widely regarded as the country's main leader, his power was largely

China in turmoil Chanting Red Guards hold aloft the "Little Red Book" of Mao Zedong during the Cultural Revolution – the "disastrous decade" as China's leadership now describes it. These self-appointed judges of who was following the correct political line created chaos – yet brought some improvements.

THE CULTURAL REVOLUTION

The Cultural Revolution was initiated in 1966 by Mao Zedong to oppose what he and his followers regarded as the erosion of China's socialism by the economic policies of Liu Shaoqi and others, who were termed "capitalist roaders" by the Maoists. A series of extreme political campaigns and bitter factional struggles at times threatened to disintegrate into civil war. "The Great Proletarian Cultural Revolution" is officially reckoned to have lasted ten years, its end being marked by the arrest of the Gang of Four shortly after Mao's death.

While it was at its height, from 1966 to 1969, schools and colleges closed and many of the students formed the famous Red Guards. They were encouraged to demonstrate against disgraced Party leaders and denounce the "enemies of socialism". Industry was disrupted, and confusion rife.

The Cultural Revolution had a divisive and wounding impact. Intellectuals were hounded, and hundreds of thousands of people illegally detained, physically attacked and tortured. Many members of China's future leadership were themselves imprisoned for long periods. Its impact on rural development was more beneficial. Irrigation systems, schools and clinics were built, and nearly 2 million "barefoot" doctors (rapidly trained paramedics) brought improved health care to the villages.

China's parallel structures State and Party organizations mirror each other at all levels. Theoretically, executive power is in the hands of the state council, elected by the National People's Congress, but day-to-day decisions of government are made by the standing committee of the 22-member Party politburo. From 1986 membership of the Party and state Central Military Commission, headed by Deng Xiaoping, was the same.

The Gang of Four The austere conformist politics pursued by Mao's wife and three other Party officials during Mao's last senile years made them widely hated.

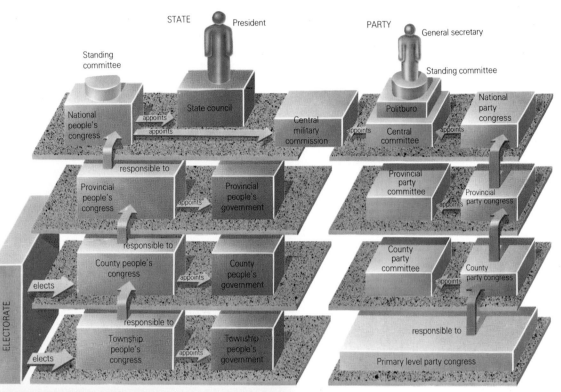

unofficial, deriving from his chairmanship of the Central Military Commission.

The Fifth Modernization

Willingness to reduce the grip of the center did not extend beyond economic reform. The individual freedom that many Chinese had hoped for after the repression of the Cultural Revolution did not materialize. The economic reforms were aimed at the Four Modernizations (of industry, agriculture, science and technology, and defense). Democracy (called the Fifth Modernization by many educated Chinese) did not figure in the CCP's list of priorities.

In the late 1970s it appeared that the CCP would tolerate more open political debate, and a loosely knit democracy movement began to emerge. In many cities unofficial magazines were circulated; posters appeared on "democracy walls" promoting different ideological viewpoints. However, the movement was to prove short-lived, and the Party soon cracked down on "dissidents".

In 1986 student demonstrations calling once again for democratization led to the dismissal of Party secretary Hu Yaobang (1915–89); his death in April 1989 caused renewed demonstrations, which turned into massive protests against corruption and poor Party leadership. Tanks entered Beijing's Tiananmen Square in June. Many of the demonstrators were killed, and Party control was reasserted.

CHINA ON THE WORLD STAGE

The largest nation on Earth was accepted into membership of the UN only in 1971: before that, the desire of the Western powers (especially the United States) to isolate the People's Republic meant that until then China's seat was occupied by the nationalist government of Taiwan. By the time China took its seat in the UN, its relations with the Soviet Union were very bad. Disagreements over ideology had led to the withdrawal of Soviet aid in 1960, and a slanging match between the two governments was carried on throughout the 1960s. In 1969 this erupted into a border conflict on the Ussuri river in the northeast of the country.

The "great power triangle"
China's changing relations with the Soviet Union and the United States reflected, and in turn influenced, the balance of power between the super-powers. Likewise, the shifts in attitude of the Soviet Union and the United States to each other led both to assume vacillating policies toward China, regarding it alternately as ally or foe.

In the hostile Cold War climate of 1950, isolated by Western embargoes, China signed a military alliance with the Soviet Union. Following China's support for North Korea in the Korean war (1950–53), the United States portrayed "Red China" as an expansionist communist regime. It consequently chose to support Chiang Kai-shek's Taiwanese government as representing all China, and attempted to "contain" China militarily.

By the time relations between China and the Soviet Union had broken down in the 1960s, hostility between the Soviet Union and the United States had evolved into an implicit recognition of each other's sphere of influence. China's arrival as a nuclear power (its first test took place in 1964) disturbed this balance. Though not a great power, in a sense it had greatness thrust upon it: for another communist country to possess the Bomb was a severe disruption of the Cold War balance of power.

The intervention of the United States in the Vietnam war, and its even more massive military presence in the western Pacific, confirmed China's need to build up a force to counterbalance those of the superpowers. It also looked for diplomatic

THE DRAGON AND THE BEAR

The intensity of China's conflicts with its communist neighbor, the former Soviet Union, astonished international observers. From the signing of a 20-year friendship treaty in 1950, relations had degenerated by the 1960s and 1970s to the point where huge armies and nuclear missiles were pointed at each other. Air raid shelters had been dug in all the cities of northern China, and the Soviet Union was using Mongolia to site major bases. Fears of a full-scale invasion had mounted in China after the Soviet intervention in Czecho-slovakia in 1968.

The Chinese resented having to play the junior partner to the Soviet Union's perception of itself as communism's international leader. To this must be added deep ideological differences.

Mao Zedong challenged the plan for economic development that the Soviet Union was thrusting at China, along with aid and advisers. In the 1960s China denounced the Soviet leadership for becoming imperialist and promoting bourgeois attitudes but despite the propaganda war the two communist giants were seen by world opinion as allies until the outbreak of hostilities on the Ussuri river in 1969. By the late 1980s tension had eased. The Soviet withdrawal from Afghanistan removed a major obstacle to the healing of the rift, and in May 1989 Soviet leader Mikhail Gorbachev visited Beijing. After the demise of the Soviet Union, relations were maintained with Russia, and President Boris Yeltsin visited Beijing in December 1992.

The Chinese army on exercise Since 1949 Chinese troops have been involved in the Korean war, in Tibet, and in border clashes with India, the Soviet Union and Vietnam.

Feting the US president Richard Nixon, seated beside Zhou Enlai, samples Chinese cuisine at a banquet in Beijing. His visit in 1972 gave China greater international prominence.

Healing the rift (*far left*) Mikhail Gorbachev's visit to Beijing in May 1989 marked a significant step in improving relations between China and the Soviet Union. It was overshadowed by the student demonstrations that were taking place in Tiananmen Square, which forced his official welcome to be moved to the airport.

means of reducing Washington and Moscow's domination of the world, welcoming discussions with other groups, such as the European Community. This strategy was largely responsible for its obtaining UN membership in 1971: in the preceding decade China had nurtured diplomatic and "people-to-people" relations with many Third World countries.

It was the United States that began to change this particular arrangement of the "great power triangle". Following the failure of its military intervention in Vietnam, it needed to secure its future in Asia. President Richard Nixon's visit to China in 1972 laid the basis for China's reincorporation into the mainstream of world politics, leading in the early 1980s to the astonishing situation whereby the United States became an arms supplier to

China and a virtual military ally, this time isolating the Soviet Union.

Improving relations
In the late 1980s, as China's own lapsed ideological purity made criticism of the Soviet Union seem ridiculous, tension on that front eased as well. Improved relations would have been quicker but for the Soviet occupation of Afghanistan (seen by China as proof of its expansionist designs), and its support for Vietnam's invasion of Cambodia (in opposition to China's support of the coalition between Prince Sihanouk and Pol Pot's Khmer Rouge forces). China several times accused Vietnam of attacking it along their disputed border, when it was more likely that China was pressuring the Vietnamese army to divert forces from its

occupation of Cambodia. There were also clashes with Vietnam over the possession of the Paracel and Spratly islands in the South China Sea – which were claimed by others as well, including the Philippines, Malaysia and Taiwan.

Since the 1980s China has sought to improve relations with its neighbors and settle remaining disputes: a treaty of friendship with Japan effectively ended the hostility that had remained after World War II; relations with Vietnam were restored in 1991. Although its suppression of the prodemocracy demonstrations in Tiananmen Square in 1989 led to the temporary imposition of international sanctions, and its attitude to negotiations over Hong Kong has proved uncompromising, China is now a full player in international trade and diplomacy.

One country, two systems

A new home, or repatriation? International conferences met to discuss the fate of the thousands of refugees from Vietnam who fled to Hong Kong in the late 1980s, and were housed in desperately overcrowded camps like this one.

As China's power has increased, it has sought to reintegrate the old province of Taiwan and the colonized enclaves of Hong Kong and Macao with the People's Republic of China.

Hong Kong island was ceded to Britain after the Opium War (1839–42), when China unsuccessfully opposed Britain's encouragement of the illegal opium trade. Kowloon Peninsula was added to it after a further defeat in 1860. Its natural deep-water harbor helped Hong Kong to grow as an important trading post, and the neighboring New Territories on the mainland were leased after a further unequal treaty in 1898. This 99-year lease expires in 1997, making the rest of the colony unviable. In the years since World War II Hong Kong has become a thriving manufacturing and financial center. Macao, on the west coast of the Pearl river estuary, has never developed the same vital commercial or military importance. It was leased to the Portuguese as a trading post in 1557, and became a Portuguese colony in 1887.

In 1984 the British and Chinese governments concluded an agreement for the return of the whole of Hong Kong in 1997. In 1987 Portugal signed a similar agreement for the return of Macao to Chinese administration in 1999.

Hong Kong and Macao will be incorporated into China as "special administrative regions" and allowed a high degree of autonomy, though foreign policy and defense issues will be in China's hands. This is the basis of the idea of "one country, two systems".

Many in Hong Kong argued that Britain had not secured adequate safeguards to ensure that the "two systems" would be recognized after 1997, nor had taken sufficient steps to promote democracy in the running of the colony beforehand. These protests reached new heights after the events in Tiananmen Square in June 1989 appeared to end hopes of democratization in China. Britain's refusal to grant rights of abode to more than a few thousand of Hong Kong's citizens wishing to leave the colony before 1997 caused further anger. China, however, continued to disregard mounting diplomatic pressure from Britain to safeguard democratic rights in Hong Kong, choosing to ignore an offer of and to build an international airport when this was made dependent on the granting of guarantees.

Reunion with Taiwan

Though only a 15 percent minority, the nationalist Chinese who took refuge on Taiwan after 1949 assumed control of the island, and they continue to govern it as the legitimate rulers of mainland China. Taiwan enjoyed a period of rapid industrial growth, and emerged as one of the strongest economies in Asia. The Chinese government announced in 1986 that the principle of "one country, two systems" could also be applied to Taiwan, with the added concession that the nationalist government there would be able to retain its own armed forces.

Both the People's Republic of China and the Taiwan government have con-

Keeping watch on the border A soldier gazes into mainland China across the high barbed-wire fence erected by the British to deter Chinese refugees from seeking the riches of the capitalist world. The "one country, two systems" arrangement guarantees to preserve Hong Kong's flourishing economy into the next century.

A colony based on trade Hong Kong's main asset is its magnificent natural harbor. Although the 99-year lease with China refers only to Hong Kong's New Territories, Britain will cede sovereignty of the entire colony as retention of Hong Kong island would be untenable without them. The island has no natural resources – even its water comes from reservoirs across the border.

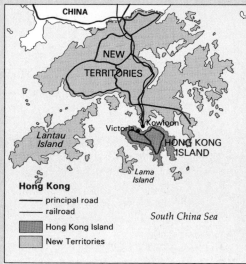

CHINA

NEW
TERRITORIES

Kowloon

Lantau
Island

Victoria

HONG KONG
ISLAND

Lama
Island

Hong Kong
——— principal road
——— railroad
  Hong Kong Island
  New Territories

South China Sea

tinued to affirm that Taiwan is part of China, and favor unity. Since the mutual hostility and armed clashes of the 1950s and 1960s, the conflict has eased considerably. The late 1980s witnessed a gradual thaw through cultural contacts and visits to their mainland homes by Taiwanese. There were even signs of a willingness to talk on the nationalist side, more intransigent than Beijing.

The offer of a "one country, two systems" deal to Taiwan's government has little attraction for the island's people, anxious about damage to their prosperity if incorporated into China's economy. It is a deal that is unlikely to be trusted until seen in operation in Hong Kong.

The struggle for democracy

4 June 1989 was one of the saddest and most shocking days in modern Chinese history. Chinese troops crushed a demonstration of students and workers. Many died as the People's Liberation Army turned its guns and tanks on its own people.

The demonstrations began with students protesting at the slowdown in Deng Xiaoping's economic reforms, and asserting vague demands for political/democratic reforms. They were joined by workers and ordinary citizens protesting at inflation and corruption. World television crews were in Beijing to witness Soviet President Mikhail Gorbachev's visit to resolve the Sino–Soviet dispute. Initially this made it difficult for the regime to crack down on the protests; then world television witnessed the death of hundreds, perhaps thousands of demonstrators and bystanders. There followed political purges in the Communist Party, and the imprisonment and execution of those accused of being central to the demonstrations.

The ruthless suppression of the prodemocracy demonstrations was an immediate indication that the regime would not tolerate political reform, while asserting that it would continue with economic reforms and not close China off from the world economy.

Monument to democracy Students from the Central Fine Arts Institute made this 10 m (33 ft) high polystyrene statue, the Goddess of Liberty and Democracy, which was erected in Tiananmen Square.

A THEATER OF WAR

NEW STATES IN TURMOIL · MANY PEOPLES, MANY REGIMES ·
CONFRONTATION IN SOUTHEAST ASIA

Three major civilizations – Buddhist, Confucian and Islamic – had shaped the cultural and political development of Southeast Asia long before the arrival of colonizers from Europe from the 16th century onward. Of the present-day states of the region, only Thailand remained free of European control. A Dutch commercial empire was based on the East Indies (Indonesia); British rule extended over Burma (Myanmar), the Malay peninsula and northern Borneo; the French established a colonial empire in Indochina (Cambodia, Laos and Vietnam); and in 1898 the United States acquired the Philippines from Spain. During World War II the entire region came under Japanese rule. It was after 1945 that its present political structure began to take shape – giving rise to civil war and extreme political instability.

COUNTRIES IN THE REGION

Brunei, Burma, Cambodia, Indonesia, Laos, Malaysia, Philippines, Singapore, Thailand, Vietnam

Disputed borders Cambodia/Thailand, Cambodia/Vietnam, Indonesia/Malaysia, Vietnam/China

STYLES OF GOVERNMENT

Republics Burma, Indonesia, Laos, Philippines, Singapore, Vietnam

Monarchies Brunei, Cambodia, Malaysia, Thailand

Federal state Malaysia

Multi-party states Cambodia, Indonesia, Malaysia, Philippines, Singapore, Thailand

One-party states Brunei, Burma, Laos, Vietnam

Military influence Burma, Indonesia, Thailand

CONFLICTS (since 1945)

Coups Burma 1962, 1992; Cambodia 1970, 1975; Indonesia 1965; Laos 1975; Thailand 1947, 1973, 1991

Revolution Philippines 1986

Civil wars Burma 1948–51; Cambodia 1970–75, 1978–91; Indonesia (East Timor) 1976–; Laos 1953–73

Independence wars Indonesia/Netherlands 1945–49; Malaysia/UK 1948–60; North Vietnam/France 1946–54

Interstate conflicts North Vietnam/South Vietnam/USA 1957–75; Cambodia/Vietnam 1978–79; Vietnam/China 1979

MEMBERSHIP OF INTERNATIONAL ORGANIZATIONS

Association of Southeast Asian Nations (ASEAN) Brunei, Indonesia, Malaysia, Philippines, Singapore, Thailand

Colombo Plan Burma, Malaysia, Philippines, Singapore, Thailand

Council for Mutual Economic Cooperation (COMECON) Vietnam

Organization of Petroleum Exporting Countries (OPEC) Indonesia

---

NEW STATES IN TURMOIL

In the postwar period Southeast Asia was the setting for a series of violent struggles instigated by communist-aligned guerrilla movements. These groups, whose origins lay in their common resistance to the Japanese, directed their activities after 1945 toward preventing a resumption of colonial rule. The first state to declare its independence was Indonesia, in 1945, though it took another four years for the Dutch to agree to a negotiated withdrawal. The Philippines became fully self-governing in 1946, and Burma gained its independence in the aftermath of the British withdrawal from India (1948). By the end of the next decade, troubled and violent years, colonial power had ceased everywhere in the region. The nationalist struggle, especially in Indochina, had far-reaching consequences.

The "domino theory"

Western observers, especially in the United States, persisted in regarding nationalist guerrilla movements as part of a concerted Soviet- and Chinese-led effort to subvert and control the region – a belief that gave rise to the "domino theory". This held that once one country fell to communist insurgency, its neighbors would also succumb, like a collapsing line of dominoes. The theory was fatally flawed, however, because it overlooked sharply defined ethnic and linguistic divisions between the different groups of insurgents, which frequently gave rise to conflict between them. It also failed to take note of the varied nature of the societies they were operating in.

In Malaya, for example, the "emergency" that began in 1948 was essentially the result of activity by communist groups that sprang from the minority Chinese community. The British strove to control them by a combination of military force with large-scale population resettlement that cut off the guerrillas from their support villages.

In Indochina, on the other hand, the communist Vietminh movement led by Ho Chi Minh (1890–1969) was deeply rooted in the nationalist resistance to the colonial order. By 1954 sustained guerrilla warfare brought about the withdrawal of the French, following their defeat at Dien Bien Phu. This left Vietnam divided at the 17th parallel of latitude, with a communist government in the north, centered on Hanoi, and a weak government in the south, supported by the United States.

Communist guerrilla activity in the south, led by the National Liberation Front, or Viet Cong, was supported by North Vietnam and China. It was seen by the United States as part of a much bigger plan for the advance of communism. Military involvement by the United States escalated throughout the 1960s: at its height 1.25 million American and South Vietnamese troops were being deployed in the war.

This ferocious application of military might failed to eradicate the guerrillas. Politicians in the United States lost the will to sustain the struggle, but a ceasefire agreement, negotiated in 1973, was breached by the North Vietnamese.

With the final withdrawal of United States troops in 1975, North Vietnam invaded and overran the south. Its capital, Saigon, was renamed Ho Chi Minh City, and in July 1976 the Socialist Republic of Vietnam was proclaimed.

In the same year the Vietnamese-backed Pathet Lao seized power in Laos, and the communist Khmer Rouge took control of Cambodia. In 1978 Vietnam invaded Cambodia, which had become increasingly hostile to it, and toppled the genocidal Khmer Rouge regime. The

A city destroyed Saigon, the South Vietnamese capital, in ruins after the final bombing offensive before it was taken by North Vietnam's armies on 30 April 1975. The former colonial capital had been laid out like a French town, with broad tree-lined boulevards and pavement cafés. Before the outbreak of war it had been a thriving center of trade. It had been the scene of a four-week battle in 1968.

Colonialism in retreat French wounded are evacuated by helicopter from the battlefield of Dien Bien Phu. Defeat at the hands of the Vietminh communist rebels dealt a devastating blow to French national pride, precipitating their withdrawal from their colonies in Indochina.

A diverse region Conflict has centered most tragically on Indochina, where the post-colonial communist nationalist movements were met by Western intervention, culminating in the Vietnam war. Elsewhere ethnic and religious diversity has proved an obstacle to national unity, and border disputes are frequent.

years of turmoil in Indochina led to the flight of over a million refugees from Cambodia, Laos and Vietnam and, in Cambodia alone, to more than a million deaths between 1975 and 1978.

Other causes of unrest

Outstanding territorial claims caused frequent clashes between states in the region in the post-colonial period. Indonesia successfully annexed West Irian (formerly Dutch New Guinea) and Portuguese Timor, though unrest continued in both; it failed to make good its claim to Sarawak and Sabah from Malaysia. There were border disputes between Thailand and Cambodia. Cambodia has a long-standing claim to Cochin China in the southern tip of the Indochina peninsula, now part of Vietnam. The Spratly and Paracel islands in the South China Sea are also the subject of contested claims.

The government of Burma (which renamed itself Myanmar in 1989) is controlled by ethnic Burmese from the lowland river valleys. It has never successfully incorporated the Kachin, Shan and Karen minorities of the border areas.

The region's island states have also been plagued by ethnic, linguistic and religious differences. In Indonesia there were a number of rebellions in the outer islands against the largely Javanese-dominated central government. There is a strong movement among Muslims in the south of the Philippines. Malaysia's population structure offers a continuing challenge to national integrity – 55 percent of Malaysians are Muslim Malays; the rest are ethnic Chinese and Indians. Malays still dominate the government, and Chinese the economy.

Southeast Asia since 1945

→ invasion
⚔ conflict
⚔ civil unrest
— Vietnam boundary 1954
1958 date of independence
colonial power
British
Dutch
French
Portuguese
US

MANY PEOPLES, MANY REGIMES

Democratic institutions play an important part in some Southeast Asian states, but there are other political systems – single-party hegemony, military dictatorship, autocratic monarchy – that also figure prominently. Diversity is the rule rather than the exception, and there is a wide range of ideologies, from Marxism to absolutist Islamic monarchy.

By the early 1990s Vietnam was one of a handful of countries in the world where the Communist Party remained in firm control; the impoverished, land-locked state of Laos was still dependent on Vietnam. Following peace agreements signed in Paris in 1991, Cambodia was administered by an interim authority sponsored by the UN, prior to elections taking place in 1993.

The power of the military

The parliamentary democracy established in Burma on independence was replaced in 1962 by a single-party state under the Burma Socialist Program Party (BSPP) headed by N Ne Win, who seized power in a military coup. It remains military in character; a civilian constitution was introduced in 1975 but without any change of personnel. In 1988 popular unrest, fueled by the stagnation produced by state control of the economy, was suppressed by the army, which remained somewhat precariously in control.

The army in Indonesia performs a more sophisticated role within a system that incorporates political parties, periodic elections and elements of a capitalist economy. After an abortive communist coup in 1965 the army assumed power under General Suharto, who ousted Indonesia's first president, Achmed Sukarno, in 1967.

Since 1971 elections to the Indonesian national parliament have been contested by a number of political parties, though in such a way as to ensure a majority for Golkar, the regime's ruling party. Parliament, which is partly composed of nominated members (mostly drawn from the military), has few powers. The official doctrine of the "dual function" gives the armed forces a political and nation-building role in addition to defense.

Although the Buddhist religious leaders and the hereditary monarch, who is formal head of state, have a significant political role in Thailand, the army has a long tradition of intervening in government. A period of "open politics" between 1973 and 1976 ended with a military coup. Although elections were later reinstated and a civilian government formed in 1983, martial law was maintained. Widespread riots followed the appointment of a general as prime minister after indecisive elections in 1992, and the king stepped in to remove him.

The legacy of British rule

With the exception of Brunei, an absolutist Islamic monarchy ruled by Sultan

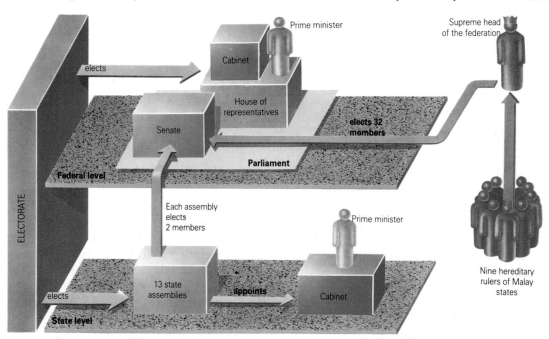

A bloodless revolution in the Philippines The supporters of Corazón Aquino call for an end to the corrupt rule of President Marcos. The nonviolent protest of the "people's power" movement eventually forced Marcos to flee the country.

An unusual federal arrangement Malaysia is a federation of 13 states, each with its own constitution, head of state and elected assembly. Nine retain their traditional rulers, the sultans. The constitution provides for these to elect one of their number to be supreme head of the federation for a term of five years, with powers similar to those of the British monarch. Effective power rests with the prime minister and cabinet drawn from the majority party or coalition in the house of representatives (Dewan Rakyar) of the two-chamber federal parliament, which is elected every five years by popular vote.

Elections in Indonesia A car decked in the colors of the Democratic Party of Indonesia (PDI). Elections to the house of representatives are held every five years, but the army-backed Golkar remains in power.

PEOPLE'S POWER IN THE PHILIPPINES

In February 1986 the dictatorship of Ferdinand Marcos (1917–89) in the Philippines was brought down by massive non-violent demonstrations of popular dissent. He had been in power since 1965. His regime, which was backed by the United States, had become increasingly repressive after the imposition of martial law in 1972. Taking charge of a potentially strong economy, he had brought it to the verge of ruin. He plundered state revenues and rewarded his clique of friends with monopolies in important industries such as sugar and coconut production.

The assassination of the popular opposition leader, Benigno Aquino, on his return from exile in the United States in 1983 was widely believed to have been carried out by Marcos' men. It provoked a clamor for political change. This reached its climax in 1986, when Marcos attempted to claim a further term as president. Despite the presence of international observers, he falsified the results of the elections, in which he had been defeated by Aquino's widow, Corazón Aquino.

A series of opposition rallies – the "people's power" movement – gathered overwhelming popular support and the backing of the Roman Catholic church. Marcos came under increasing pressure to stand down, particularly from the United States. Once it became clear that the army would not back him Marcos fled into exile, and Corazón Aquino became president.

Hassanal Bolkiah, the constitutions of the former British colonies in Southeast Asia preserved elements of the British style of parliamentary government.

The most important political party in the Federation of Malaysia is the United Malays' National Organization (UMNO). It was formed in 1946 to fight against colonial rule, and came to lead the multi-party Barisan National (National Front) coalition, which ruled all federal and most state administrations after 1974.

Singapore, an independent republic since it left the Malaysian federation in 1965, also adopted British parliamentary institutions. The dominant party since 1959 has been the People's Action Party (PAP) led by Lee Kuan Yew. Strong social and political controls asserted by the government have constrained the activities of opposition parties.

On gaining independence in 1946 the Philippines adopted the political institutions of the former ruling power, the United States. These have never worked well in a peasant society still dominated by a relatively small number of wealthy families. In 1972 President Ferdinand Marcos (1919-89) weakened these institutions still further by proclaiming martial law (lifted in 1981).

President Corazón (Cory) Aquino, who replaced his unpopular personal dictatorship in 1986, introduced a more conciliatory style of government. In 1992 Mrs. Aquino endorsed her former defense minister Fidel Ramos for the presidency and he won the election to succeed her. It remained troubled by disloyalty from pro-Marcos elements in the armed forces, and by continued resistance from Marxist and Muslim insurgents.

CONFRONTATION IN SOUTHEAST ASIA

With the exception of Burma, which maintains a policy of strict nonalignment and is closed to foreigners (though a recipient of Japanese and other international aid), the states of Southeast Asia divide into two main groups. On the one hand are the formerly Soviet-aligned states of Indochina, and on the other the island states of the Pacific. These remain to a greater or lesser degree within the United States' orbit of influence.

Following the North Vietnamese take-over of the south, Chinese and Vietnamese interests in Southeast Asia came into ever-increasing conflict. China developed a close relationship with the Khmer Rouge regime in Cambodia; its claim to exclusive control of the Paracel and Spratly islands in the South China Sea was vigorously contested by Vietnam. A series of border clashes with the Khmer Rouge led Vietnam to seek closer ties with the Soviet Union. These were strengthened when Vietnam invaded Cambodia to expel the Khmer Rouge in 1978. The following year China led a raid into Vietnam in retaliation.

The government of Hun Sen that replaced the Khmer Rouge in Cambodia was installed and kept in place with massive Vietnamese support. After 1975, when it helped the communist Pathet Lao to seize power in Laos, Vietnam maintained a considerable influence there as well, leading to fears of further Vietnamese expansion in the region.

By the late 1980s Soviet President Mikhail Gorbachev's reshaping of the Soviet Union's foreign policy objectives had brought a substantial reduction of military and economic aid to Vietnam. As a result, its troops were gradually withdrawn from Cambodia between 1987 and 1989. All-party peace agreements in Paris in 1991 saw the abandonment of communism in Cambodia, and later Vietnam resolved its differences with China.

Regional cooperation
Political cooperation to withstand Marxist influence in the region became an objective of the states that belonged to the Association of Southeast Asian Nations

Soviet-made tanks in Ho Chi Minh City The Soviet Union's support for Vietnam after the takeover of the south exacerbated existing bad relations with China, particularly after the Soviet Union also provided massive aid to Vietnam to maintain Hun Sen's puppet government in Cambodia.

The victims of war Refugees from the turmoil in Indochina have sometimes had to endure years of detention in conditions that can be appalling. Over 125,000 Cambodians fled from the fighting between Khmer Rouge guerrillas and the invading Vietnamese army in 1978–79 to camps like this one on the Thai border.

(ASEAN): Indonesia, Malaysia, the Philippines, Singapore and Thailand and, after 1984, Brunei. ASEAN, established in 1967, was originally intended to foster social, cultural and economic relations between its members. Moves toward economic integration did not develop significantly, though member states did act together in some international trade negotiations. The organization had greater success in achieving concerted political action. After Vietnam's invasion of Cambodia, ASEAN (together with China and the United States) provided political and economic aid to Khmer Rouge groups on the Thai border.

Despite these common efforts, there were clear differences in the foreign policies and the international relations separately pursued by the ASEAN states. After a brief flirtation with China in the early 1960s, Indonesia was careful to follow a course of nonalignment. Thailand and the Philippines were original signatories to the Manila Pact of 1954, which inaugurated the Southeast Asia Treaty Organization (SEATO). This was established by Western powers, including Britain, France, Pakistan and the United States, after the Korean war (1950–53) to resist possible communist aggression in the southern hemisphere. After 1971 the security of both Malaysia and Singapore was underwritten by the Five-Power Defense Arrangement with Australia, Britain and New Zealand. Brunei retained a special defensive relationship with Britain.

By the end of the century, the United States' reappraisal of its international security role seemed likely to lead to realignments within the region. In 1991 the Philippines' senate voted to urge the withdrawal of all United States' forces, and did not renew the lease of Subic Bay naval base. The ASEAN states were set to gain greater prosperity within the expanding Pacific economy.

THE FLOOD OF REFUGEES

In 1989 the number of people displaced from their country of birth by war, famine, political upheaval, repression and persecution reached more than 14 million worldwide. While greater movements of people have taken place elsewhere – more than 7 million have been displaced in Africa alone – the desperate plight of refugees caught up in the catastrophe of the Vietnam war and other conflicts in Indochina caused worldwide concern. The problem of resettlement of such vast numbers continued to demand a concerted political response, including the determination to tackle the root causes of displacement, which the developed world seemed unable to find.

In the decade after the end of the Vietnam war more than 1.6 million people fled the country, many of them taking to the seas in tiny fishing boats. One million of these began the painful process of rebuilding their lives in another country. Others did not have the chance. In the late 1980s thousands of Vietnamese refugees remained in camps throughout Southeast Asia, and the unstemmed flight of further numbers was placing enormous pressure on desperately meager resources. Western governments were unwilling to resettle Southeast Asians without clear proof of persecution, and the refugees faced forced repatriation or lengthy detention.

The tragedy of Cambodia

The area now known as Cambodia was occupied between the 6th and 15th centuries by the Khmer empire, a civilization that left the temple complex of Angkor Wat as its most durable legacy. On its break-up its territories were fought over for the next four centuries by the armies of Thailand and Vietnam, until it was brought under French control in the mid-19th century.

Cambodia became an independent state in 1953, following the collapse of French colonial power in Indochina. It was ruled by Norodom Sihanouk, first as king and then, after his abdication in 1955, as prime minister. He endeavored to remain neutral throughout the Vietnam war, maneuvering a course between the United States and the communist powers, until he was overthrown in 1970 by a military revolt while he was absent from the country on a trip to Moscow; a pro-US government, led by Marshal Lon Nol, was installed. Sihanouk formed a government in exile, which allied itself with the Cambodian communist movement (the Khmer Rouge), backed by Vietnam and China. The Lon Nol government became increasingly dependent on US military aid to fight the guerrillas and maintain its tenuous authority. Intensive US bombing of the countryside shattered the already fragile economy, and drove many people into crowded cities or into the ranks of the guerrillas.

In April 1975 Lon Nol's government fell and the Khmer Rouge occupied the capital, Phnom Penh. It looked as if the whole of Indochina would come under allied Marxist rule, but once in power the Khmer Rouge become dominated by a clique of intellectuals who had been students together in France. The most notorious of them, Pol Pot and Ieng Sary, had already begun to purge the movement of pro-Vietnamese elements, and they now ruthlessly suppressed and removed all other factions and groups within the Khmer Rouge. Sihanouk, though at first titular head of state, was imprisoned in his residence in the capital, and many of his family were killed.

The "killing fields" of Pol Pot
The regime emptied the cities, setting their populations to work on massive agricultural programs. Between 1975 and 1978 more than a million Cambodians died, either as a result of forced labor or in purges aimed at destroying the educated class of society, which it was feared might form a center of opposition to the regime. In the countryside mass graves ("killing fields") are testimony to the brutality of Pol Pot's regime.

Not content with remaking Cambodian society, the Khmer Rouge also sought to restore the ancient glory of the Khmer people through the recovery of territory lost to Vietnam in the 18th and 19th centuries. Border incursions provoked tension between the two countries, and the Pol Pot regime developed close links with China as the Vietnamese drew nearer to the Soviet Union.

In December 1978 Vietnam invaded Cambodia and replaced the Khmer Rouge with a client government led by Hun Sen. Thousands of refugees fled to the Thai border. Armed by China and aided by the

Sihanouk – an indefatigable politician As king of Cambodia Norodom Sihanouk, born in 1922, declared his country's independence from France. Overthrown in 1970, he allied himself with the Khmer Rouge and was reappointed head of state in 1975, until Pol Pot turned against him. He later formed an uneasy coalition with the Khmer Rouge and one other resistance group against the Vietnamese occupation.

ASEAN states and the United States, the remnants of the Khmer Rouge slowly gained power among them, and mounted attacks against the new regime.

Without international backing for its government – in the late 1980s the Khmer Rouge held on to Cambodia's seat in the United Nations – Hun Sen's regime made only slow progress in recovery and reconstruction. The cost of supporting the regime placed an enormous strain on Vietnam's economy, and in 1988-89, encouraged by the Soviet Union, it began to withdraw its troops. A UN peace plan was signed in 1991 and UN-supervised elections in May 1993 resulted in the formation of a coalition government headed by the nationalist FUNCINPEC and the Cambodian People's Party (PCC). Prince Sihanouk returned to reclaim his crown but peace was marred by the guerilla activities of the Khmer Rouge who refused to participate in the elections.

Testimony to horror A mass grave in one of the killing fields of Cambodia. Many victims of Pol Pot's regime died as a result of famine and disease; others were murdered in mass purges of "intellectuals".

A brutal regime A propaganda poster erected by the pro-Vietnamese government displays the stark brutalities of Pol Pot's drive to create a self-reliant peasant society.

PROSPERITY OUT OF DEFEAT

BREAKING WITH THE PAST · DEMOCRACIES UNDER STRAIN · JAPAN'S PLACE IN A CHANGING WORLD

For most of their history Japan and Korea have been politically distinct, and frequently hostile to each other, despite a similar cultural and linguistic heritage. The Japanese imperial dynasty can be traced back to the 7th century; in the 12th century real power was assumed by the shoguns, or military leaders, who preserved the country from invasion and imposed rule through a rigid social hierarchy. Foreigners were almost entirely excluded until 1853, when Japan's self-imposed isolation was broken by the United States' navy. For almost all the last thousand years Korea has been a single country. It was annexed by Japan in 1910 and ruled as a colony until 1945 when, as a consequence of Japan's surrender at the end of World War II, the country was divided north and south at the 38th parallel of latitude.

BREAKING WITH THE PAST

Scarcely more than a decade after Japan had been opened up to trading relations with the West the power of the shogunate was destroyed (1868). Direct imperial rule was restored in the name of the Meiji emperor (Mitsuhito, 1868-1912), and the leaders of the new regime, ruling from Tokyo, set about modernizing Japan's economic, social and political systems.

Feudal practices were abolished, and by 1890 a form of constitutional government had been adopted that was based on Western models. Although they were built around the institution of the emperor, these did allow some degree of elected popular representation. At the same time, rapid industrialization took place, the army was modernized, and a powerful navy was also founded.

The demand for space and raw materials to support Japan's growing population and industries led to a surge of colonial expansion. Once its hold over the northern island of Hokkaido and the islands to the south had been established, Japan's energies turned toward the mainland.

Here the goal of its ambitions was Korea, but first other rivals in the region had to be overcome. China and then Russia were defeated in wars that gave Japan control of Taiwan, then known as Formosa (1895), and the southern half of Sakhalin island (1905), as well as extensive rights in southern Manchuria on the Chinese mainland. Korea, which had earlier been made a Japanese protectorate, was annexed in 1910.

COUNTRIES IN THE REGION
Japan, North Korea, South Korea

Disputed territories   Japan/Russia (Kuril Islands and south Sakhalin)

STYLES OF GOVERNMENT
Republics   North Korea, South Korea

Monarchy   Japan

Multi-party states   Japan, South Korea

One-party state   North Korea

One-chamber assembly   North Korea, South Korea

Two-chamber assembly   Japan

CONFLICTS   (since 1945)
Internal unrest   South Korea 1960, 1979, 1987

Coup   South Korea 1961

Interstate conflicts   North Korea/South Korea (with Chinese and UN involvement) 1950–53

MEMBERSHIP OF INTERNATIONAL ORGANIZATIONS
Colombo Plan   Japan

Organization for Economic Cooperation and Development (OECD)   Japan

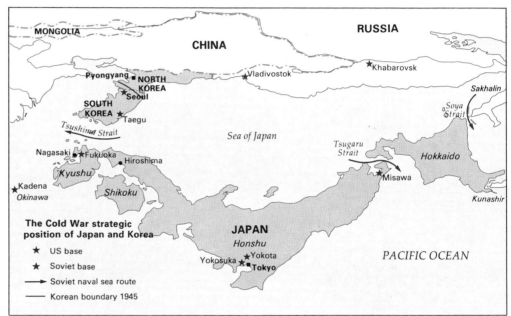

The Cold War strategic position of Japan and Korea
- ★ US base
- ★ Soviet base
- → Soviet naval sea route
- — Korean boundary 1945

The balance of power in northeast Asia Japan's proximity to military installations in the east of the former Soviet Union gave it a key role in Western defense strategies. Major United States air and sea bases are maintained on Japan, and in South Korea. The demilitarized zone that separates the two Koreas has become one of the most heavily fortified borders in the world.

Hiroshima's terrible legacy On 6 August 1945 the world's first atomic bomb was dropped on Hiroshima. By the end of the year more than 140,000 people had died from the effects of radiation. This first dreadful step into the nuclear age – which despite all the suffering it caused sparked off the nuclear arms race of the 1950s and 1960s – still haunts the Japanese people.

Old and new in Japan Mourners at the funeral of Emperor Hirohito, who died in 1989. After Japan's defeat in World War II he formally rejected belief in the divinity of the emperor, paving the way for the country's adoption of a democratic constitution.

This first period of Japanese expansionism was followed by a second in the 1930s, when the army seized from the civilian government the initiative in foreign policy. Military success in Manchuria and in eastern and central China led to Japanese control of these areas. With domestic dissent effectively stifled and the civilian administrators subordinate, continued expansionism led to conflict with Western powers. Japan entered World War II on 7 December 1941 by launching a devastating attack on the United States navy at Pearl Harbor, Hawaii, and in a succession of victories overran Indochina, the Philippines, the Malay Peninsula, Burma and the Dutch East Indies.

The years of democratization
Following defeat in 1945 Japan was occupied by Allied forces, mainly from the United States, until 1952, when full sovereignty was restored. Although the Soviet Union had declared war on Japan only days before the surrender, it acquired control over most of the islands to the north of Hokkaido, including south Sakhalin and the Kuril Islands. Four of these (Etorafu, Kunashiri, Shikotan and the Hobomai group) are claimed by Japan, preventing the conclusion of a peace treaty with the Soviet Union's successor states. In the south, the Okinawa islands remained under United States control until 1972: the United States still maintains military bases there, as well as on the main islands of Japan.

In 1947 a new parliamentary constitution was adopted. It allowed the emperor no part in government, but made him "the symbol of the state and of the unity of the people, deriving his position from the will of the people". At the same time a comprehensive program was undertaken under United States supervision to make Japanese institutions more liberal and democratic. Reforms focussed in particular on land ownership, education, the police force, and the domination of the economy by huge industrial and commercial business monopolies. Not all were equally successful – giant corporations such as Mitsui and Mitsubishi still dominate the economy – but the 1947 constitution provided the political framework within which Japan has prospered in the modern world.

Korea – a partitioned country
The Soviet Union, whose troops occupied Korea north of the 38th parallel of latitude in 1945, installed a Soviet-trained communist government. In 1948 it declared North Korea a Democratic People's Republic, with Kim Il Sung as president. In the south, by contrast, a conservative pro-Western regime emerged under the leadership of Syngman Rhee (1875–1965).

Growing antagonism between the two led to the Korean war (1950–53). A North Korean offensive across the 38th parallel was repulsed from the south by a United Nations force. Led by the United States, it then occupied most of the north until North Korean forces, with massive Chinese support, drove it back. A long, bloody stalemate ensued as truce negotiations made sluggish progress. The armistice agreed in June 1953 between the communist and United States armies was not superseded by a formal peace treaty.

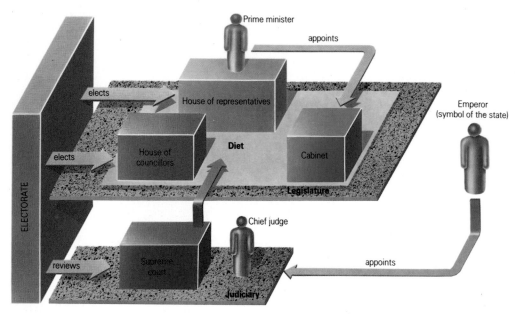

DEMOCRACIES UNDER STRAIN

Japan's 1947 constitution vests executive power in the prime minister, who selects a cabinet that is collectively responsible to the two-chamber parliament (Diet). This is made up of the House of Councillors (the upper house) and of the more powerful House of Representatives.

Following the model of the United States constitution, the judiciary was made a distinct branch of government. The Supreme Court controls the activities of all the lesser courts. Its power of judicial review gives it the right to declare any piece of legislation or administrative action unconstitutional, but in practice it has used this power very sparingly.

The 252 members of the House of Councillors are elected for six-year terms, with half the membership standing for election every three years. A total of 152 seats are elected from constituencies whose boundaries coincide with Japan's 47 prefectures; the rest are chosen by a form of proportional representation, with the whole country acting as one constituency. The House of Councillors has no effective voice with regard to budget decisions or agreeing treaties, and the House of Representatives is able to override any of its enactments provided that a two-thirds majority is gained.

The 512 members of the House of Representatives are elected from 130 electoral districts for a maximum of four years. Very few parliaments live out their

Japan's two-chamber parliamentary system The American authors of the 1947 constitution envisaged only a single popularly elected chamber, but the Japanese insisted on adding an elected upper House of Councillors. Every bill passes through a complex committee system in each house before being approved by a full session of both houses together.

full terms as the cabinet (in practice the prime minister) can dissolve the house at any time, and must do so if a vote of no confidence is passed, as happened in 1980. Each district returns between three and five members. Electors have a single vote, which they cast by writing the candidate's name on the ballot paper.

Political reform

The Liberal Democratic Party (LDP) held power in Japan from 1955, when it was formed, until 1993 when it lost its overall parliamentary majority and was replaced by a seven-party coalition which ranged from left-wing to center right.

Public dissatisfaction had been growing in the 1980s with the level of corruption in the political system, this was attributed to the country's electoral system of multimember constituencies. Members had to compete against each other as well as for election and this had encouraged the growth of factions within the ruling party.

In the late 1980s and early 1990s the LDP was riven by corruption scandals and internal dissension. As a result they suffered a massive drop in support and unprecedently lost control of the upper house in July 1989. When it became

apparent that political reform was not forthcoming LDP members voted for a motion of no-confidence in the government precipitating a general election, held in July 1993, in which the LDP was defeated. LDP rebels formed two new parties, the Japan Renewal Party and the New Party Harbinger which became part of the new coalition government. The other parties were the Democratic Socialist Party (DSP); the Japan National Party (JNP); the Social Democratic Party of Japan; the Clean Government Party and the United Social Democratic Party.

The new government set about introducing political reform and, after a struggle in both houses, a new constitution

Alternative voices A campaigner for a right-wing minority party airs his views in a Tokyo street. Despite the presence in Japan of numerous political parties of all persuasions, the LDP's domination of the Diet stretched back more than three decades. By the 1980s its unlimited monopoly appeared to be coming into question. Women, too, were beginning to demand a greater part in Japan's overwhelmingly male-dominated politics.

A powerful faction leader Yasuhiro Nakasone held several ministerial posts before establishing his own faction within the LDP in the 1960s. As prime minister (1982–87) he narrowly escaped involvement in the 1983 Lockheed scandal, which had some similarities with the more damaging Recruit scandal that took place in 1988–89.

JAPAN'S MONEY-POLITICS

One consequence of single-party rule is that it compels politicians at every level to depend on financial donations from business companies. Election campaigns in multi-member constituencies are extremely expensive, with the result that even in a non-election year a member of the Diet is expected to spend large sums of money on his constituents. For example, a system perfected by Kakuei Tanaka (prime minister 1972–74) ensures that all couples getting married and every person over 80 celebrating a birthday in his constituency receives a card signed personally by the member, and that a thick candle with his name on it is burned at every funeral ceremony.

A member's income covers only a fraction of such expenses, and to ensure reelection he needs to raise substantial sums elsewhere. His main source of cash will be his faction: many "tunnel organizations" exist to pipe money into faction funds. In 1988–89 investigations into the activities of the Recruit Cosmos business empire revealed a web of corruption that extended throughout the LDP and even the opposition parties; one of the ways the company had made donations was to hand out to politicians cheap shares that soon afterward shot up in value.

Only a thorough reform of Japan's political system, including the creation of single-member constituencies and the abolition of the factions, would bring an end to Japan's money-politics. There was condemnation of the system within Japan following these revelations, which resulted in the resignation of two prime ministers.

---

was agreed early in 1994. This created 300 single-seat constituencies and 200 seats to be filled from party lists by proportional representation. The electoral map was redrawn to correct disproportionate voting power in rural districts and politicians were subjected to a contribution limit of 500,000 yen from any one company in one year. Provision was made for the future abolition of such donations.

Autocracies in the Korean peninsula
When he died in 1994, Kim Il Sung, North Korea's "great leader", had been in power longer than any other head of state in the world. After the systematic elimination of his main rivals in the 1960s, he remained in control of the ruling Workers' Party. The industrialization of North Korea was extremely rapid up to the 1960s but was not sustained at the same pace, and by the late 1980s its rate of economic growth had fallen far behind that of South Korea.

Between 1945 and 1979 South Korea was ruled by two autocratic rulers. Syngman Rhee was overthrown in 1960 by a student-led revolution; his successor, General Park Chung Hee (1917-79), created a "civilianized" military government that retained the outer appearances of a representative democracy, but denied citizens many of their basic rights. He pursued an economic strategy that transformed South Korea into one of the fastest-growing economies in the world. Assassinated in October 1979, he was succeeded by another ambitious general, Chun Doo Hwan.

After widespread demonstrations in the summer of 1987, Chun agreed to hold a presidential election, which was won by his protégé Roh Tae Woo. The first president for 30 years to achieve power without a coup, Roh promised further progress toward democracy. In elections in 1988 the ruling Democratic Justice party lost its overall majority in the national assembly. It joined with two minor parties to form the Democratic Liberal Party (DLP), but continued to lose ground in elections in 1992.

JAPAN'S PLACE IN A CHANGING WORLD

Japan's changing status in the world since World War II is most clearly seen in the shifts of its relationship with the United States. The terms of the treaty that returned full sovereignty to Japan in 1952 contained a separate security agreement. In this Japan consented to the deployment of United States forces "in and about Japan so as to deter armed aggression upon Japan" and even "to put down full-scale internal riots". Some of these details were revised in 1960, and again in 1970, amid massive anti-government rallies, but Japan remained the junior partner in the relationship – one part of the United States' security network in the Pacific – until at least the early 1970s.

By this time Japan's economic strength had grown to a point where its leaders felt entitled to a more equal voice in policy making. Their continuing refusal to open diplomatic relations with communist China, despite enormous domestic pressure to do so, was therefore regarded as a considerable act of loyalty toward the United States. To the Japanese, President Richard Nixon's announcement in July 1971, without prior consultation, of his forthcoming visit to Beijing was, to say the least, disconcerting. A month later a series of economic measures aimed at making Japanese exports less competitive in the United States market compounded the outrage.

These twin "Nixon shocks" generated an intense debate in Japan on the wisdom of remaining a military satellite of the United States. A sudden increase in the price of oil and other raw materials later in the decade served to underline Japan's considerable economic vulnerability, and further encouraged the construction of a more flexible and more independent foreign policy.

Forging new ties

In 1978 a treaty of peace and friendship was concluded with China. By the 1980s Japan was China's biggest trade partner, and a considerable investor in projects aimed at modernizing China's economic base. The unresolved dispute over ownership of the islands north of Japan has inhibited the improvement of relations with the Soviet Union. Fears of Soviet military power in the region were reinforced in 1983 when a Korean airliner that had strayed into Soviet airspace over Sakhalin island was shot down, though

JAPAN'S DEFENSIVE FORCES

Article 9 of Japan's new constitution renounced "the threat or use of force as the means of settling international disputes": Japan would in future have no armed forces. The Korean war, when the Japanese islands provided an essential base for United States troops, changed all that. As a part of the United States strategy to contain communism in Asia, Japan in 1954 was permitted to form a modest "self-defense force". The JSP and other left-wing parties have since then consistently protested against this decision.

Japan's military budget steadily increased until, in 1988, it stood at $29 billion – the third highest in the world after the Soviet Union and United States. This brought it to more than 1 percent of GNP, the level at which it had until then been set.

Japan rearmed Its fleet of 60 destroyers is surpassed only by those of the USA and USSR. Britain has 37 destroyers, France 17.

Heightened perceptions of the military threat presented by the Soviet Union in the early 1980s hardened Japan's defense policies. Three specific roles were ascribed to its land and sea forces – to defend Japan from limited land attacks, to block the three straits through which the Soviet navy must pass to reach the Pacific, and to protect the sea lanes to a distance of 1,800 km (1,000 nautical mi) from Japan. Defense budgets were set to grow by over 6 percent a year to carry out these roles. Then, in 1992, the Diet voted to allow Japanese troops to be sent overseas on "peace missions" – regarded by many as a reversal of the constitution pledge.

Anti-war demonstrations Members of a students' revolutionary group act out the horrors of nuclear war. The JSP and other left-wing parties have spearheaded opposition to Japan's growing defense commitment, which they argue is a betrayal of the non-militaristic clauses of the constitution.

relations later improved. Talks with Russia about the future of the northern islands seemed close by the early 1990s.

A resumption of normal relations with South Korea proved more delicate, as the government there was reluctant to enter into negotiations with its former colonial rulers. The signing of a treaty with General Park's government in 1965 provoked massive demonstrations in both Japan and South Korea. Diplomatic relations remained volatile and subject to stress, but economic ties strength ened. In the 1970s Japan became the largest foreign investor in South Korea's economic

boom, and now found itself threatened by an economic rival in the region.

As Japan's prosperity grew it increased its profitable trading relations throughout Southeast Asia, despite residual resentment of its former imperialist role there. It joined the Organization of Economic Cooperation and Development (OECD) – the "rich nations club" – in 1964. The approach of the single market in Europe in 1992 aroused Japanese fears of a "fortress Europe" from which they would be excluded. By the late 1980s Japan had become the largest supplier of aid in the world – $11 billion in 1988.

Toward a new Asia
It seems likely that by the end of the century a new equilibrium will have emerged in Asia. As new relationships are formed Japan's role will be crucial. In

the late 1980s its alliance with the United States was under strain because of differences over trade and defense, but the consequences of a complete break were too dire for either side to contemplate.

There were signs of significant shifts from former positions in the Korean peninsula. Under President Roh, South Korea expanded its trade and diplomatic links with communist states. It is possible that the United States will at some future date reduce its sizeable military presence, which was the target of frequent hostile demonstrations during the 1980s.

North Korea's relations with China and the Soviet Union, its major allies, were more complex. In the climate of internal reform neither China nor the Soviet Union was fully enthusiastic in its support of Kim Il Sung's conservative but nonetheless strategically vital regime.

Two Koreas or one?

Korea is one of several states that have been partitioned in the 20th century as the result of war and political pressure from outside powers – Germany, and Vietnam between 1954 and 1975 are the other examples. Both sides recognize each other as part of the same nation, and reunification is officially proclaimed by both regimes as a desirable and attainable political objective.

During the early 1970s the rival Korean regimes competed with each other to win international support for their respective reunification plans. The North favored a confederation of the two existing regimes, which would become autonomous subsystems in a single state ("one country – two systems"). The South put forward the formula of "cross-recognition", modeled on the precedent of East and West Germany, whereby the interested outside powers – in this case China, Japan, the Soviet Union and the United States – would recognize both Korean governments as legitimate. Both would become full members of the United Nations (North Korea had observer status only). A second proposal suggested that national elections should be held, and that a new constitution for a unitary Korean state and government ("one country – one

system") should be agreed.

Neither side was able to gain clear support from the world's governments for these proposals. The North retained the support of the communist world and membership of the nonaligned movement, but most governments adopted a nonpartisan approach to the question of Korean reunification, favoring a vaguely defined "peaceful resolution".

A new phase in negotiations

During the 1980s the overriding role of ideology in international affairs receded as the importance of economic relations increased, strengthening the position of South Korea. In 1988 the successful hosting of the 24th Olympic Games in the South Korean capital, Seoul, coincided with a new, legitimately elected government, the achievement of a massive trade surplus, and the opening of closer political relations with communist states. South Korea initiated a new phase of negotiations with the North.

Discussions between the two rivals took several forms, but once again continuing military tension – represented by the annual Team Spirit exercise carried out by South Korean and United States troops – proved a stumbling block to

Student riots in South Korea A succession of authoritarian rulers ensured that political opposition was vigorously and sometimes brutally suppressed. However, student demonstrations like this one in the summer of 1987 helped to secure the holding of democratic presidential elections as South Korea sought to improve its international image before the 1988 Seoul Olympics.

progress. The conservative ruling elite in South Korea reacted harshly to the opposition parties seeking nongovernmental links with the North, and to workers attempting to establish militant and independent trade unions.

Outside powers all supported a relaxation of tension in Korea, but none was willing to sacrifice vital security interests in the region for the sake of Korean reunification; they were limited in the

A divided nation South Korean troops patrol the border with the North – the line of the 1953 ceasefire. It is estimated that more than 10 million families remain split as a result of the partition.

The "great leader" A massive bronze statue of North Korea's Kim Il Sung (1912–1994) towers above passers by. His personality cult has outstripped those of Joseph Stalin and Mao Zedong this century.

degree to which they could influence the attitude of either of the Korean regimes.

At that time the path toward further democratization in the South could by no means be considered assured. The regime in the North had not even set foot upon that path. It remained one of the most conservative of all communist states, unaffected by the internal reforms that were taking place elsewhere within the communist bloc. Although his son Kim Jong-Il is his chosen successor, it is impossible to predict what will happen following Kim Il Sung's death in 1994. However, South Korea's growing economic, diplomatic, political and military advantages made it almost certain that North Korea would have to undergo considerable restructuring if it hoped to become able to compete.

The reunification of Korea remains a hope for the future. The Armistice of 1953, to which South Korea was not a signatory, has not been replaced by a formal peace treaty, but in 1991 a non-aggression agreement and nuclear pact was signed between North and South Korea, signaling a move in that direction. In the same year North Korea joined the United Nations. Its signing of the Nuclear Safeguard Agreement, allowing inspection of its nuclear facilities, indicated more open relations.

PACIFIC STATES

COMMONWEALTH DOMINIONS · PARLIAMENTARY RULE · BREAKING THE IMPERIAL LINK

The eleven independent sovereign states of Australasia are all former territories of the British empire. By far the largest two states, Australia and New Zealand, were colonized only in the last 200 years. They are today both stable liberal democracies within the Commonwealth. Of the widely scattered island groups in the region nine achieved their independence between 1962 (Western Samoa) and 1980 (Vanuatu). The largest, Papua New Guinea, was administered by Australia until independence in 1975. Nauru (independent 1968) is the smallest state in the world, with a population of only 8,000. The rest are dependencies: Australia and New Zealand administer a number of island groups; France has three overseas territories (French Polynesia, New Caledonia and the Wallis archipelago), and the United States ten.

COUNTRIES IN THE REGION

Australia, Fiji, Kiribati, Nauru, New Zealand, Papua New Guinea, Solomon Islands, Tonga, Tuvalu, Vanuatu, Western Samoa

Island territories Cocos Islands, Christmas Island, Norfolk Island, Heard and McDonald Islands (Australia), Cook Islands, Niue (New Zealand)

Dependencies of other states American Samoa, Guam, Johnston Atoll, Midway Islands, Northern Marianas, US Trust Territory of the Pacific Islands (Marshall Islands, Micronesia, Palau), Wake Island (USA); French Polynesia, New Caledonia, Wallis and Futuna Islands (France)

STYLES OF GOVERNMENT

Republics Fiji, Kiribati, Nauru, Vanuatu

Monarchies Australia, New Zealand, Papua New Guinea, Solomon Islands, Tonga, Tuvalu, Western Samoa

Federal state Australia

Multi-party states Australia, Fiji, New Zealand, Papua New Guinea, Solomon Islands, Vanuatu, Western Samoa

States without parties Kiribati, Nauru, Tonga, Tuvalu

CONFLICTS (since 1945)

Nationalist movements Australia (Aborigines); New Caledonia (Kanaks); New Zealand (Maoris)

Coups Fiji 1987

MEMBERSHIP OF INTERNATIONAL ORGANIZATIONS

Colombo Plan Australia, Fiji, New Zealand, Papua New Guinea

Organization for Economic Cooperation and Development (OECD) Australia, New Zealand

South Pacific Forum Australia, Cook Islands, Fiji, Kiribati, Micronesia, Nauru, New Zealand, Niue, Papua New Guinea, Solomon Islands, Tonga, Tuvalu, Vanuatu, Western Samoa

Notes: Parts of the continent of Antarctica are claimed by Argentina, Australia, Chile, France, New Zealand, Norway and the UK.

COMMONWEALTH DOMINIONS

The Dutch explorer Abel Tasman (1603–59) circumnavigated Australia as early as 1642, but permanent European settlement did not begin until 1788, when a British penal colony was established in Botany Bay (Sydney). Eighteen years earlier Captain James Cook (1728–79) had claimed the surrounding land as the British territory of New South Wales, and it was here, on the east coast, that the earliest colonies were set up, beginning with Hobart in Van Diemen's Land in 1804 (renamed Tasmania in 1853). Within the next three decades coastal settlements were established to the north at Moreton Bay (Brisbane), on the west coast at Perth, and on the south coast at Melbourne and Adelaide. By 1862 boundaries had been drawn around the hinterlands of these original settlements, and these define Australia's six states today.

In 1901 the states came together to form the federal Commonwealth of Australia, which became the second self-governing dominion (after Canada in 1867) in the British empire. Formal independence was recognized by the Statute of Westminster (1931). To prevent Sydney, already Australia's largest city, from dominating the federation, Melbourne was selected as the first capital city. The federal government took over administration of the sparsely settled Northern Territory and, in 1926, created the Australian Capital Territory, an enclave halfway between Melbourne and Sydney, where a new capital, Canberra, was built.

The fate of the Maoris

Some time before the end of the 14th century New Zealand had been occupied by Maoris from Polynesia. By the time Tasman sailed round New Zealand (1642) they were hunting and fishing over both islands, having settled mainly on North Island. By the end of the 18th century New Zealand's natural harbors were being used by European whalers- for refitting and refurbishing sailing vessels; exploitation of its natural resources, especially timber, followed. Traders and missionaries from Australia began to establish permanent settlements.

At first the British government was content that these settlements should be informally linked with the colony of New South Wales, but in 1840 the Treaty of

Waitangi was signed with Maori chiefs. In it they ceded "absolutely and without reservation all rights and powers of sovereignty to Britain". In return the British guaranteed the Maoris "the full exclusive and undisturbed possession of their lands and estates, forests, fisheries and other properties which they may collectively or individually possess so

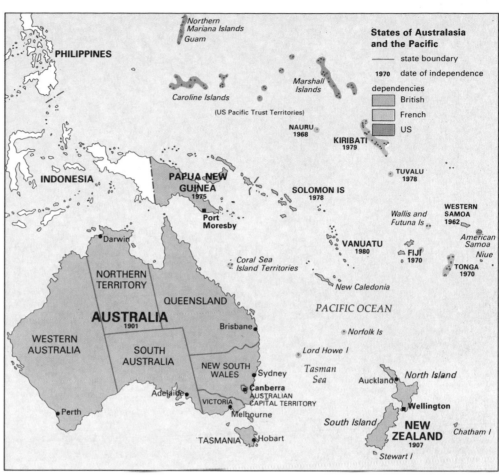

The demand for Aboriginal rights European settlers drove the Aborigines off their land into the remote outback areas of Australia. Recent protest has centered on establishing Aboriginal rights even to this land, now threatened with mineral exploitation.

Island states Recent moves to keep the South Pacific free of northern hemisphere interference have given its scattered island states a sense of regional identity. French Polynesia lies some distance east of Samoa.

long as it is their wish and desire to retain the same in their possession".

Soon afterward, sheep farming was introduced from Australia, and large-scale agriculture rapidly developed. A series of "land wars" fought by the Maoris in the 1860s led to concessions, including representation in parliament. New Zealand became self-governing in 1853 and was granted dominion status in the British empire in 1907; full independence was granted in 1931.

From a population of 250,000 in 1840 the Maoris had dropped to just 45,000 in 1900, and their extinction seemed likely. Their numbers have, however, returned to the precolonial level. Since the 1960s there has been a resurgence of Maori cultural and political awareness. In 1975 the Waitangi Tribunal was established to hear Maori claims to wrongful dispossession of their lands by the *pakeha* (the Maori term for outsiders), and to assess compensation. After a century and a half, however, such claims are almost impossible to resolve.

In Australia the Aborigines fared even worse. In 1788 they numbered about 300,000; today there are about 160,000, with some 20,000 of mixed ancestry. They were exterminated in Tasmania. Their rights and interests have been almost entirely disregarded: they did not even appear on voting lists until 1962 or on the census until 1971. Despite a strong movement for recognition of Aboriginal rights, there is still unofficial segregation in housing, education and jobs. Not surprisingly, most Aborigines refused to celebrate the Australian bicentenary of 1988.

Ethnic conflict in the Pacific

Elsewhere in the region there has been conflict between indigenous peoples and European or other ethnic groups. In the French overseas territory of New Caledonia a referendum in September 1987 showed a clear majority of people (one third of French descent) in favor of remaining a dependency, but the minority Kanak population continued to press for independence. In Fiji the largest population group are Indians brought in by the British to work in the sugar plantations. In 1987, following electoral defeat, indigenous Fijians engineered a military coup that enabled them to retain political control of the island.

...AND WE ASSERT BY VIRTUE OF THE CONSTITUTION THAT ALL POWER BELONGS TO T
PEOPLE --- ACTING THROUGH THEIR ELECTED REPRESENTATIVES..

PARLIAMENTARY RULE IN THE SOUTH PACIFIC

British rule has left a strong legacy throughout Australasia and Oceania in the form of popularly elected, multi-party assemblies based on the Westminster parliamentary model. These are found in all the former territories of the British empire apart from Tonga, Tuvalu and Western Samoa, where government is under the control of the traditional ruling families. The British monarch, who as head of the Commonwealth is the formal head of state, is represented in Australia, New Zealand, Papua New Guinea, Tuvalu and the Solomon Islands by a governor-general; Fiji, Kiribati, Nauru and Vanuatu have elected presidents, Tonga and Western Samoa have hereditary heads of state.

Australia's complex form of democratic government has borrowed many features from the British system. It has a two-chamber parliament, consisting of a

Stating the principles of democracy This richly decorated mosaic embellishes the front of the parliament building at Port Moresby. The inscription proudly continues: "Parliament may make laws having effect within and without the country for the peace, order and government of Papua New Guinea and the welfare of the people."

Demanding to be heard Eva Rickard, a Maori leader, speaks at Waitangi, where the treaty ceding sovereignty to Britain was signed in 1840. Though there has been a considerable drift to the cities of the North Island, most Maoris retain strong rural roots, and have preserved much of their social organization. Since the 1960s a movement to redress the perceived wrongs of the treaty has developed inside and outside parliament. In 1981 a parliamentary party – Mana Motuhake – was formed to advance a tougher nationalist line. It was led by a former government minister, Matiu Rata, who was disillusioned with the Labor Party's lack of commitment for the Maori cause. It now forms the major opposition party in all four Maori constituencies.

POLITICAL CRISIS IN FIJI

Fiji is the only former British colony in the Pacific to have left the Commonwealth. This came about in 1987 as the result of internal ethnic conflict. It stemmed from the fact that indigenous Fijians make up only 43 percent of the population, but own 80 percent of the land; Indians form 51 percent. In the 1970 constitution, 22 seats in the house of representatives were assigned to indigenous Fijians, 22 to Indians and 8 to others. The Alliance Party, led by Sir Kamisese Mara, formed the first government after independence. For the next 17 years it continued to exercise power largely on behalf of the minority Fijians. The regime was the only one in the Commonwealth that maintained friendly relations both with South Africa and with the white rebel regime in Southern Rhodesia.

The Alliance Party lost many votes at the 1982 election, and was finally voted out of office in 1987 by a coalition of the Indian-dominated National Federation Party and Labor Party. Timoci Bavadra, who became prime minister, favored a foreign policy of nonalignment, and indicated an intention to sign the Rarotonga Treaty supporting the South Pacific nuclear-free zone. Shortly afterward Colonel Sitiveni Rambuka led an attack on parliament and arrested all 28 members of the cabinet. A new government of indigenous Fijians was formed, which immediately made clear its hostility to the non-Fijian population: Indians were invited to "choose any other country to live in". In October 1987 the Queen accepted the resignation of the governor-general, and Fiji ceased to be a member of the Commonwealth.

senate and a house of representatives, and a prime minister and cabinet who are chosen from the majority party of the lower house. Its federal system, however, owes something to the United States model: each of its six states is headed by a governor and has its own executive, legislature and judiciary. The most distinctive features of the Australian system lie in its unusual electoral arrangements.

Australia's voting system

Voting in elections to both chambers is compulsory. Those to the 148-seat house of representatives take place every three years by a form of voting known as the alternative vote system. This requires voters to place in order of preference all candidates standing in a constituency. The candidate who wins more than 50 percent of the first preference votes is elected: if no candidate reaches that threshold, lower preferences are counted until a majority candidate emerges.

Three main parties compete for government: the Australian Labor Party (ALP), the Liberal Party and the National Party. The voting system has frequently worked against the ALP. Although it has won more votes than the other two parties in nearly every election since 1945, it has gained a majority of seats in less than half of them. In constituencies where its candidate fails to get 50 percent of first preference votes, it is usually overtaken by either the Liberal or National candidate when the lower preferences have

all been counted in. Consequently, a Liberal–National alliance may often win a majority of seats even when it has failed to gain more than 50 percent of the vote.

For elections to the senate the single transferable vote system is used. Each state is a multi-member constituency returning 12 senators. Normally half the senate is elected every three years. If, however, the entire senate is dissolved, 12 seats have to be contested in each state at the next election. Voters have to place the candidates in order of preference, and the winners finally emerge in a complicated counting procedure.

Candidates may number as many as 80 in a state. A few voters, rather than pay a

fine for failing to vote, have indulged in a practice known as "donkey voting" – they simply list the candidates in order from the top of the ballot sheet to the bottom. Nevertheless, this system of voting produces results close to proportional representation, and allows the smaller parties, notably the Democratic Labor Party, to gain representation in the senate.

Normally no party wins a majority of seats in senate elections. In 1975, when the ALP government headed by Gough Whitlam did not have control of the senate, the opposition parties blocked the passage of the government's finance bills through the senate. Judging this to be a constitutional crisis, the governor-general dismissed the government and invited the leader of the opposition to form a caretaker coalition government – the only time that such action has been taken by a Commonwealth governor-general. The crisis fueled demands in some quarters for a republican constitution.

Electing New Zealand's government

New Zealand's single-chamber house of representatives is elected by the simple plurality or "first-past-the-post" system. Two main parties, the Labor Party and the National Party, compete for 95 seats. A unique feature is that four constituencies are allocated to Maoris. People of 50 percent or more Maori descent vote in them; those of less than 50 percent descent may choose to be included in the Maori or the general electoral roll. Since 1967 Maoris have been able to stand in general constituencies and non-Maoris in the Maori constituencies.

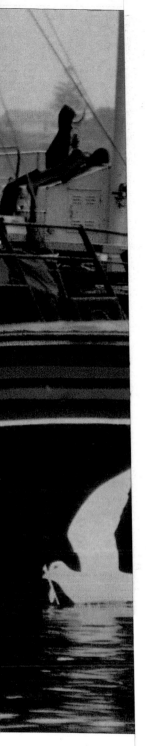

From the 1950s
northern hemispl
relatively unpopt
prime area for
tests. This is why
important here.
Between 1952
Minister Robert
without cabinet c
roval, made avail
South Australia, f
nuclear tests; 9
1957–58, plus 4 at
mid-Pacific and 6
The United State
Bikini Atoll and E
Marshall Islands,

Australia's federal system As the representative of the British sovereign, the governor-general is head of state, but power lies with the cabinet, headed by the prime minister. The six state parliaments enjoy considerable autonomy.

Prime minister

Governor-general

Cabinet

advises

House of representatives

elects

some drawn from

Executive council

Senate

Federal parliament

elects

Federal level

Appointed by British monarch

ELECTORATE

State parliament

elects

State level

Antarctica: a unique political experiment

Antarctica is the only world landmass of continental extent where large tracts are not claimed by any sovereign state. Claims to sovereignty over parts of Antarctica are held by a number of states, but nationals of any state are allowed to move freely over them as long as their purpose is peaceful, and states have come to an agreement to retain the area as a nonmilitarized zone.

Antarctica's political uniqueness is a reflection of its extremely inhospitable environment, which offers no inducement to human settlement other than for reasons of scientific research. It has no permanent residents, and the number of people working there at any one time is extremely small – no more than a few hundred in mid-winter and a few thousand in mid-summer.

Initial interest in Antarctica was in the marine resources of its coastal waters. Many countries in the 19th century established shore stations for whaling, sealing and similar activities. It later became the focus for exploration, culminating in the "race to the Pole" between British and Norwegian teams in 1911.

The first territorial claim was made in 1908 by Britain; it covered the islands now known as the Falkland Islands Dependencies (South Georgia, South Orkney, South Sandwich and South Shetlands) plus the Graham's Land peninsula south of latitude 60°, which was renamed the British Antarctic Territory in 1962. In addition, in 1923 Britain claimed the Ross Dependency (which was to be administered by New Zealand) as a triangular segment of territory extending to the South Pole. Argentina, Australia, Chile, France and Norway have also established similar claims.

During the Cold War period of superpower hostility after 1945 both the Soviet Union and the United States indicated that they reserved the right to lodge territorial claims of their own. Each increased its activity on the continent, especially in 1958, when Antarctica was the center of intense scientific study during International Geophysical Year.

The Antarctic Treaty

A unique cooperative regime has been established in Antarctica. The Antarctic Treaty signed in Washington DC in 1959 recognized the freedom of the 12 signatory states to undertake peaceful scientific work anywhere on the continent. Six more states later obtained consultative status. Antarctica was to remain what it had always been – a nonmilitarized zone. All territorial claims, including those of mineral exploitation, were rescinded while the treaty remained extant.

The cost of developing Antarctica's mineral resources as yet far outweighs the

Laying claim to Antarctica A lone Chilean flag flies over the barren landscape of the Antarctic Peninsula. The peninsula extends north toward the southern tip of South America, and is also claimed by both Argentina and Britain.

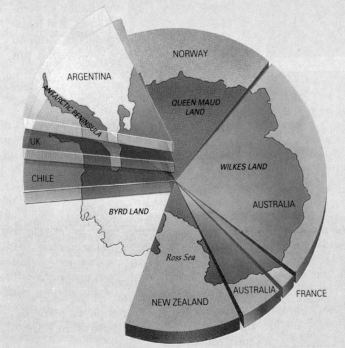

Slices of the cake Seven states have made territorial claims to Antarctica, dividing it up between them like a giant cake. In several places the slices are overlapping. Under the Antarctic Treaty of 1959 these claims are held in abeyance in the interests of scientific research and also (since 1988) the conservation of resources.

value of what would be produced. These resources include large quantities of low-grade coal, found in the Transantarctic Mountains near the Ross Ice Shelf in 1908, and a substantial iron ore deposit in the Prince Charles Mountains of the Australian sector. When, and if, it becomes economic to exploit these resources, the

sovereignty claims will assume far greater importance. This will put the Antarctic Treaty to a major test. However, in 1991 a 50-year ban on mining activity was secured, giving reason for greater optimism about the future.

In recent years, attention has switched to conservation of the natural life of the region. This led to a separate Convention on the Conservation of Antarctic Marine Living Resources (CCAMLR) in 1980. In November 1988 the terms of the 1959 Antarctic Treaty came up for renewal; a further convention was signed, covering nature conservation in the context of resource exploitation. A total of 35 countries are now signatories to the articles of the Antarctic Treaty. These measures show that states are able to take collective action without conflict or under military threat. Much of their success undoubtedly arises from recognition of the fragile nature of the Antarctic's ecosystems, and limited resources.